Beetle–Pathogen Interactions in Conifer Forests

APPLIED BOTANY AND CROP SCIENCE
Series Editors: R.W. Snaydon, University of Reading, UK;
J.M. Barnes, Co-operative State Research Service, United States Department of Agriculture, Washington, USA

Other books in the series:
Physiological Ecology of Forest Production *J.J. Landsberg*
Weed Control Economics *B.A. Auld, K.M. Menz and C.A. Tisdell*
Improving Vegetatively Propagated Crops *A.J. Abbott and R.K. Atkin (Eds)*
Beetle–Pathogen Interactions in Conifer Forests *T.D. Schowalter and G.M. Filip (eds)*

Beetle–Pathogen Interactions in Conifer Forests

Edited by

T.D. SCHOWALTER AND G.M. FILIP
Oregon State University, Corvallis, USA

ACADEMIC PRESS
Harcourt Brace & Company, Publishers
London San Diego New York
Boston Sydney Tokyo Toronto

ACADEMIC PRESS LIMITED
24–28 Oval Road
LONDON NW1 7DX

United States Edition published by
ACADEMIC PRESS INC.
San Diego, CA 92101

This book is printed on acid-free paper

A catalogue record for this book is available from the British Library

ISBN 0–12–628970–0

Typeset by Jenny England, Woking, Surrey
Printed and bound in Great Britain by The University Printing House, Cambridge

Contents

Contributors

Fred A. Baker	Department of Forest Resources, Utah State University, Logan, UT 84322, USA
C. Wayne Berisford	Department of Entomology, University of Georgia, Athens, GA 30602, USA
Catalino A. Blanche	Department of Plant and Soil Sciences, Southern University and A & M College, Baton Rouge, LA 70813, USA
Donald E. Bright	Biosystematics Research Institute, Canada Agriculture, K.W. Neatby Bldg., Ottawa, Ontario K1A 0C6, Canada
Bov. B. Eav	USDA Forest Service, Methods Application Group, 3825 E Mulberry, Rm. 228, Fort Collins, CO 80524, USA
Donald L. Dahlsten	Division of Biological Control, University of California, Berkeley, CA 94720, USA
P. Fenn	Department of Plant Pathology, University of Arkansas, Fayetteville, AR 72701, USA
Gregory M. Filip	Department of Forest Science, Oregon State University, Corvallis, OR 97331, USA
Donald J. Goheen	USDA Forest Service, Pacific Northwest Region, P.O. Box 3623, Portland, OR 97208, USA
Susan Hagle	USDA Forest Service, Northern Region, P.O. Box 7669, Missoula, MT 59802, USA
Everett M. Hansen	Department of Botany and Plant Pathology, Oregon State University, Corvallis, OR 97331, USA
Thomas C. Harrington	Department of Plant Pathology, Iowa State University, Ames, IA 50011, USA
John D. Hodges	Department of Forestry, Mississippi State University, Mississippi State, MS 93762, USA
Peter L. Lorio Jr.	USDA Forest Service, Southern Forest Exp. Stn., 2500 Shreveport Hwy., Pineville, LA 71360, USA
John C. Moser	USDA Forest Service, Southern Forest Exp. Stn., Pineville, LA 71360, USA (retired)
T. Evan Nebeker	Department of Entomology, Mississippi State University, Mississippi State, MS 93762, USA
Timothy D. Paine	Department of Entomology, University of California, Riverside, CA 92521, USA
Thomas W. Phillips	USDA Agricultural Research Service, Department of Entomology, University of Wisconsin, Madison, WI 53706, USA
Kenneth F. Raffa	Department of Entomology, University of Wisconsin, Madison, WI 53706, USA
Scott M. Salom	Department of Entomology, Virginia Polytechnic Institute and State University, Blacksburg, VA 24061, USA
Timothy D. Schowalter	Department of Entomology, Oregon State University, Corvallis, OR 97331, USA
Richard Schmitz	USDA Forest Service, Intermountain Research Station, Forestry Sciences Laboratory, 507 25th Street, Ogden, UT 84401, USA
C.G. Shaw III	USDA Forest Service. Rocky Mountain Research Station, Fort Collins, CO 80524, USA
Frederick M. Stephen	Department of Entomology, University of Arkansas, Fayetteville, AR 72701, USA

Preface

The purpose of this book is to synthesize the enormous volume of work on the importance of interactions among bark beetles, pathogens and trees to productivity of conifer forests in North America. Bark beetles and pathogens are credited with causing substantial losses of timber. They also affect values of other forest resources. Tree mortality caused by these organisms in dense conifer forests often fuels devastating fires, such as those in Yellowstone National Park in the western US during the summer of 1988. Direct and indirect interactions among bark beetles and pathogens require an integrated approach to forest management. The traditional approach of managing individual pest species as damage is recognized may accomplish reduction of the target pest but inadvertently promote other pest species.

This synthesis comes at a time of renewed interest in bark beetle–pathogen interactions in conifer forests. The prospect of global warming and drying, in addition to growing problems of air pollution and industrial forest management, has raised concern that bark beetle and pathogen activity will increase in forests made vulnerable by these stresses. Prevention of future pest problems will require changes in our perspective of forest ecosystems and in our approach to managing forest pests.

Despite the importance of bark beetles and pathogens to forest productivity and forest management, no single book has addressed the various aspects of interaction among trees, insects and pathogens. Various books have addressed bark beetle or pathogen ecology and management, insect–fungal interactions, forest patterns and processes, or forest management. We believe that this synthesis will improve appreciation for the complex interactions among bark beetles, pathogens and trees as these influence productivity of conifer forests.

We also note that no texts currently available for the integrated forest insect and disease management courses now offered and required at most forestry schools in North America integrate information on insect–pathogen interactions. Although a textbook for such a course would have a broader scope than that of this book, this volume synthesizing interactions among bark beetles, pathogens and conifers provides the basic integration for a major set of interacting components in North American forests and addresses interactions of these components with other insects and pathogens.

We have endeavoured to produce an integrated synthesis rather than a collection of project summaries. Our objective is to focus on patterns and processes central to bark beetle and pathogen epidemiology in conifer forests rather than on particular interactions that have been widely reported through technical journals and other outlets. This should increase the value of this book as a textbook, as well as a reference, for forest entomologists, pathologists, ecologists and managers, from students in forest protection courses through researchers and administrators.

A large number of individuals and agencies have contributed to the research and preparation of this book. We thank the National Science Foundation, the US Department of Agriculture (USDA) Competitive Grants Program, Agricultural Research Service and Forest Service, and the many universities, agricultural experiment stations, state forestry departments, and private timber organizations for research support. We especially acknowledge the support of USDA Regional Research Project W-110 for supporting research and providing the mechanism for coordinating the many research projects synthesized in this book. W.W. Allen, J.M. Barnes, A.W. Richford and R.W. Snaydon provided invaluable advice and assistance in the preparation of this book. R.L. Edmonds and D.L. Wood provided a technical review of the book. A.A. Berryman, R.A. Blanchette, J.N. Borden, J.P. Dunn, F. Guinn, F.P. Hain, V.L. Harper, J.L. Hayes, P.J. Kramer, T.D. Paine, J.R. Parmeter, Jr., K.F. Raffa, F.M. Stephen, P. Turchin, K.E. Ward, D.L. Wood and J.J. Worrall provided technical reviews of various chapters.

Timothy D. Schowalter and Gregory M. Filip

PART I
Introduction

–1–

Bark Beetle–Pathogen–Conifer Interactions: an Overview

T.D. SCHOWALTER[1] and G.M. FILIP[2]
Departments of [1]Entomology and [2]Forest Science, Oregon State University, Corvallis, OR, USA

1.1 INTRODUCTION

In North America, as in other parts of the world, forest management historically has been equated with timber management. A narrow focus on maximizing productivity of commercially valuable tree species precluded recognition of the importance of associated species or their interactions for sustained forest productivity and stability. Any factor (biotic or abiotic) which interfered with the goal of maximum timber yields was considered a "pest" to be vigorously suppressed (Stark and Waters, 1985). Only recently have interactions among the various forest species, including "pests," become recognized both as indicators of forest health and as contributors to long-term forest health. Accumulating scientific evidence now supports a view of forests as integrated ecosystems in which species interactions respond to changes in forest condition and maintain nutrient cycling and soil fertility critical to forest productivity and stability. This view, supported by increasing insect and pathogen problems in stressed managed forests and increasing public demand for non-timber amenities, together with recent legislation requiring environmental impact assessment and protection of endangered species, is revolutionizing forestry and, consequently, our approach to forest protection.

1.2 THE ECOSYSTEM VIEW OF FORESTS

1.2.1 The importance of interactions

The complexity of forest systems and the need to manage forests for productivity and water yield over long time periods led to development of an ecosystem perspective during the

BEETLE–PATHOGEN INTERACTIONS IN CONIFER FORESTS
ISBN 0-12-628970-0

1950s and 1960s. Forested watersheds were ideal ecosystems with relatively distinct boundaries. Various instruments were used to measure amounts of materials added to the ecosystem (input) through bedrock weathering, precipitation, atmospheric filtering and biomass accumulation and lost from the ecosystem (output) through evapotranspiration, leaching, streamflow and harvest (see Bormann and Likens, 1979; Edmonds, 1982; Swank and Crossley, 1987; Perry *et al.*, 1989). Eventually, as part of the International Biological Program (IBP) and Long Term Ecological Research (LTER) Program, studies of species diversity led to consideration of the consequences of mutualistic, competitive and predatory interactions among species within integrated communities (Edmonds, 1982). The emerging view of forests as ecosystems with internal mechanisms for maintaining productivity and stability, in the face of constant change in environmental conditions, is changing our approach to managing forest resources and the organisms often viewed as "pests."

Natural forests are typically composed of more than one tree species or age class, many other plant and animal species, and various structures such as snags (dead standing trees) and decomposing logs. These components function in various ways to maintain nutrient cycles (critical to soil fertility) and long-term forest productivity and stability (Franklin *et al.*, 1989) and influence regional and global climate. Forests reduce solar heating of the soil surface and contribute to cloud formation, moderating regional climates, and regulate fluxes of atmospheric oxygen and carbon dioxide. Plants and autotrophic bacteria capture water, carbon dioxide and nutrients from the atmosphere and soil and incorporate these resources into biomass. Herbivores consume plants and plant parts, influencing nutrient requirements by vegetation and controlling the flow of energy and nutrients to the forest floor. Saprophagous invertebrates, bacteria and fungi release nutrients from biomass and control soil fertility. Mycorrhizal fungi infuse soil and decomposing litter and provide nutrients to host trees in return for necessary photosynthates. The activities and interactions of these components prevent bottlenecks in nutrient supply that could limit forest productivity and reduce stability to environmental fluctuations (Mattson and Addy, 1975; Tilman, 1982, 1988; Seastedt, 1984; Schowalter *et al.*, 1986).

Selective consumption of particular plants and plant parts by herbivores and pathogens not only maintains nutrient cycling but also provides for continuous turnover and rejuvenation of vegetative material. Such turnover is necessary to prevent stagnation and reduced photosynthetic efficiency and to facilitate adaptation to environmental change. Replacement of stressed plants by plants more tolerant of prevailing conditions could enhance long-term productivity and the stability of forest function.

1.2.2 Importance of non-commercial components

The functions of many forest components remain unknown. However, experience has shown that inconspicuous or unappreciated components often are critical to maintenance of forest health and long-term productivity. Temple (1977) noted that the age (300–400 years) of the last 13 surviving *Sideroxylon sessiliflorum* (=*Calvaria major*), a tree that once covered the South Pacific island of Mauritanius, coincided with the extermination of the dodo bird in 1680. When *S. sessiliflorum* seeds were force-fed to turkeys (approximately the size of a dodo), the seed coats were sufficiently abraided during passage through the bird's gut to permit germination, thereby indicating the dodo's contribution to reproduction of *S. sessiliflorum.* A major tree species was nearly lost because of the extermination of its unappreciated, but necessary, associate.

Other components of natural forests also have been found to promote long-term productivity and stability. Loss or suppression of these components can disrupt necessary interactions and ecological processes. Bark beetles and pathogens interact with many of these components to enhance forest health, as in the examples below.

1.2.2.1 Non-commercial plants

Non-commercial trees and shrubs often have been regarded as "weeds," competing with crop trees for limiting resources. However, many of these tree and shrub species also contribute to soil fertility (often through association with nitrogen-fixing bacteria), soil retention on steep slopes, moisture retention through soil shading, and interruption of fire, insect and pathogen transmission (Binkley *et al.*, 1982; Perry, 1988; Borchers and Perry, 1990). For example, interspersed hardwoods or other non-hosts disrupt *Dendroctonus frontalis* infestations in southern pine forests (Schowalter and Turchin, 1993) and *Leptographium wageneri* infection in northwestern conifer forests (Hansen *et al.*, 1988; Chapter 11). Disruption of infestations by non-host vegetation may be due to the physical interruption of dispersal from host to host or to confusing of attractive host compounds by non-attractive or repellent non-host compounds in the forest aerosol (Visser, 1986; Hansen *et al.*, 1988). Although competition may slow growth of crop trees in the short term, these compensatory effects of plant associates may contribute to the survival and growth of crop trees over the longer period of forest development.

1.2.2.2 Dead trees and decomposing logs

Dead trees and decomposing logs have been viewed as wasted biomass, fire hazards, sources of destructive insects and pathogenic fungi, or impediments to reforestation efforts. Consequently, these structures have been zealously removed or destroyed. Recent research has demonstrated that dead trees and decomposing logs provide habitat for insectivorous birds and vectors of mycorrhizal fungi; stabilize soils, especially on steep slopes; contribute to soil development and fertility; and store water during dry periods (Harmon *et al.*, 1986; Carpenter *et al.*, 1988; Edmonds and Eglitis, 1989; Zhong and Schowalter, 1989; Schowalter *et al.*, 1992). Roots and mycorrhizae infuse logs and transport nutrients into growing tissues. Models indicate that forest health and productivity are enhanced by decomposing logs and may decline if these structures are removed (Harmon *et al.*, 1986). Healthy forests are more resistant to pest outbreaks than are forests stressed by resource limitation (Mattson and Haack, 1987; Chapters 4 and 5). Consequently, retention of woody residues has become a management priority in many North American forests. Bark beetles and wood-boring insects initiate and stimulate this process of nutrient turnover by channelizing wood and introducing nitrogen-fixing bacteria and saprophytic fungi necessary for nutrient turnover (Bridges, 1981; Edmonds and Eglitis, 1989; Schowalter *et al.*, 1992).

1.2.2.3 Fire

Fire historically has been suppressed at any cost, leading to changes in tree species composition or density, imbalances in nutrient availability for tree growth, and predisposition of stressed forests to insect and pathogen outbreaks. *Pinus* forests, in particular, depend on frequent fires to mineralize nutrients and remove competing fire-intolerant vegetation. Frequent fires maintain open, park-like forests that are resistant to crown fires and pest spread (Wright and Heinselman, 1973; Schowalter *et al.*, 1981a; Perry, 1988). Fire suppression can contribute to development of dense conifer forests stressed by competition for resources, susceptible to spread of bark beetles, root pathogens, and dwarf mistletoe and, inevitably, vulnerable to catastrophic crown fire (Fig.1.1). Fire suppression in the southern US contributes to eventual replacement of pines by shade-tolerant hardwoods, in large part through pine mortality to *D. frontalis* (Fig. 1.2). Fire suppression, together with selective harvest of *Pinus* and *Larix*, in western North America has favored forest succession to mixed *Pseudotsuga* and *Abies* forests

Fig. 1.1. Effects of wildfire in mixed-conifer forest in Yellowstone National Park, Wyoming. Note distribution and reduced density of surviving trees.

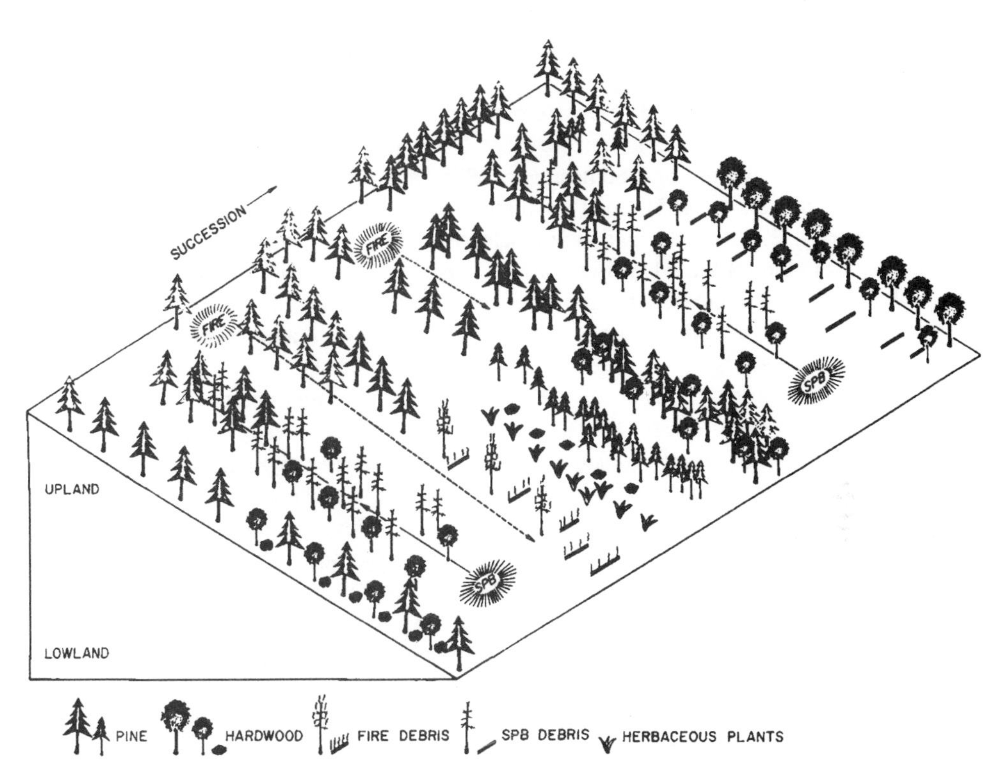

Fig. 1.2. Diagrammatic representation of the coniferous forest (upland and lowland) in the southeastern US, illustrating interaction between *Dendroctonus frontalis* (SPB) and fire. Successional transformations resulting from fire and *D. frontalis* extend from left to right; dotted arrows indicate direction of movement. Fire, a regular feature of the generally dry, well-drained uplands, invades generally moist, poorly drained lowlands where drought or *D. frontalis* create favorable conditions. *D. frontalis* depends upon fire for regeneration of pine stands. The hardwood climax forest (far right lowland) results from freedom from fire and can be reduced by fire (from Schowalter *et al.* (1981a), copyright Entomological Society of America).

now subject to widespread mortality to a variety of insects, including *Choristoneura occidentalis* and *Orgyia pseudotsugata*, and pathogens, including *Armillaria* (Hadfield, 1988). In the absence of fire or other disturbances that maintain particular tree species composition, insects and pathogens selectively remove fire-tolerant tree species stressed by increased competition and decreased nutrient turnover and often accelerate succession to a forest of shade-tolerant species (Geiszler *et al.*, 1980; Schowalter *et al.*, 1981a). Although politically controversial, periodic fire is necessary to preserve the health and composition of many natural forests. Fire now is used in many forested regions as a management tool to reduce fuel accumulation and enhance forest productivity, especially in the southern US, and to reduce susceptibility to insects and pathogens (Wade and Ward, 1976; Miller, 1979).

1.2.2.4 Insects and Pathogens

Insects and pathogens also are integral components of forest ecosystems. These organisms contribute to forest health and long-term productivity in several ways.

First, insects and fungi select particular host species on the basis of their biochemical suitability and attractiveness (Rosenthal and Janzen, 1979; Cates and Alexander, 1982;

Wood, 1982; Coley *et al.*, 1985). Insect and pathogen outbreaks are not random events, but reflect conditions that promote population growth and spread, e.g. conditions that increase abundance of hosts with limited defensive capabilities or that stress trees and weaken their chemical defenses (Rosenthal and Janzen, 1979; Cates and Alexander, 1982; Kareiva, 1983; Coley *et al.*, 1985; Waring and Schlesinger, 1985; Schowalter *et al.*, 1986; Mattson and Haack, 1987; Chapters 4 and 5). Such selection tailors tree density and species composition to site conditions. For example, *D. frontalis* selectively removes pines and favors hardwoods on wetter sites in the southern US (Fig. 1.2). Increased plant diversity resulting from mortality to abundant hosts increases the ability of the forest to maintain canopy conditions and forest functions under a wider range of environmental conditions and reduces the likelihood of future population outbreaks.

Second, nutrient turnover from foliage and wood increases nutrient availability for more efficient plants and plant tissues (Mattson and Addy, 1975). Schowalter *et al.* (1991) found that 20% defoliation in young *Pseudotsuga menziesii* doubled the amount of precipitation and litter reaching the forest floor and increased nitrogen, potassium, and calcium turnover 20–30%, compared to non-defoliated trees. These nutrients became available for new foliage production. Forest productivity can be enhanced for decades following defoliator outbreaks (Wickman, 1980; Alfaro and MacDonald, 1988), provided that defoliation does not trigger bark beetle or pathogen outbreaks (Berryman and Wright, 1978). Similarly, mining by bark beetles and wood-boring insects rapidly fragments the nutrient-rich phloem and provides entry and food resources for nitrogen-fixing bacteria and saprophytic microorganisms that mineralize and release bound nutrients for uptake by roots and mycorrhizae (Carpenter *et al.*, 1988; Edmonds and Eglitis, 1989; Zhong and Schowalter, 1989; Schowalter *et al.*, 1992).

Finally, tree deformation, mortality and canopy opening create habitat and food resources for other species with roles equally critical to forest productivity and stability (Franklin *et al.*, 1989). For example, *Tsuga* and other trees germinate best on the nutrient-rich substrates provided by decaying logs. Many cavity-nesting birds and other important predators of forest insects depend on habitats in snags and heart-rotted trees.

Recognition of the importance of species interactions to the productivity and stability of forest ecosystems has emerged only over the past two decades (cf. Mattson and Addy, 1975; Edmonds, 1982; Franklin *et al.*, 1989). Progress has been limited by the difficulties of organizing and supporting the multidisciplinary research necessary to evaluate the effects of interactions, and by scientific methodologies that focus on effects of single, or a few, factors rather than on interactions among many factors. Nevertheless, our appreciation for the complex effects of these organisms on forest productivity and stability is fundamental to our approach to managing forest insects and pathogens.

Although tree-injuring outbreaks may continue to be undesirable from a timber management perspective, their occurrence may provide clues to problems of forest health that require remedy. Suppression of pests without regard for the factors triggering outbreaks may be counterproductive.

1.3 ADVANCES IN FOREST PEST MANAGEMENT

1.3.1 The management context

Our approach to managing forest resources and pests is determined by our perception of forests as integrated ecosystems or loosely organized species assemblages. A traditional view of forests as collections of independent species justified removal of non-commercial species from the forest and led to a forestry in which crop species were managed as tree farms. Declining forest health and pest problems resulting from this approach, and the emergence of the ecosystem perspective, encourage a broader management approach that addresses critical ecological processes, including nutrient cycling and species interactions, and their consequences for forest health.

Management of forest resources presents a number of problems distinct from those of agricultural systems. First, forests are managed over long time periods (decades to centuries), allowing for cumulative effects of various factors affecting resource values. Second, management often must balance competing demands for forest resources, such as timber and fiber, watersheds, recreation, rangeland and fish and wildlife (Leuschner and Berck, 1985). This balance applies especially in the USA and other countries where multiple use of forest resources is mandated by federal legislation. Identification of the management goals for particular forest tracts is a prerequisite for effective management of multiple resources, hence for management of pests.

Management of North American forests has become controversial as vast areas of undisturbed forests have been replaced rapidly by cities, farms and younger, managed forests over the past century, with serious consequences for pest activity. At a time when little forested land in North America remains in its natural condition, subject to natural disturbances, accumulating evidence indicates that current management practices are promoting insect and pathogen outbreaks (Fig. 1.3; Franklin *et al.*, 1989; Chapter 11). Fire suppression results in dense, unhealthy stands of conifers with retarded nutrient cycling and

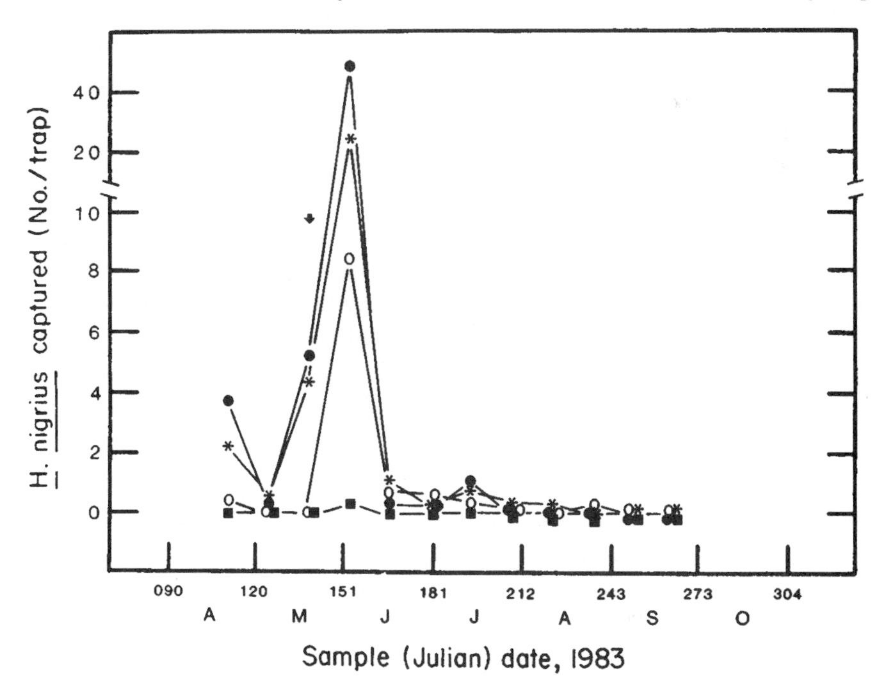

Fig. 1.3. Mean number of *Hylastes nigrinus* caught per trap on sticky traps during 1983 in precommercially thinned and unthinned 2-ha plots in two 12-year-old plantation of Douglas-fir (n = 8 plots per treatment). Stars designate plots thinned in September 1982; closed circles, plots thinned in January 1983; open circles, plots thinned in May 1983; closed squares, unthinned (control). The arrow indicates the time of thinning in May. Elevated populations of *H. nigrinus* increase the probability of introducing black-stain root disease fungus, *Leptographicus wageneri* (from Witcosky *et al.* 1986, copyright National Research Council of Canada).

little resistance to potential pests, especially where fire naturally occurred frequently. Young, rapidly growing monocultures favored by forest managers soon become vulnerable to pests because of selective breeding for rapid growth at the expense of resistance to pests, inadequate nutrient supply by impoverished soils, exposure to extreme environmental conditions prior to canopy development, and elimination of natural barriers (present in diverse stands and landscapes) to pest spread (Franklin et al., 1989). Air pollution and climate change exacerbate forest susceptibility to pests (Cobb *et al.*, 1968; Smith, 1981; Franklin *et al.*, 1992).

The responsiveness of forest insects and pathogens to changes in forest conditions is ominous in view of the regional scale of forest conversion to managed forests, increased atmospheric pollution, and potential global climate change (Graham *et al.*, 1990; Gear and Huntley, 1991). Forest researchers and managers must adopt a broader view of forest ecosystems and the importance of species interactions if our current activities are to provide the diversity and flexibility necessary for continued productivity in future forests.

1.3.2 Emerging views

Major advances have occurred over the past decade in our approach to managing forest pests (Waters and Stark, 1980; Coulson and Stark, 1982; Waters *et al.*, 1985). The traditional approach focused on suppression of individual pest species as plant injury was detected. Chemical pesticides were the favored control option because of their low cost and apparent effectiveness. Reliance on pesticides resulted in pesticide resistance (due to inherent ability of pests to detoxify plant defenses) and environmental degradation (due to disruption of community structure and function). During the 1970s two emerging views began to change the way in which forests and forest "pests" are managed.

1.3.2.1 Recognition of species interactions

Species interactions have a key role in promoting or mitigating destructive population trends. The ability of most healthy trees to resist colonization by insects and pathogens by means of physical barriers, chemical toxins and feeding deterrents has become a major focus of research (Rosenthal and Janzen, 1979; Cates and Alexander, 1982; Harborne, 1982; Coley *et al.*, 1985; Bazzaz *et al.*, 1987; Chapters 5 and 8). However, chemical defense is subject to resource availability and plant metabolic priorities (Tuomi *et al.*, 1984; Haukioja *et al.*, 1985; Lorio *et al.*, 1990; Chapter 5). Community diversity also dictates the ease by which herbivores can identify and reach acceptable hosts (Kareiva, 1983; Rausher, 1983; Hunter and Aarssen, 1988) and the combination of predators and parasites available to prevent herbivore population growth (Dahlsten, 1982).

Furthermore, tree injury or death no longer can be attributed simply to attack by the most conspicuous species. Rather, multiple-species pest complexes, interacting with soil and climatic factors that impair tree resistance, drive a process of gradual decline and replacement of susceptible trees and stands (Berryman and Wright, 1978; Goheen and Filip, 1980; Filip and Goheen, 1982; Whitney, 1982; Schowalter, 1985; Chapters 4 and 9). Populations of adapted insects and pathogens can grow on dense populations of relatively healthy plants. Such species and environmental factors that stress a tree predispose it to other species that depend on stressed hosts. For example, defoliators, especially when aided by moisture limitation, can predispose conifers to bark beetles (Berryman and Wright, 1978, Chapter 4). Associated species can have profound effects on the success of tree-killing species

(Dahlsten, 1982; Chapter 7). For example, tarsonemid mites phoretic on *D. frontalis* are responsible for the transmission of pathogenic *Ophiostoma minus* into pines attacked by the bark beetle (Bridges and Moser, 1986; Chapter 7).

1.3.2.2 Integrated pest management

Emergence of the ecosystem and multiple-pest concepts resulted in integrated pest management (IPM), or integrated forest protection (IFP), methodology (e.g. Coulson and Stark, 1982; Stark and Walters, 1985). This methodology addresses the need to manage pest species within the framework of an integrated ecosystem. Assessment of various impacts of pests on other species and on key ecosystem processes is particularly important within the context of multiple-use forest management. IPM is based on four premises: (1) pest management must contribute to the resource management goals; (2) pest management must be consistent with ecological principles; (3) the target of pest management is maintenance of pest populations below economic thresholds; and (4) pest management is accomplished through a combination of treatment strategies. These premises are intended to maintain a resource focus, with pest management applied only as necessary to prevent unacceptable losses of resource values to the target pest(s), and to protect key ecological processes and species interactions that limit further damage by target or non-target pests.

Studies of interactions within integrated forest communities have increased our understanding of the contributions of these interactions to processes fundamental to long-term forest productivity and stability. At the same time, these studies provide the practical information necessary to improve our management of forest resources and forest pests. Multidisciplinary research on bark beetles and pathogens in conifer forests over the past 15 years has been both contributor to and beneficiary of concepts of forest ecosystems, multiple-species interactions, and IPM.

1.4 BARK BEETLES, PATHOGENS AND CONIFERS

1.4.1 Nature of interactions

Bark beetles and pathogenic fungi have been recognized as major interacting factors affecting the productivity and ecological succession of conifer forests. Traditionally, these organisms have been viewed as enemies by forest managers. The ability of bark beetles and fungal pathogens to kill or deform conifers over extensive areas is beyond dispute (Coulson and Stark, 1982; Filip and Goheen, 1982, 1984; Waters *et al.*, 1985; Hansen *et al.*, 1988; Chapter 9). Considerable effort and expense have been directed toward suppression, usually of individual species as tree injury or mortality became apparent.

With the emergence of new perspectives outlined above, these organisms have been recognized as integral components of forest ecosystems. Rather than threatening forests, these "pests" have contributed largely to the forest structure and productivity appreciated by early settlers and forest managers. Bark beetles and fungal pathogens, interacting with fire, have been instrumental in maintaining healthy conifer forests through natural thinning, nutrient cycling, and selection for site-adapted trees (Schowalter *et al.*, 1981a; Chapter 9).

Conifers display a variety of adaptations to prevent or tolerate injury by fire, pathogens or bark beetles under normal conditions. These adaptations include fire-resistant bark or meristems, and chemical defenses such as terpenes and phenols (Fig. 1.4). In fact, destructive wild-

Fig. 1.4. *Dendroctonus ponderosae* removing pitch from entrance bored into *Pinus contorta*. Oleoresin flow is a major physical and chemical barrier to bark beetle or pathogen invasion of conifers with a primary oleoresin system. Oleoresin flow depends on tree condition.

fires typically occur where fuels have accumulated as a result of extensive tree mortality in dense stands (Perry, 1988). Bark beetles and pathogenic fungi normally are restricted in occurrence to scattered weakened or dead trees which have impaired chemical and physical defenses. By removing these trees gradually, bark beetles and pathogens reduce tree crowding and prevent massive fuel accumulation. Populations of insects or pathogens remain inocuous as long as suitably injured or weakened trees are scattered across forest landscapes. Tree-killing epidemics occur when populations reach critical thresholds in stands that are composed predominantly of host trees (Kareiva, 1983; Rausher, 1983) and/or stressed by storm damage, drought, air pollution, overstocking, prior defoliation, mistletoe, fungal infection, etc. (Berryman and Wright, 1978; Mattson and Haack, 1987; Schowalter *et al.*, 1981b; Chapters 4, 5 and 11). Under such conditions, sheer numbers of beetles or fungal propagules assure that widely distributed resources are colonized and increase the likelihood that resistant hosts, as well as more susceptible hosts, will succumb to mass attacks (Schowalter, 1985).

Outbreaks of bark beetles and pathogens usually are associated because bark beetles transport fungal spores into attacked trees, and pathogens create newly stressed hosts attractive to bark beetles (Whitney, 1982; Chapters 6, 8 and 9). The association of bark beetles and fungal pathogens with dead and dying conifers has, over evolutionary time, led to interdependence, characterized by life history synchronization (Dowding, 1984), nutritional enhancement of wood by fungi and other microorganisms (Bridges, 1981; Whitney et al., 1987; Chapter 7), fungal adaptation for transport by beetles (Dowding, 1984; Beaver, 1988; Chapter 3), specialized structures (mycangia) for transporting fungal propagules (Whitney, 1982; Chapter 2), and beetle orientation to diseased trees (Cobb *et al.*, 1968, 1974; Witcosky *et al.*, 1987; Chapters 4 and 9). The interaction between bark beetles and pathogenic fungi, acting in concert to exploit suitable resources, requires that they be managed as a unit. Management of one without regard for the other will be ineffective in protecting forest resources.

Interactions among trees and their associated organisms create unlimited opportunities for collaboration among entomologists, pathologists, ecologists, tree physiologists, chemists and forest managers for the purpose of understanding and managing these interactions. Clearly, appreciation of the nature of interactions among the beetles, pathogens and conifer hosts will contribute to better management of conifer ecosystems. While these interactions are the focus of this book, interactions with other components of forest ecosystems, such as fire, mistletoes, defoliators, etc., also will be considered as these affect forest productivity and management goals.

1.4.2 Scope of this book

The interactions highlighted in this work involve bark beetles of the family Scolytidae (and some weevils, family Curculionidae), root pathogenic fungi belonging to several taxa, and their conifer hosts, primarily in the family Pinaceae. Conifer forests dominate large portions of North America, including the boreal region that spans the continent between 45° and 68° north latitude, the coastal plain forests from the mid-Atlantic states south and west across the Gulf of Mexico (Fig. 1.5), and the western montane forests (Fig. 1.6). These forests range in structure and complexity from young single-species plantations to the complex temperate rainforests of the Pacific Northwest, the most massive conifer forests in the world with tree diameters often exceeding 2 m and heights exceeding 70 m. Bark beetles and pathogenic fungi occur throughout these regions, accounting for extensive tree growth loss and mortality (Leuschner and Berck, 1985; Chapter 9). Global warming and drying are expected to increase the incidence and severity of tree mortality to these organisms (Franklin *et al.*, 1992; Chapter 4).

The considerable ecological and economic effects of these interacting organisms has warranted a concerted effort over the past 15 years to improve our understanding and management of these interactions. The resulting work is among the best examples of successful collaboration between scientists and managers on interactions among trees and associated organisms. This book represents a synthesis of information derived from this effort and is organized around four major problem areas fundamental to understanding and managing interactions among bark beetles, pathogenic fungi, and conifers.

1.4.2.1 Systematics

First, the systematics of the taxa involved must be understood. Pest species must be distinguishable from closely related non-pests. Such distinction is necessary for proper ascription of tree injury or death, and for assessment of future injury potential and feasibility of management options. New techniques for distinguishing species or biotypes are being evaluated. This topic is addressed in Chapters 2 and 3.

1.4.2.2 Factors influencing interactions

Second, while insects and pathogen propagules may land randomly on suitable and unsuitable plants, neither insects nor pathogens colonize trees at random. Rather, insects and pathogens most often colonize trees that are susceptible as a result of impaired defensive

Fig. 1.5. Coastal plain *Pinus* forest in the southern US (Louisiana).

capabilities or of rapid growth at the expense of defenses (Coley *et al.*, 1985; Lorio *et al.*, 1990). However, in diverse forests, limited life span or dispersal capability often prevents discovery of suitable hosts (Kareiva, 1983; Rausher, 1983). Therefore, factors influencing host susceptibility, host distribution across forest landscapes, and host apparency (perception by dispersing beetles) largely determine bark beetle and pathogen epidemiologies. Furthermore, bark beetles and pathogens show complex interactions that often facilitate each other's population growth and spread, but which under some circumstances may be antagonistic. Associated species that may or may not achieve pest status also are involved in

Fig. 1.6. Montane coniferous forest in the western US (Oregon).

the epidemiologies of the pests (Dahlsten, 1982). Various invertebrates compete with the bark beetles or feed on fungi. Microorganisms influence the nutritional quality of the wood resource for beetles and pathogens and/or compete with pathogenic fungi. These factors are discussed in Chapters 4–7.

1.4.2.3 Effects of interactions

Effects of bark beetle–pathogen interactions are the topics of Chapters 8 and 9. Effects on trees begin with cellular disruption and blockage of water transport by pathogens and proceed through tree death. Effects on stands are variable, depending on stand age and condition and the severity of tree suppression or mortality. Stands can be thinned, or subjected to waves of mortality that can require unscheduled salvage and replanting. Both stands and landscapes become more diverse, in terms of both species composition and age structure, thereby tending to provide natural barriers to continued pest activity.

1.4.2.4 Management of interactions

Finally, we must be concerned with managing forest resources effectively and efficiently. Models have become powerful tools for predicting future pest activity as well as assessing the need for and efficacy of suppression strategies. Models, however, represent state-of-the-art information availability and are necessarily based on assumptions with regard to parameters with insufficient data. Management strategies are available to manage pests pursuant to established resource management goals. However, given that current pest problems often reflect past management strategies, we must assess carefully the scientific basis for management decisions and anticipate responses of non-pests within the forest community. Models are the subject of Chapter 10; management options are addressed in Chapter 11.

Interactions among bark beetles, pathogens and conifers are sufficiently complex and well known to warrant their emphasis in this book. However, our approach to studying and managing these relationships is applicable to other interactions or ecosystem types. Therefore, our synthesis may serve as a model for understanding and managing other pest complexes.

REFERENCES

Alfaro, R.I. and MacDonald, R.N. (1988). Effects of defoliation by the western false hemlock looper on Douglas-fir tree-ring chronologies. *Tree-ring Bull.* **48**, 3–11.

Bazzaz, F.A., Chiariello, N.R., Coley, P.D. and Pitelka, L.F. (1987). Allocating resources to reproduction and defense. *BioScience* **37**, 58–67.

Beaver, R.A. (1988). Insect–fungus relationships in the bark and ambrosia beetles. *In* "Insect–Fungus Interactions" (N. Wilding, N.M. Collins, P.M. Hammond, and J.F. Webber, eds), pp. 121–143, Academic Press, London.

Berryman, A.A. and Wright, L.C. (1978). Defoliation, tree condition, and bark beetles. *In* "The Douglas-fir Tussock Moth: A Synthesis" (M.H. Brookes, R.W. Stark and R.W. Campbell, eds), pp. 81–87, USDA Forestry Serv. Tech. Bull. 1585, USDA Forest Serv., Washington, DC.

Binkley, D., Cromack, K., Jr. and Fredriksen, R.L. (1982). Nitrogen accretion and availability in some snowbrush ecosystems. *Forest Sci.* **28**, 720–724.

Borchers, S.L. and Perry, D.A. (1990). Growth and ectomycorrhiza formation of Douglas-fir seedlings grown in soils collected at different distances from pioneering hardwoods in southwest Oregon clear-cuts. *Can. J. For Res.* **20**, 712–721.

Bormann, F.H. and Likens, C.E. (1979). "Pattern and Process in a Forested Ecosystem." Springer-Verlag, New York. 253pp.

Bridges, J.R. (1981). Nitrogen-fixing bacteria associated with bark beetles. *Microb. Ecol.* **7**, 131–137.

Bridges, J.R. and Moser, J.C. (1986). Relationship of phoretic mites (Acari: Tarsonemidae) to the bluestaining fungus, *Ceratocystis minor*, in trees infested by southern pine beetle (Coleoptera: Scolytidae). *Environ. Entomol.* **15**, 951–953.

Carpenter, S.E., Harmon, M.E., Ingham, E.R., Kelsey, R.G., Lattin, J.D. and Schowalter, T.D. (1988). Early patterns of heterotroph activity in conifer logs. *Proc. Royal Soc. Edinburgh* **94B**, 33–43.

Cates, R.G. and Alexander, H. (1982). Host resistance and susceptibility. *In* "Bark Beetles in North American Conifers: A System for the Study of Evolutionary Biology" (J.B. Mitton and K.B. Sturgeon, eds), pp. 212–263. University of Texas Press, Austin, TX.

Cobb, F.W., Jr., Wood, D.L., Stark, R.W. and Parmeter, J.R., Jr. (1968). Theory on the relationships between oxidant injury and bark beetle infestation. *Hilgardia* **39**, 141–152.

Cobb, F.W., Jr., Parmeter, J.R., Jr., Wood, D.L. and Stark, R.W. (1974). Root pathogens as agents predisposing ponderosa pine and white fir to bark beetles. *In* "Proc. Fourth International Conference on *Fomes annosus*", pp. 8–15. USDA Forest Serv., Washington, DC.

Coley, P.D., Bryant, J.P. and Chapin, F.S., III. (1985). Resource availability and plant antiherbivore defense. *Science* **230**, 895–899.

Coulson, R.N. and Stark, R.W. (1982). Integrated management of bark beetles. *In* "Bark Beetles in North American Conifers: A System for the Study of Evolutionary Biology" (J.B. Mitton and K.B. Sturgeon, eds), pp. 315–349. University of Texas Press, Austin, TX.

Dahlsten, D.L. (1982). Relationships between bark beetles and their natural enemies. *In* "Bark Beetles in North American Conifers: A System for the Study of Evolutionary Biology" (J.B. Mitton and K.B. Sturgeon, eds), pp. 140–182. University of Texas Press, Austin, TX.

Dowding, P. (1984). The evolution of insect–fungal relationships in the primary invasion of forest timber. *In* "Invertebrate Microbial Interactions" (J.M. Anderson, A.D.M. Rayner and D.W.H. Walton eds), pp. 133–153. Cambridge University Press, Cambridge.

Edmonds, R.L. (1982). "Analysis of Coniferous Forest Ecosystems in the Western United States." Hutchinson Ross Publ. Co., Stroudsburg, PA, 419 pp.

Edmonds, R.L. and Eglitis, A. (1989). The role of the Douglas-fir beetle and wood borers in the decomposition and nutrient release from Douglas-fir logs. *Can. J. For. Res.* **19**, 853–859.

Filip, G.M. and Goheen, D.J. (1982). Tree mortality caused by root pathogen complex in Deschutes National Forest, Oregon. *Plant Dis.* **66**, 240–243.

Filip, G.M. and Goheen, D.J. (1984). Root diseases cause severe mortality in white and grand fir stands in the Pacific Northwest. *Forest Sci.* **30**, 138–142.

Franklin, J.F., Perry, D.A., Schowalter, T.D., Harmon, M.E., McKee, A. and Spies, T.A. (1989). Importance of ecological diversity in maintaining long-term site productivity. *In* "Maintaining Long-term Productivity of Pacific Northwest Forest Ecosystems" (D.A. Perry, B. Thomas, R. Meurisse, R. Miller, J. Boyle, P. Sollins and J. Means, eds), pp. 82–97. Timber Press, Portland, OR.

Franklin, J.F., Swanson, F.J., Harmon, M.E., Perry, D.A., Spies, T.A., Dale, V.H., McKee, A., Ferrell, W.K., Means, J.E., Gregory, S.V., Lattin, J.D., Schowalter, T.D. and Larson, D. (1992). Effects of global climatic change on forests in Northwestern North America. *In* "Global Warming and Biological Diversity" (R.L. Peters and T.E. Lovejoy, eds), pp. 244–257. Yale University Press, New Haven, CT.

Gear, A.J. and Huntley, B. (1991). Rapid changes in the range limits of Scots pine 4000 years ago. *Science* **251**, 544–547.

Geiszler, D.R., Gara, R.I., Driver, C.H., Gallucci, V.F. and Martin, R.E. (1980). Fire, fungi and beetle influences on a lodgepole pine ecosystem in south-central Oregon. *Oecologia* **46**, 239–243.

Goheen, D.J. and Filip, G.M. (1980). Root pathogen complexes in Pacific Northwest forests. *Plant Dis.* **64**, 793–794

Graham, R.L., Turner, M.G. and Dale, V.H. (1990). How increasing CO_2 and climate change affect forests. *BioScience* **40**, 575–587.

Hadfield, J.S. (1988). Integrated pest management of a western spruce budworm outbreak in the Pacific Northwest. *Northw. Environ. J.* **4**, 301–312.

Hansen, E.M., Goheen, D.J., Hessburg, P.F., Witcosky, J.J. and Schowalter, T.D. (1988). Biology and management of black-stain root disease in Douglas-fir. *In* "Leptographium Root Diseases on Conifers" (T.C. Harrington and F.W. Cobb, Jr., eds), pp. 63–80. American Phytopathological Society Press, St. Paul, MN.

Harborne, J.B. (1982). "Introduction to Ecological Biochemistry", 2nd edn. Academic Press, London.

Harmon, M.E., Franklin, J.F., Swanson, F.J., Sollins, P., Gregory, S.V., Lattin, J.D., Anderson, N.H., Cline, S.P., Aumen, N.G., Sedell, J.R., Lienkaemper, G.W., Cromack, K., Jr. and Cummins, K.W. (1986). Ecology of coarse woody debris in temperate ecosystems. *Adv. Ecol. Res.* **15**, 133–302.

Haukioja, E., Niemela, P. and Siren, S. (1985). Foliage phenols and nitrogen in relation to growth, insect damage, and ability to recover after defoliation in the mountain birch, *Betula pubescens* ssp. *tortuosa*. *Oecologia* **65**, 214–222.

Hunter, A.F. and Aarssen, L.W. (1988). Plants helping plants. *BioScience* **38**, 34–40.

Kareiva, P. (1983). Influence of vegetation texture on herbivore populations: resource concentration and herbivore movement. *In* "Variable Plants and Herbivores in Natural and Managed Systems" (R.F. Denno and M.S. McClure, eds), pp. 259–289. Academic Press, New York.

Leuschner, W.A. and Berck, P. (1985). Impacts on forest uses and values. *In* "Integrated Pest Management in Pine-Bark Beetle Ecosystems" (W.E. Waters, R.W. Stark and D.L. Wood, eds), pp. 105–120. John Wiley & Sons, New York.

Lorio, P.L., Jr., Sommers, R.A., Blanche, C.A., Hodges, J.D. and Nebeker, T.E. (1990). Modeling pine resistance to bark beetles based on growth and differentiation balance principles. *In* "Process Modelling of Forest Growth Responses to Environmental Stress" (R.K. Dixon, R.S. Meldahl, G.A. Ruark and W.G. Warren, eds), pp. 402–409. Timber Press, Portland, OR.

Mattson, W.J. and Addy, N.D. (1975). Phytophagous insects as regulators of forest primary production. *Science* **190**, 515–522.

Mattson, W.J. and Haack, R.A. (1987). The role of drought in outbreaks of plant-eating insects. *BioScience* **37**, 110–118.

Miller, W.E. (1979). Fire as an insect management tool. *Bull. Entomol. Soc. Am.* **25(2)**, 137–140.

Perry, D.A. (1988). Landscape pattern and forest pests. *Northw. Environ. J.* **4**, 213–228.

Perry, D.A., Thomas, B., Meurisse, R., Miller, R., Boyle, J., Sollins, P. and Means, J., eds (1989). "Maintaining Long-term Productivity of Pacific Northwest Forest Ecosystems." Timber Press, Portland, OR.

Rausher, M.D. (1983). Ecology of host-selection behavior in phytophagous insects. *In* "Variable Plants and Herbivores in Natural and Managed Ecosystems" (R.F. Denno and M.S. McClure, eds), pp. 223–257. Academic Press, New York.

Rosenthal, G.A. and Janzen, D., eds (1979). "Herbivores: Their Interaction with Secondary Plant Metabolites." Academic Press, New York.

Schowalter, T.D. (1985). Adaptations of insects to disturbance. *In* "The Ecology of Natural Disturbances and Patch Dynamics" (S.T.A. Pickett and P.S. White, eds), pp. 235–252. Academic Press, New York.

Schowalter, T.D. and Turchin, P. (1993). Southern pine beetle infestation development: interaction between pine and hardwood basal areas. *Forest Sci.* (in press).

Schowalter, T.D., Coulson, R.N. and Crossley, D.A., Jr. (1981a). Role of southern pine beetle and fire in maintenance of structure and function of the southeastern coniferous forest. *Environ. Entomol.* **10**, 821–825.

Schowalter, T.D., Pope, D.N., Coulson, R.N. and Fargo, W.S. (1981b). Patterns of southern pine beetle (*Dendroctonus frontalis*) infestation enlargement. *Forest Sci.* **27**, 837–849.

Schowalter, T.D., Hargrove, W.W. and D.A. Crossley, Jr. (1986). Herbivory in forested ecosystems. *Annu. Rev. Entomol.* **31**, 177–196.

Schowalter, T.D., Sabin, T.E., Stafford, S.G. and Sexton, J.M. (1991). Phytophage effects on primary production, nutrient turnover, and litter decomposition of young Douglas-fir in western Oregon. *For. Ecol. Manage.* **42**, 229–243.

Schowalter, T.D., Caldwell, B.A., Carpenter, S.E., Griffiths, R.P., Harmon, M.E., Ingham, E.R., Kelsey, R.G., Lattin, J.D. and Moldenke, A.R. (1992). Decomposition of fallen trees: effects of initial conditions and heterotroph colonization rates. *In* "Tropical Ecosystems: Ecology and Management" (K.P. Singh and J.S. Singh, eds), pp. 373–383. Wiley Eastern Ltd., New Delhi.

Seastedt, T.R. (1984). The role of microarthropods in decomposition and mineralization processes. *Annu. Rev. Entomol.* **29**, 25–46.

Smith, W.H. (1981). "Air Pollution and Forests." Springer-Verlag, New York, 379 pp.

Stark, R.W. and Waters, W.E. (1985). Concept and structure of a forest pest management system. *In* "Integrated Pest Management in Pine-Bark Beetle Ecosystems" (W.E. Waters, R.W. Stark and D.L. Wood, eds), pp. 49–60. John Wiley & Sons, New York.

Swank, W.T. and Crossley, D.A., Jr., eds (1987). "Forest Hydrology and Ecology at Coweeta." Springer-Verlag, New York.

Temple, S.A. (1977). Plant–animal mutualism: coevolution with dodo leads to near extinction of plant. *Science* **197**, 885–886.

Tilman, D. (1982). "Resource Competition and Community Structure." Princeton University Press, Princeton, NJ.

Tilman, D. (1988) "Plant Strategies and the Dynamics and Structure of Plant Communities." Princeton University Press, Princeton, NJ.

Tuomi, J., Niemela, P., Haukioja, E., Siren, S. and Neuvonen, S. (1984). Nutrient stress: an explanation for plant anti-herbivore responses to defoliation. *Oecologia* **61**, 208–210.

Visser, J.H. (1986). Host odor perception in phytophagous insects. *Annu. Rev. Entomol.* **31**, 121–144.

Wade, D.D. and Ward, D.E. (1976). Prescribed use of fire in the South – a means of conserving energy. *Proc. Tall Timbers Fire Ecol. Conf.* **14**, 549–558.

Waring, R.H. and Schlesinger, W.H. (1985). "Forest Ecosystems: Concepts and Management." Academic Press, New York.

Waters, W.E. and Stark, R.W. (1980). Forest pest management: concept and reality. *Annu. Rev. Entomol.* **25**, 479–509.

Waters, W.E., Stark, R.W. and Wood, D.L., eds (1985). "Integrated Pest Management in Pine–Bark Beetle Ecosystems." John Wiley & Sons, New York, 256 pp.

Whitney, H.S. (1982). Relationship between bark beetles and symbiotic organisms. *In* "Integrated Pest Management in Pine–Bark Beetle Ecosystems" (W.E. Waters, R.W. Stark and D.L. Wood, eds), pp. 183–211. John Wiley & Sons, New York.

Whitney, H.S., Bandoni, R.J. and Oberwinkler, F. (1987). *Entomocorticium dendroctoni* gen. et sp. nov. (Basidiomycotina), a possible nutritional symbiote of the mountain pine beetle in lodgepole pine in British Columbia. *Can. J. Bot.* **65**, 95–102.

Wickman, B.E. (1980). Increased growth of white fir after a Douglas-fir tussock moth outbreak. *J. Forestry* **78**, 31–33.

Witcosky, J.J. Schowalter, T.D. and Hansen, E.M. (1986). The influence of time of precommercial thinning on the colonization of Douglas-fir by three species of root-colonizing insects. *Can. J. For. Res.* **16**, 745–749.

Witcosky, J.J., Schowalter, T.D. and Hansen, E.M. (1987). Host-derived attractants for the beetles, *Hylastes nigrinus* (Coleoptera: Scolytidae) and *Steremnius carinatus* (Coleoptera: Curculionidae). *Environ. Entomol.* **16**, 1310–1313.

Wood, D.L. (1982). The role of pheromones, kairomones and allomones in host selection and colonization behavior of bark beetles. *Annu. Rev. Entomol.* **27**, 411–446.

Wright, H.E., Jr. and Heinselman, M.L. (1973). The ecological role of fire in natural conifer forests in western and northern North America. *Quart. Res.* **3**, 319–328.

Zhong, H. and Schowalter, T.D. (1989). Conifer bole utilization by wood boring beetles in western Oregon. *Can. J. For. Res.* **19**, 943–947.

PART II
Systematics

–2–
Systematics of Bark Beetles

DONALD E. BRIGHT
Biosystematics Research Institute, Canada Agriculture, Ottawa, Ontario, Canada

2.1 INTRODUCTION

The entire field of bark-beetle–fungus relationships is a relatively new and rapidly advancing field of study. A close relationship between insects and fungi was recognized early, and a true symbiotic relationship demonstrated (Baumberger, 1919). Different components of the bark beetle–fungus relationship have been reviewed by Baker (1963), Francke-Grossmann (1966, 1967), Graham (1967), Morris (1979), Whitney (1982) and Wilding *et al.* (1989). Barras and Perry (1975) presented an annotated bibliography of the literature treating this subject for the period 1965–1974. In this bibliography, they recorded 244 references. A similar bibliography covering the period from 1975 to the present would undoubtedly increase the number of references by a factor of two or more.

The causes of this accelerated research are several, among which is the increasing realization that fungal associates are important factors in predisposing trees to bark beetle attack, in causing tree death, or in making the microhabitat more amenable to bark beetle survival. Much of the early literature focused on the fungus as a secondary effect of bark beetle tunneling in the cambium or as food for the larvae. Recent literature has tended to focus more on the role of the fungus as one more player in the complex system of bark beetle attack and the successful colonization of the host tree.

In all biological studies, systematics provides the basic framework for evaluating and integrating biological information. Systematics also provides the system for transmission of this biological information. This chapter addresses the present status and future trends of the systematics of bark beetles belonging to the family Scolytidae (Fig. 2.1). These beetles excavate characteristic mines in the phloem of conifers during oviposition and larval development (Fig. 2.2) The systematics of this family remains an important area of study, especially as we re-evaluate bark beetle phylogeny in view of the relationships with fungi.

BEETLE–PATHOGEN INTERACTIONS IN CONIFER FORESTS
ISBN 0-12-628970-0

Fig. 2.1. A representative scolytid, *Dendroctonus pseudotsugae*. (Photo by T.D. Schowalter.)

2.2 STATUS OF BARK BEETLE SYSTEMATICS

The museum taxonomist who works on insects, particularly those perceived to be a "highly visible" group, does not often have time to reflect on future directions or the relationship between systematics and field biology. Much time is spent identifying and curating specimens and developing taxonomic aids. It is necessary to pause periodically and reflect on the status of the taxonomy of a particular group and what directions it might take in the future. In this section, I shall consider the future of bark beetle systematics based on where we are now and the possibilities for future direction.

2.2.1 Present state

Numerous authors have noted in recent years that taxonomy, as a scientific discipline, is losing support, despite its importance to field biology and pest management. Upon retiring, taxonomists on university staffs often are replaced by scientists in more glamorous fields. Governments, in the face of budget deficits and declining resources, often focus more directly on mission-oriented or short-term problem-solving research, at the expense of basic research. Thus, the pool of taxonomists available to provide research and identification is shrinking rapidly.

Opportunities for systematic work on bark beetles remain, in spite of the tremendous growth in taxonomic knowledge over the past few decades. Interest in bark beetles has increased dramatically because of recognition of their ecological and economic importance. Several devastating outbreaks during the past few decades have indicated an immense need for accelerated research in bark beetle biology. When *Dendroctonus frontalis* began to cause serious damage in the southeastern United States in the early 1970s, there was a great demand for new and improved ways to deal with this problem. This resulted in approval and funding

Fig. 2.2. Representative galleries of bark beetles; left to right, vertical parent gallery and radiating larval mines of *Dendroctonus pseudotsugae* in *Pseudotsuga menziesii*, winding parent gallery, short larval mines and pupal cells of *Dendroctonus frontalis* in *Pinus taeda*, parent galleries extending from a central nuptial (mating) chamber, and perpendicular larval mines of *Ips pini* in *Pinus ponderosa*, and transverse parent galleries extending in opposite directions from a central nuptial chamber, and perpendicular larval mines of *Scolytus ventralis* in *Abies grandis*. The grain of the phloem runs vertically in all cases. (Photo by T.D. Schowalter.)

of the Expanded Southern Pine Beetle Research and Application Program in 1974 (Waters *et al.*, 1985). Almost $12 million was made available over the 6-year program. The result was a massive increase in literature dealing with almost all aspects of *D. frontalis* bionomics, including papers dealing with the systematic status of various populations of the beetle (Vité *et al.*, 1974; Anderson *et al.*, 1979; Lanier *et al.*, 1988). How much of this money made its way to taxonomists is not known, but the interest developed in *D. frontalis* and other bark beetles undoubtedly had a positive effect when funding for recent taxonomic projects, such as the world catalog (Wood and Bright, 1987, and in preparation), was being considered.

For North America (and probably Europe), the basic alpha-level taxonomy of the Scolytidae is essentially complete. New species are likely to be sibling species in currently recognized species complexes, that are indistinguishable using traditional morphological techniques, or species from atypical hosts. The work that remains is the refining of basic classification, clarification of evolutionary and ecological relationships, application of cladistics theory, and use of new tools such as molecular systematics. For the remainder of the world, basic taxonomic knowledge is inadequate at almost all levels, as discussed below.

2.2.1.1 Geographic and faunal studies

Faunal studies, addressing distributions and biological relationships among related species and populations, have been almost exclusively the activity of systematists or informed amateurs or other professional entomologists with an interest or special knowledge of a particular geographic area or taxonomic group. This will remain the trend well into the future. Regional works and faunal studies are the main provider of identification keys and species descriptions. Notwithstanding the allure and enthusiasm involved in using the newer taxonomic techniques, such as cladistics, molecular genetics or morphometrics, good regional taxonomic revisions of bark beetle genera or tribes will be required well into the 21st century. Of the 215 valid genera of Scolytidae currently recognized world-wide (Wood, 1986), slightly more than half have been monographed adequately either for a limited area or for the entire fauna during the past 25 years. Of these, 96 genera are included in one comprehensive monograph (Wood, 1982).

There are few regional monographs at present, reflecting the scarcity of bark beetle systematists. Only North and Central America (Wood, 1982), the United Kingdom (Duffy, 1953), Denmark (Hansen, 1956), Middle Europe (Schedl, 1981), Czechoslovakia (Pfeffer, 1989), Europe (Grune, 1979), China (Yin *et al.*, 1984), and Korea (Choo, 1983) have been adequately addressed. The Japanese fauna is well known, but no recent comprehensive monograph is available. No comprehensive work has been published on the fauna of the region formerly encompassed by the USSR since Stark's (1952), which is now out of date. The faunas of Africa, Southeast Asia, South America, India, Australia and New Guinea are virtually unknown. What is available from these regions is restricted to scattered species descriptions, lists etc., and the occasional generic treatment.

The last, and presently only, world catalog for the Scolytidae was published by Hagedorn (1910) and contains references for about 2000 species. The world fauna is now estimated to be in excess of 6000 species, and the Hagedorn catalog is clearly out of date. The problem of a world catalog will be resolved by the soon-to-be-published catalog of the Scolytidae and Platypodidae (Wood and Bright, in press). This publication will contain a listing of all 6000+ species of scolytids and platypodids in the world, along with basic taxonomic information on types, type localities, synonyms, distribution, host plants and references to all aspects of each species' biology, management, etc. This publication should do much to update scolytid taxonomy, but will not eliminate the need for future study.

2.2.1.2 Research collections

It is obvious that progress in taxonomy and ecology of bark beetles and fungi will require an abundance of carefully collected, well-preserved, and well-labeled specimens, especially specimens that include host data. Taxonomists must continue to work with museum specimens, at least for the foreseeable future, although biological data, cytological experiments, and other techniques using live material, will add much to museum-based studies.

No single museum currently has an adequate collection of Scolytidae, but useful work can be pursued when specimens from various collections are combined by loan or exchange. Too often, the growth of a collection depends on the enthusiasm and energy of a devoted individual. When that person retires or departs, the collection invariably suffers, and growth slows dramatically. Between the 1930s and 1950s, the collection of Scolytidae at the United States National Museum of Natural History underwent rapid expansion due to the work of M.W. Blackman and W.H. Anderson. After their retirements, growth of the collection almost ceased. In similar fashion, the Canadian National Collection in Ottawa grew from 75 drawers of Scolytidae in 1966 to its present crowded condition in 270 drawers.

Major holders of Scolytidae collections are few. In North America, the US National Museum of Natural History, the Canadian National Collection and the private collection of

S.L. Wood, all with world-wide coverage, are especially notable. The Natural History Museum, London (formerly the British Museum of Natural History) also has a world-wide collection and holds many primary types, but growth is relatively slow. The collections of W.F.H. Blandford (mainly Central America) and F.G. Browne (Central Africa and Southeast Asia) are both in the Natural History Museum, London. The Natural History Museum in Vienna, Austria contains the extensive collection of K.E. Schedl. This collection is valuable for its type material but is of limited research use because almost all species in the collection are represented by few specimens. The remainder of the world's collection of Scolytidae, many of which contain valuable type material, regional collections or specimens with historical value, cannot serve alone as sources of research material, although each is valuable when combined with other collections.

Voucher specimens representing populations identified in ecological or biological studies are an important source of material for systematic study and provide for verification of published identifications. Voucher specimens provide a permanent standard of reference for determining the precise identity of species on which published information is based. All researchers working on the biology of scolytids, or any other group, should be encouraged to deposit voucher specimens documenting their research populations in a major museum. This practice is not widely used at present. Francoeur (1982) stated that many of the published data in entomological journals is reduced in value or rendered useless because of the evolving taxonomic base on which identification of study species is made. Robinson (1975) stated that biologists are responsible for using valid scientific names and documenting their research and identifications with voucher specimens. He further notes that researchers are responsible for instilling the concepts of good taxonomy and use of voucher specimens in their students. Voucher specimens become increasingly important as studies of bark beetle associations with fungal complexes forces us to re-examine bark beetle systematics.

2.2.2 Future state

2.2.2.1 Systematic services

Although many universities offer graduate instruction in systematics, the trend is away from natural history and systematics studies and toward the high-tech sciences, such as molecular biology. Although tremendous advances in biology have been made in these emerging fields, all branches of biology depend on systematics to identify insects for biodiversity, biogeographic and pest management studies, and to pursue generic revisions and other activities that indicate relationships among species and underlie the choice of species for biological study. In Canada, only three universities train the majority of insect systematics in that country, and all three have lost, or soon will lose, the taxonomist responsible for the systematics program. Two of these universities did not replace the retirees with another taxonomist. As of March 1991, the third had not made a decision.

Few, if any, graduate students currently are pursuing study of scolytid systematics at any institution. Those who have graduated with an interest in scolytid systematics have been forced to work in a different area of research after failing to find a permanent position as a systematist. Yet the US and Canadian governments and various forest industries continue to commit large sums of money annually to research projects designed to improve pest management systems for bark beetles because of the substantial losses of forest products to these insects. The challenge for the future is to train systematists with an interest in the Scolytidae and to provide them with resources to pursue systematics research.

2.2.2.2 Future approaches to bark beetle systematics

In the past, the taxonomy of bark beetles has relied almost exclusively on the study of external morphological features. This method has served well and will undoubtedly continue to be the main source of information for taxonomic discrimination well into the future. However, taxonomists now have at their disposal an exciting array of non-morphological techniques, each with its own individual capabilities, but all with a fascinating potential for future advances. These non-morphological techniques include morphometrics (actually morphological but considered here as an extension beyond the usual sense of morphology), cytogenetic studies and molecular studies (including isozyme analysis, cuticular hydrocarbons, DNA sequencing, etc.). These methods can be reviewed here only briefly, but I anticipate that their potential will be recognized and their use encouraged.

Morphometrics involves the measurement of numerous morphological features and the development of non-variable ratios that can be used to characterize species or populations. Only two morphometric studies have involved a group of bark beetles. Sturgeon and Mitton (1986) measured and analyzed 15 characters on *Dendroctonus ponderosae* from five populations on three different hosts (*Pinus ponderosa, P. contorta*, and *P. flexilis*) in Colorado. Beetles from *P. flexilis* were more similar to those from *P. ponderosa* than to those from *P. contorta*. Sexual dimorphism also was demonstrated. Cane *et al.*, (1990) studied the *Ips grandicollis* group, a set of seven very similar species. They measured 23 morphological characters, but were unsuccessful in locating loci or macromorphological measurements that were useful for absolute discrimination among the seven species. They reported that variation in individual size and possible allometry clearly confounded interspecific morphological comparison and resisted statistically acceptable transformations.

In a study with implications for future studies of bark beetle systematics, Foottit and Mackauer (1980) used multivariate morphometric analysis to compare 18 populations of *Adelges piceae* from all known infestation areas in North America. They separated the various populations into three distinct groups that reflected the adelgid's colonization history in North America. Identification functions based on the morphometric analysis of adult characters enabled 85% of the specimens to be assigned correctly to one of the groups. These groups later were recognized as subspecies (Foottit and Mackauer, 1983). This example could serve as a model for future morphometric studies of bark beetles.

The pioneering studies of bark beetle cytogenetics by Lanier have shown that this tool for systematics research holds much promise as a means of species discrimination. In a series of papers from 1966 to 1991 (see G.N. Lanier in Wood and Bright, 1987), Lanier used breeding experiments, cross matings and karyology, combined with pheromone responses and general biology, to assess the taxonomic validity of species in the genera *Ips* and *Dendroctonus*. New synonymy was proposed, three undescribed species of *Ips* (*apache, paraconfusus* and *hoppingi*) were discovered, previously established synonymy was verified, and the systematics of these two genera was placed on a more secure foundation.

The value of cytogenetic studies can be demonstrated by two examples. In a generic revision of *Dendroctonus*, Wood (1963) placed *D. jeffreyi*, known only from *Pinus jeffreyi*, in synonymy with *D. ponderosae*, known from many *Pinus* species in western North America, based on their morphological similarity, geographic sympatry, similar host plants, and similar biology. Thomas (1965) concurred with the synonymy based on his studies of larval and pupal morphology. However, doubts about the synonymy continued to be expressed. Differences in attack patterns (Eaton, 1956), in sizes of beetles emerging from different pine species (Hopkins, 1909; Wood, 1963), and in responses to resin vapors from their own and each other's host tree (Smith, 1965) indicated that two sibling species actually might be represented. Lanier and Wood (1968) demonstrated that specimens from *P. jeffreyi* (*D. jeffreyi*) were completely unable to produce fertile offspring when mated with specimens from other pines (*D. ponderosae*). This incompatibility, along with differences in survival of laboratory broods, variation in egg incubation time, differences in karyotypes, and previously unrecognized morphological differences confirmed the separate species status of *D. ponderosae* and *D. jeffreyi*.

Conophthorus banksianae was described by McPherson *et al.* (1970) when they recognized that beetles feeding in twigs seemingly displayed different biological characteristics from the typical cone-feeding *C. resinosae*. The two species are impossible to distinguish morphologically and can be considered sibling species. Wood (1989) placed *C. banksianae* in synonymy with *C. resinosae* based on the absence of different morphological characters. However, Mattson (1989) maintained that the two species are distinct because of their unique host acceptance behavior. De Groot and Ennis (1990) investigated the cytotaxonomy of these species and showed that the karyotypic formulae did not support the designation of *C. banksianae* as a distinct species.

Similar research needs to be conducted with other genera of Scolytidae. Our understanding and management of bark beetles in the genus *Scolytus*, species of which cause serious mortality in *Abies* and related tree species and act as vectors of several phytopathogenic fungi, *Hylurgops* and *Hylastes*, also vectors of various root pathogens, and a number of other genera with troublesome systematics would benefit from this research.

Molecular techniques provide a number of opportunities for systematic research. One of the most exciting and relatively unexplored areas of scolytid taxonomy is cuticular hydrocarbon analysis. Museum taxonomists have hoped for a technique by which museum specimens can be analyzed for their biochemical properties without destroying the specimens. Presently, cuticular hydrocarbon analysis comes closest to fulfilling this need. The advantage of this technique is that specimens can be returned to the collection undamaged after their cuticular waxes have been removed and analyzed. Page *et al.* (1990a) extracted the cuticular waxes from pinned museum specimens of *D. ponderosae* collected in 1914; all specimens were structurally unaltered when returned to their museum drawers. The cuticular waxes of these specimens evidently were identical to those of newly collected specimens.

Only two studies of bark beetle cuticular hydrocarbons have been published (Page *et al.*, 1990a,b). Page *et al.* (1990b) analyzed the cuticular hydrocarbons of eight species of *Conophthorus*. The results were preliminary but provided strong evidence for the continued synonymy of *C. banksianae* with *C. resinosae* and the possible existence of a sibling species in the *C. ponderosae* complex. These studies demonstrate the value of cuticular hydrocarbon analysis in elucidating the evolution and systematics of bark beetles.

Enzyme electrophoresis is another area of molecular systematics that has received attention. Higby and Stock (1982) compared enzymes of *D. ponderosae* and *D. jeffreyi* and confirmed Lanier and Wood's (1968) recognition of two species. Stock *et al.* (1979) assessed the genetic differences between two populations of *D. pseudotsugae* using isozyme analysis and showed that the two populations (coastal and inland) were clearly in the process of speciation. Of the 13 gene loci compared in the Idaho (inland) and Oregon (coastal) populations, eight were virtually identical, four were different, and one showed an intermediate amount of differentiation. This degree of genetic difference is characteristic of populations differentiated to or beyond the race or subspecies level. Populations of *D. frontalis* likewise were analyzed by Anderson *et al.* (1979). They showed statistically significant differences in gene frequency among the five populations tested. *Dendroctonus frontalis* populations in Mexico and

Arizona are genetically distinct from each other and from the southeastern US population. Anderson *et al.* (1979) concluded that populations in Mexico and Arizona had diverged from the main body of the species, providing information on the speciation process in this species.

A number of similar studies have been conducted recently on other species or genera of Scolytidae. Ritzengruber and Fuhrer (1986) showed that *Pityogenes chalcographus* populations from six different regions of northern and central Europe each showed characteristic electrophoretic patterns. Stock *et al.* (1987) compared the isozymes of the European *D. micans* with 10 North American species in the genus and showed distinct differences. However, the closely related (and possibly synonymous) North American species, *D. punctatus,* was not available for comparison. Cane *et al.* (1990) used electrophoretic data in their study of the *Ips grandicollis* complex and concluded that the species *I. lecontei*, previously included in the complex, was not related to the remainder of the included species and should be removed from the group, and that the questionable species *I. hoppingi* was a distinct species.

The potential value of recombinant DNA analysis for bark beetle systematics has not attracted attention as yet. The techniques have been used in aphid systematic studies (Foottit, 1990). The results, to date, have been encouraging, but the full potential cannot be assessed. This technique holds promise for helping to elucidate the systematics and phylogeny of the Scolytdiae.

2.3 PRESENT CLASSIFICATION OF THE SCOLYTIDAE

Bright and Stock (1982) reviewed the history of scolytid taxonomic investigation. The contributions of early researchers such as Hopkins, Swaine, Eggers, Eichhoff, Blackman, Schedl, and others, were outlined. We concluded our review with comments concerning the then-unpublished monograph of the North and Central American Scolytidae (Wood, 1982). Following that monograph, Wood (1986) presented a reclassification of the genera of Scolytidae for the world. In that paper, Wood divided the family into two subfamilies, 25 tribes and 215 genera. Wood's classification has done much to stabilize the classification of the Scolytidae and, although refinements are certain, the basic structure of his classification likely will be followed well into the next century.

2.3.1 Phylogeny

The taxonomic position of the Scolytidae and Platypodidae within the Coleoptera has been the subject of considerable debate. The two groups usually are considered to be two distinct families within the superfamily Curculionoidea (Morimoto, 1962; Wood, 1973, 1982; Bright, in press) or as subfamilies within the Curculionidae, the family including weevils and related groups (Crowson, 1967, 1981; Lawrence, 1982; Lawrence and Newton, 1982). Wood (1973) presented his argument for distinct families based mainly on morphological evidence. Kuschel (1966) pointed out the similarity in habits of the Araucariini (Curculionidae: Cossoninae).

Most authorities adhering to the subfamily concept of the Scolytidae have used the similarity in body structure and biology of the Cossoninae and Scolytidae as support for their position. In these interpretations, the Scolytidae evolved somewhere in the region of the Cossoninae or other specialized Curculionidae and are thus a relatively recent evolutionary product. Wood (1973) contends (and I agree) that the Scolytidae (and Platypodidae) arose as a monophyletic unit early in the evolution of the Curculionoidea before some of the major structural and biological features characterizing the Curculionidae developed.

Crowson (1981) states that the first unquestioned scolytid fossil dates from the Baltic amber (Oligocene, 40 million years BP). However, Wood (1982) argues that engravings on Cretaceous coniferous bark from Europe reported by Brongniart (1877) almost certainly were made by phloeophagous Scolytidae. Blair (1943) also described characteristic subcortical insect borings in Early Cretaceous coniferous wood. These borings are described as very similar to those made by some modern Scolytidae, consisting of a series of expanding "larval" mines radiating out from a more or less straight "parent" gallery. Crowson (1981) contends that it is unlikely that true scolytids were present that early. Other previous reports of Eocene fossils are of doubtful authenticity. The Baltic amber scolytids all represent still living tribes, and many extant genera in both currently recognized subfamilies, although the more primitive Hylesininae are more abundantly represented.

2.3.2 Co-evolution with fungi

Primitive scolytids probably arose in or near the Cretaceous Era from anthribid-like ancestors living in close proximity to various fungi in decaying or dead plant tissue. Competition for resources in this environment would favor adaptations that ensure rapid detection and utilization of dead wood. Ethanol, produced in stressed trees and derived from fermentation of living tissue by bacteria and some yeasts (Kramer and Kozlowski, 1979), is a primary attractant for some bark beetles (Border, 1982; Berryman, 1989). The ability of early scolytids to orient toward sources of this volatile substance would facilitate the discovery and utilization of decomposing host resources. In this habitat, beetles would interact with various microorganisms, including phytopathogenic fungi, that would be associated with dead or dying trees.

Co-evolution of the scolytids and fungi in this environment would be an advantage for both (Whitney, 1982; Dowding, 1984). The beetles would carry with them an agent that would assist in breaking down plant tissue and facilitate efficient utilization of the host. The fungi would benefit by having a means of dispersal directly into susceptible plant material. At this point, one can visualize the development of spore-carrying modifications, mycangia, on the beetle and the utilization of alcohols produced by fermentation or by transformation of defensive chemicals of living trees, e.g. the synthesis by fungi of verbenol from tree-produced alpha-pinene. Elaboration of this scheme would lead to closer ties between beetles and fungi and eventual development of mycangia. At the present time, probably all Scolytidae have some degree of dependence on fungi, but relatively few have developed mycangia, or perhaps mycangia have not yet been discovered for many species. Most species evidently depend on passive transmission of the pathogenic fungi, such as those transmitting *Ophiostoma ulmi* (Webber and Gibbs, 1989) or *Leptographium wageneri* (Witcosky *et al.*, 1986).

Table 2.1 shows the association of mycangia in the various tribes of the Scolytidae. These data are derived from Beaver (1989) with additional personal observations. No clear phylogenetic sequence can be detected in the type or presence of mycangia. However, no mycangia have been reported in the primitive tribe Hylastini. Spores are carried in punctures on the body or lodged in other surface irregularities. However, in the closely related primitive tribe Hylesinini, fully developed mycangial pits are found on the proepisternum of females of *Phloeoborus* and *Bothrosternus*. These pits are lined with setae and contain a waxy substance that evidently sustains fungal spores during transport.

Table 2.1 Present higher classification of Scolytidae (from Wood, 1986) with type of mycangia found (from Beaver, 1989)

Subfamily Hylesininae

Tribe Hylastini (3 genera)
  No mycangia recorded. Spores carried on body surface
Tribe Hylesinini (12 genera)
  Phloeoborus — pit on proepisternum (♀) (all spp.)
  Dactylipalpus — transverse crevice on pronotum (♀) (all spp.)
Tribe Tomicini (14 genera)
  Dendroctonus — Various locations: maxillary pouches (♀, ♂) (*ponderosae*); below anterior margin of pronotum (♀) (several species); lateral callus on pronotum (♀) (*brevicomis, frontalis* and related species); externally on body (♀, ♂) (*pseudotsugae*)
  Tomicus — External on body and in sutural groove on elytra and on metanotum (♀, ♂) (*minor*)
Tribe Phrixosomini (1 genus)
  No mycangia recorded
Tribe Hyorrhynchini (3 genera)
  No mycangia recorded
Tribe Diamerini (7 genera)
  No mycangia recorded
Tribe Bothrosternini (5 genera)
  Eupagiocerus — Pubescent patches on proepisternum (♀) (spp.?) (personal observation)
Tribe Phloeotribini (2 genera)
  No mycangia recorded
Tribe Phloeosinini (12 genera)
  No mycangia recorded
Tribe Hypoborini (8 genera)
  No mycangia recorded
Tribe Polygraphini (8 genera)
  No mycangia recorded

Subfamily Scolytinae

Tribe Scolytini (4 genera)
  Scolytus — In punctures on vertex of head (♀, ♂) (*ventralis*); in elytral punctures (♀, ♂) (*scolytus*)
Tribe Ctenophorini (4 genera)
  No mycangia recorded
Tribe Scolytoplatypodini (1 genus)
  Scolytoplatypus — Median pit on pronotum (♀, ♂) (many spp.)
Tribe Micracini (13 genera)
  No mycangia recorded
Tribe Cactopinini (1 genus)
  No mycangia recorded)
Tribe Carphodicticini (3 genera)
  No mycangia recorded
Tribe Ipini (6 genera)
  Ips — Paired mandibular pouches (♀) (*acuminatus*); pronotal punctures (♀, ♂) (*sexdentatus*)
  Pityogenes — Pit(s) on frons (♀) (most spp.) (personal observation)
Tribe Dryocoetini (18 genera)
  Dryocoetes — Paired mandibular pouches (♀, ♂) (*confusus*)
Tribe Crypturgini (6 genera)
  No mycangia recorded
Tribe Xyloterini (3 genera)
  Trypodendron — Pit on proepimeron (♀) (all spp. (?))
  Xyloterinus — Pit on proepimeron (♀) and paired mandibular pouches (♀, ♂) (*politus*)
  Indocryphalus — Pit on proepimeron (♀) (*intermedius*)

Table 2.1 continued

Tribe Xyleborini (24 genera)
Xyleborus — Paired mandibular pouches (♀) (many spp.); intersegmental pouches between pro- and mesonotum (♀) (*dispar*)
Xyleborinus — Pouches in elytral bases (♀) (many spp.)
Euwallacea — Paired mandibular pouches (♀) (*fornicatus*)
Cryptoxyleborus — Pouches in elytral bases (♀) (*naevus* and related spp.)
Xylosandrus — Intersegmental pouch between pro- and mesonotum (♀)(many species)
Eccoptopterus — Spiral pit in mesonotum (♀) (all spp.)
Tribe Xyloctonini (5 genera)
No mycangium reported
Tribe Cryphalini (24 genera)
Hypothenemus — Shallow lateral depressions on pronotum (♀) (*curtipennis*)
Tribe Corthylini (29 genera)
Pityoborus — Pubescent lateral patches on pronotum (♀) (all spp.)
Gnathotrichus — Enlarged procoxal cavities (♀) (most spp.); internal cavities on pronotal base (*primus*) (personal observation)
Monarthrum — Enlarged procoxal cavities (♀) (many spp.); mentum-pregular pouch (♀) (*bicallosum*)
Corthylus — Paired tubes opening into precoxal cavity (♀) (several spp.)
Microcorthylus — Paired tubes opening into precoxal cavity (♀) (*castaneus*)

It is interesting to note that very few true Curculionidae have developed a mutualistic association with fungi. Consequently, none have developed mycangia or similar modifications for transporting fungi. Evidently, primitive Curculionidae did not require fungi to aid in host discovery and utilization and followed a different evolutionary pathway. The Anthribidae, which resemble Scolytidae in many basic morphological features and have a mutualistic association with fungi, are probably in the ancestral line of primitive scolytids. However, any resemblance of Scolytidae to the specialized Curculionidae is likely a recent phenomenon and due, no doubt, to convergent evolution.

2.4 CONCLUSIONS

This brief review of bark beetle systematics has examined the present status of systematics of the Scolytidae and has anticipated some directions for the future. Clearly, abundant opportunities remain for meaningful taxonomic research in this fascinating group of beetles. While some groups of scolytids, such as *Dendroctonus* and *Ips*, are subjects of state-of-the-art taxonomic investigation, others are still in the era of species description and alpha-level taxonomy.

The ecological and economic importance of this group of beetles has supported extensive research on biology and management and shown the importance of systematics in clarifying patterns of host use and responses to pest management practices. Further advances in bark beetle systematics will be necessary to explain how problems of incipient speciation or sibling species may be confounding our understanding of ecological relationships among the beetles, their hosts and associated fungi, and thereby limiting our ability to manage bark beetles and their fungal associates effectively.

REFERENCES

Anderson, W.W., Berisford, C.W. and Kimmich, R.H. (1979) Genetic differences among five populations of the southern pine beetle. *Ann. Entomol. Am.* **72**, 323–327.

Baker, J.M. (1963). Ambrosia beetles and their fungi, with particular reference to *Platypus cylindrus* Fab. *Symp. Soc. Gen. Microbiol.* **13**, 232–265.

Barras, S.J. and Perry, T. (1975). Interrelationships among microorganisms, bark or ambrosia beetles, and woody host tissue: an annotated bibliography, 1965–1974. USDA Forest Serv. Gen. Tech. Rpt. SO-10, USDA Forest Serv., Southern Forest Exp. Stn., New Orleans, LA, 34 pp.

Baumberger, J.P. (1919). A nutritional study of insects with special reference to microorganisms and their substrata. *J. Exp. Zool.* **28**, 1–81.

Beaver, R.A. (1989). Insect–fungus relationships in the bark and ambrosia beetles. *In* "Insect–fungus Interactions" (N. Wilding, N.M. Collins, P.M. Hammond and J.F. Webber, eds), pp. 121–143. Academic Press, New York.

Berryman, A.A. (1989). Adaptive pathways in scolytid–fungus associations. *In* "Insect–fungus Interactions" (N. Wilding, N.M. Collins, P.M. Hammond and J.F. Webber, eds), pp. 145–159. Academic Press, New York.

Blair, K.G. (1943). Scolytidae from the Wealden Formation. *Entomol. Mon. Mag.* **62**, 59–60.

Borden, J.H. (1982). Aggregating pheromones. *In* "Bark Beetles in North American Conifers" (J.B. Mitton and K.B. Sturgeon, eds), pp. 74–139. University of Texas Press, Austin, TX.

Bright, D.E. (in press). Review of the weevils of Canada and Alaska. I. (Coleoptera: Curculionoidea, except Scolytidae and Curculionidae). Canada Dept. Agric., Ottawa.

Bright, D.E. and Stock, M.W. (1982). Taxonomy and geographic variation. In "Bark Beetles in North American Conifers" (J.B. Mitton and K.B. Sturgeon, eds), pp. 46–73. University of Texas Press, Austin, TX.

Brongniart, C.J.E. (1877). Note sur des perforations observees dans deux morceaux de bois fossile. *Ann. Soc. Entomol. Fr.* **7**, 215–220.

Cane, J.H., Stock, M.W., Wood, D.L. and Gast, S.J. (1990). Phylogenetic relationships of *Ips* bark beetles (Coleoptera: Scolytidae): electrophoretic and morphometric analysis of the *grandicollis* group. *Biochem. Syst. Ecol.* **18**, 359–368.

Choo, H.Y. (1983). "Taxonomic Studies on the Platypodidae and Scolytidae (Coleoptera) from Korea." Seoul National University, Seoul, Korea., 128 pp.

Crowson, R.A. (1967). "The Natural Classification of the Families of Coleoptera." N. Lloyd, London, 178 pp.

Crowson, R.A. (1981). "The Biology of the Coleoptera." Academic Press, New York, 802 pp.

de Groot, P. and Ennis, T.J. (1990). Cytotaxonomy of *Conophthorus* (Coleoptera: Scolytidae) in eastern North America. *Can. Entomol.* **122**, 1131–1135.

Dowding, P. (1984). The evolution of insect–fungal relationships in the primary invasion of forest timber. *In* "Invertebrate Microbial Interactions" (J.M. Anderson, A.D.M. Rayner and D.W.H. Walton, eds), pp. 133–153. Cambridge University Press, Cambridge.

Duffy, E.A.J. (1953). "Handbook for the Identification of British Insects: Coleoptera, Scolytidae and Platypodidae," Vol. 5, pp. 1–20, Royal Entomol. Soc., London.

Eaton, C.B. (1956). Jeffrey pine beetle (*Dendroctonus jeffreyi*). USDA Forest Serv Forest Pest Leafl. 11, USDA Forest Serv. Washington, DC, 7 pp.

Foottit, R.G. (1990). Technological developments in aphid systematics. *In* "Proc. 36th Annual Meeting of the Canadian Pest Management Society," pp. 141–145. Can. Pest Manage. Soc.

Foottit, R.G. and Mackauer, M. (1980). Morphometric variation between populations of the balsam woolly aphid, *Adelges piceae* (Ratzburg) (Homoptera: Adelgidae), in North America. *Can. J. Zool.* **58**, 1494–1503.

Foottit, R.G. and Mackauer, M. (1983). Subspecies of the balsam woolly aphid, *Adelgas picaea* (Homoptera: Adelgidae), in North America. *Ann. Entomol. Soc. Am.* **76**, 299–304.

Francke-Grossmann, G. (1966). Uber Symbiosen von xylomycetophagen und phloeophagen Scolyoidea mit holzbewohnenden Pilsen. *Beih. Mater. Org.* **1**, 503–522.

Francke-Grossmann, G. (1967). Ectosymbiosis in wood-inhabiting insects. *In* "Symbiosis," Vol. 2 (S.M. Henry, ed), pp. 141–205. Academic Press, New York.

Francoeur, A. (1982). The need for voucher specimens in behavioral and ecological studies. *Bull. Entomol. Soc. Can.* **8(2)**, 23.

Graham, K. (1967). Fungal–insect mutualism in trees and timber. *Annu. Rev. Entomol.* **12**, 105–126.
Grune, S. (1979). "Brief Illustrated Key to European Bark Beetles." Verlag Schaper, Hanover, 182 pp.
Hagedorn, J.M. (1910). Ipidae. *In* "Coleopterum Catalogus" (S. Schenkling, ed), pp. 1–134. W. Junk, Berlin.
Hansen, V. (1956). "Biller. XVIII Barkbiller. Med et biologisk afsnit ved B. Beier Peterson." Danmarks Fauna, Copenhagen Bd. 62, 196 pp.
Higby, P.K. and Stock, M.W. (1982). Genetic relationships between two sibling species of bark beetles (Coleoptera: Scolytidae), the Jeffrey pine beetle and the mountain pine beetles, in northern California. *Ann. Entomol. Soc. Am.* **75**, 668–674.
Hopkins, A.D. (1909). Contributions toward a monograph of the scolytid beetles. I. The genus *Dendroctonus*. USDA Bureau Entomol. Tech. Bull. 17(1), USDA Bureau Entomol., Washington, DC, 164 pp.
Kramer, P.J. and Kozlowski, T.T. (1979). "Physiology of Woody Plants." Academic Press, New York.
Kuschel, G. (1966). A cossonine genus with bark beetle habits, with remarks on the relationships and biogeography. *N.Z. J. Sci.* **9**, 3–29.
Lanier, G.N. and Wood, D.L. (1968). Controlled mating, karyology, morphology and sex-ratio in the *Dendroctonus ponderosae* complex. *Ann. Entomol. Soc. Am.* **61**, 517–526.
Lanier, G.N., Hendrichs, J.P., and Flores, J.E. (1988). Biosystematics of the *Dendroctonus frontalis* complex. *Ann. Entomol. Soc. Am.* **81**, 403–418.
Lawrence, J.F. (1982). Coleoptera. *In* "Synopsis and Classification of Living Organisms" (S.P. Parker, ed), pp. 482–553. McGraw-Hill, New York.
Lawrence, J.F. and Newton, A.F. (1982) Evolution and classification of beetles. *Annu. Rev. Ecol. Syst.* **13**, 261–290.
Mattson, W.J. (1989). Contributions to the biology of the jack pine tip beetle, *Conophthorus banksianae* (Coleoptera: Scolytidae), in Michigan. *In* "Proc. 3rd Cone and Seed Insects Working Party Conf, IUFRO Working Party Group S2.07-01" (G.E. Miller, ed), pp. 117–132. Canadian Forestry Serv., Pacific Forest Res. Ctr., Victoria, BC.
McPherson, J.E., Jr., Stehr, F.W. and Wilson, L.F. (1970). A comparison between *Conophthorus* shoot-infesting beetles and *Conophthorus resinosae* (Coleoptera: Scolytidae). II Reciprocal host and resin toxicity tests: with description of a new species. *Can. Entomol.* **102**, 1016–1022.
Morimoto, K. (1962). Comparative morphology and phylogeny of the superfamily Curculionoidea of Japan. (Comparative morphology, phylogeny and systematics of the Curculionoidea of Japan I.). *J. Fac. Agric. (Kyushu Univ.)* **11**, 331–373.
Morris, D.M. (1979). The mutualistic fungi of xyleborine beetles. *In* "Insect–Fungus Symbiosis" (L.R. Batra, ed), pp. 53–63. Halsted Press, Chichester.
Page, M., Nelson, L.J., Haverty, M.I. and Blomquist, G.J. (1990a). Cuticular hydrocarbons as chemotaxonomic characters for bark beetles: *Dendroctonus ponderosae, D. jeffreyi, D. brevicomis,* and *D. frontalis* (Coleoptera: Scolytidae). *Ann. Entomol. Soc. Am.* **83**, 892–901.
Page, M., Nelson, L.J., Haverty, M.I. and Blomquist, G.J. (1990b). Cuticular hydrocarbons of eight species of North American cone beetles, *Conophthorus* Hopkins. *J. Chem. Ecol.* **16**, 1173–1198.
Pfeffer, A. (1989). "Kurovcoviti Scolytidae a jadrohlodroviti *Platypodidae*." Academia, Praha, 137 pp.
Ritzengruber, O. and Fuhrer, E. (1986). Isoenzymanalyse verschiedener Populationen von *Pityogenes chalcographus* L. (Coleoptera: Scolytidae). I. Methodenanpassung, Enzympolymorphismus. *J. Appl. Entomol.* **101**, 187–194.
Robinson, W.H. (1975). Taxonomic responsibilities of non-taxonomists. *Bull. Entomol. Soc. Am.* **21**, 157–159.
Schedl, K.E. (1981). Familie: Scolytidae (Borken- und Ambrosiakafer). *In* "Die Kafer Mitteleuropas, Band 10" (H. Freude, K. Harde and G. Lohse, eds), pp. 34–99, 280–295. Goecke and Evers, Krefeld.
Smith, R.H. (1965). A physiological difference among beetles of *Dendroctonus ponderosae* (=*D. monicolae*) and *D. ponderosae* (=*D. jeffreyi*). *Ann. Entomol. Soc. Am.* **58**, 440–442.
Stark, V.N. (1952). Koredy. Fauna SSSR. Zhestkokrylye 31. Akademiia Nauk SSSR, Zoologicheskii Institut (N.S.) 49, 462 pp.
Stock, M.W., Pitman, G.B. and Guenther, J.D. (1979). Genetic differences between Douglas-fir beetles (*Dendroctonus pseudotsugae*) from Idaho and coastal Oregon. *Ann. Entomol. Soc. Am.* **72**, 394–397).

Stock, M.W., Gregoire, J.C. and Furniss, M.M. (1987). Electrophoretic comparison of European *Dendroctonus micans* and ten North American *Dendroctonus* species (Coleoptera; Scolytidae). *Pan-Pacif. Entomol.* **63**, 353–357.

Sturgeon, K.B. and Mitton, J.B. (1986). Allozyme and morphological differentiation of mountain pine beetles *Dendroctonus ponderosae* Hopkins (Coleoptera: Scolytidae) associated with host tree. *Evolution* **40**, 290–302.

Thomas, J.B. (1965). The immature stages of Scolytidae: the genus *Dendroctonus* Erichson. *Can. Entomol.* **97**, 374–400.

Vité, J.P., Islas, F., Renwick, J.A.A. and Kliefoth, R.A. (1974). Biochemical and biological variation of southern pine beetle populations in North and Central America. *J. Appl. Entomol.* **75**, 422–435.

Waters, W.E., Stark, R.W. and Wood, D.L., eds (1985). "Integrated Pest Management in Pine-Bark Beetle Ecosystems." John Wiley & Sons, New York, 256 pp.

Webber, J.F. and Gibbs, J.N. (1989). Insect dissemination of fungal pathogens of trees. *In* " Insect–Fungus Interactions" (N. Wilding *et al.*, eds), pp. 161–193. Academic Press, New York.

Whitney, H.S. (1982). Relationships between bark beetles and symbiotic organisms. *In* "Bark Beetles in North American Conifers" (J.B. Mitton and K.B. Sturgeon, eds), pp. 183–211. University of Texas Press, Austin, TX.

Wilding, N., Collins, N.M., Hammond, P.M. and Webber, J.F., eds (1989). "Insect–Fungus Interactions." Academic Press, New York, 344 pp.

Witcosky, J.J., Schowalter, T.D. and Hansen, E.M. (1986). *Hylastes nigrinus* (Coleoptera: Scolytidae), *Pissodes fasciatus*, and *Steremnius carinatus* (Coleoptera: Curculionidae) as vectors of black-stain root disease of Douglas-fir. *Environ. Entomol.* **15**, 1090–1095.

Wood, S.L. (1963). A revision of the bark beetle genus *Dendroctonus* Erichson (Coleoptera: Scolytidae). *Great Basin Nat.* **23**, 1–117.

Wood, S.L. (1973). On the taxonomic status of the Platypodidae and Scolytidae (Coleoptera). *Great Basin. Nat.* **33**, 77–90.

Wood, S.L. (1982). The bark and ambrosia beetles of North and Central America (Coleoptera: Scolytidae), a taxonomic monograph. Great Basin Nat. Mem. 6, 1359 pp.

Wood, S.L. (1986). A reclassification of the genera of Scolytidae (Coleoptera). Great Basin Nat. Mem. 10. 126 pp.

Wood, S.L. (1989). Nomenclatural changes and new species of Scolytidae (Coleoptera), part IV. *Great Basin Nat.* **49**, 167–185.

Wood, S.L. and Bright, D.E. (1987). A catalog of Scolytidae and Platypodidae (Coleoptera), part 1: bibliography. Great Basin Nat. Mem. 11, 685 pp.

Wood, S.L. and Bright, D.E. (in press). "A catalog of Scolytidae and Platypodidae (Coleoptera), part 2: taxonomic index. Great Basin Nat. Mem. 13.

Yin, H. Huang, F. and Li, Z. (1984). "Economic Insect Fauna of China, Fasc. 29. Coleoptera: Scolytidae." Science Press, Beijing, 205 pp.

–3–

Biology and Taxonomy of Fungi Associated with Bark Beetles

THOMAS C. HARRINGTON
Department of Plant Pathology, Iowa State University, Ames, IA, USA

3.1 INTRODUCTION

The interrelationships among bark beetles and microorganisms are complex, varied and only partly understood. Microbes can be commensal, mutualistic or parasitic symbionts of bark beetles. The range of associations has been discussed by Whitney (1982). This chapter provides an overview of the fungal symbionts and their biology, emphasizing associates of North American bark beetles.

Two problems in elucidating relationships between microbes and bark beetles are our limited knowledge of the diversity of microorganisms involved and a slowly evolving taxonomic scheme for fungi. Re-evaluation of generic limits frequently results in new combinations for fungal species, which results in understandable frustration for entomologists and plant pathologists. The taxonomy of two recently revised genera, *Ophiostoma* and *Leptographium*, will be discussed in detail.

Aside from the fungal symbionts, other fungi important to bark beetles provide weakened host material for beetle breeding (Chapter 4). This chapter focuses on the biology and taxonomy of the most important group of these tree pathogens, the root disease fungi. For four of these pathogens, new taxa or host-specialized variants with distinct biologies have been recognized recently, and the taxonomic status of these variants will be discussed.

3.2 FUNGAL SYMBIONTS

Whitney (1982) reviewed the subject of symbioses between microorganisms and bark beetles, and ambrosial-type fungi were discussed by Beaver (1989). Others (Francke-

BEETLE–PATHOGEN INTERACTIONS IN CONIFER FORESTS
ISBN 0-12-628970-0

Grosmann, 1967; Graham, 1967; Dowding, 1984; Harrington, 1988) have also reviewed the fungal symbionts with an emphasis on *Ophiostoma* and related fungi . Whitney (1982) gave an extensive list of microorganisms associated with bark beetles attacking conifers; most of these microbes were fungi, though bacteria and protozoans also were listed and discussed. Here, fungal associates are emphasized, and bacteria are mentioned only briefly.

Bacteria are present in high numbers in the gut of bark beetles and in phloem tissue around beetle galleries, though they are not present in as great a number as fungi (Moore, 1971; Whitney, 1982; Bridges *et al.*, 1984). The role of these bacteria in the biology of bark beetles is not well known, but some are believed to be important insect pathogens (Moore, 1971). In phloem tissue, bacteria may alter the nutrients available to the beetles. Nitrogen-fixing members of *Enterobacter* are associated with bark beetles (Bridges, 1981). Bacteria also may metabolize compounds in the gut of the beetle and produce beetle pheromones (Brand *et al.*, 1975), though some pheromones can be produced by axenic beetles (Conn *et al* ., 1984; Hunt and Borden, 1989).

3.2.1 Basidiomycotina

This subdivision has not been associated as commonly with bark beetles as have the other higher fungi (Ascomycotina and Deuteromycotina). Whitney *et al.* (1987) recently reviewed the literature on basidiomycetous associates and described *Entomocorticium dendroctoni* from galleries of *Dendroctonus ponderosae*. This fungus is not pathogenic to the beetle and evidently does not decay wood, but it may be nutritionally beneficial to the beetle (Whitney *et al.*, 1987).

Two unidentified basidiomycetes (which may prove to be the same species) are mycangial inhabitants of *D. frontalis* and *D. brevicomis* (Barras and Perry, 1972; Whitney and Cobb, 1972). The *D. frontalis* associate is beneficial to the development of the beetle brood, perhaps by altering the C/N ratio of the phloem or by competing with *Ophiostoma minus*, a blue-stain fungus that is detrimental to beetle brood development (Barras and Hodges, 1969; Bridges, 1983; Bridges and Perry, 1985). Volatiles produced by the basidiomycete also may enhance attractiveness of beetle pheromones (Brand and Barras, 1977).

Decay by *Fomitopsis (Fomes) pinicola*, *Cryptoporus (Polyporus) volvatus* and other fungi in recently-killed trees usually develops more quickly in the vicinity of bark beetle galleries than in other areas of the bole, but the reason for this is not clear (Harrington *et al.*, 1981). Monokaryotic isolates (presumably from single basidiospores) of wood decay fungi have been obtained from in-flight bark beetles but not from beetles freshly emerged from brood galleries (Harrington *et al.*, 1981; Pettey and Shaw, 1986). Other insects also acquire propagules of wood decay fungi (Helton *et al.*, 1988). Because these fungi all produce wind-disseminated basidiospores (Harrington, 1980), it has been hypothesized that the beetles somehow acquire spores during their flight period (Harrington *et al.*, 1981). These wood decay fungi are not dependent upon the beetles for dissemination (Dowding, 1970, 1973). Once inside the gallery, these fungi are slow to develop in the phloem and xylem and may not affect the developing beetle brood.

Heterobasidion annosum, a wood decay fungus and an important root pathogen, also has been isolated from bark beetles, but in this case it would appear that conidiophores produced in the galleries of the beetles are important in acquisition (Himes and Skelly, 1972). Such vectoring may be a very rare event, however. Hunt and Cobb (1982) suggested that *H. annosum* is pathogenic to *D. valens*.

3.2.2 Ascomycotina and Deuteromycotina

Among the Ascomycotina, the Ophiostomatales, with ascocarps and ascospores adapted for insect dispersal, are the best known associates of bark beetles and will be discussed separately. Other ascomycetes, such as *Pyxidiophora*, also produce ascospores suitable for dispersal by animals, especially by mites phoretic on bark beetles (Blackwell *et al.*, 1988). Still other species of ascomycetes (e.g. *Pezizella chapmanii*) are common in beetle galleries, but their means of dispersal are not clear (Whitney *et al.*, 1984).

Most members of the Deuteromycotina (the Fungi Imperfecti) are believed to be asexual states of Ascomycotina, but in many cases the connections between the anamorphs (imperfect or asexual state) and the teleomorphs (perfect or sexual state) are unknown, or the teleomorph has been lost. Many imperfect fungi produce conidia that are splashed by rain or carried in the wind (dry-spored). Sticky conidia suitable for insect dispersal are relatively uncommon in the Deuteromycotina. Yet, imperfect fungi with dry spores are isolated commonly from the phloem and xylem tissue attacked by bark beetles and from the beetles themselves (Moore, 1971; Whitney, 1971; Carpenter *et al.*, 1988). Apparently, the spores adhere to the exoskeleton of the bark beetles or their animal associates (e.g. phoretic mites or nematodes) and are vectored to new galleries. However, these fungi are primarily wind-disseminated and their activities are not dependent upon beetle vectoring (Verrall, 1941; Dowding, 1970, 1973). *Penicillium* and *Trichoderma* species are dry-spored, imperfect fungi commonly found in beetle galleries and isolated from beetles, but we know surprisingly little about their effect on beetle biology. Recently, Fox *et al.* (1990, 1991) have proposed that the pitch canker fungus, *Fusarium subglutinans*, can be vectored by bark, cone and twig beetles (mostly Scolytidae). *Beauveria bassiana* and some of the other dry-spored fungi are pathogens of bark beetles (Moore, 1971; Whitney *et al.*,1984).

Some fungi colonizing coniferous logs or lumber cause a blue to gray stain. In general, the Ophiostomatales cause a deep stain throughout the sapwood, but the stain caused by many of the other fungi is more superficial. Melanins in the hyphae appear to be responsible for the staining (Zink and Fengel, 1988). Use of the term "blue-stain fungi" for the beetle-associated Ophiostomatales is inappropriate because many of the members of this order (e.g. *Ophiostoma nigrocarpum*) do not cause stain (Whitney and Cobb, 1972), and not all blue stain is caused by the Ophiostomatales. For instance, *Alternaria alternata* (Zink and Fengel, 1988) and other wind-disseminated Deuteromycotina can cause blue stain (Verrall, 1941).

Yeasts are very numerous in and on beetles and in phloem tissue around their galleries. Common genera include *Pichia* (especially *P. pini*), *Hansenula* (e.g. *H. capsulata* and *H. holstii*) and *Candida* (Lu *et al.*, 1957; Shifrine and Phaff, 1956; Whitney, 1971; Bridges *et al.*, 1984; Leufvén and Nehls, 1986; Carpenter *et al.*, 1988). Yeasts are sometimes found in mycangia (Whitney and Farris, 1970), although spores are disseminated more commonly on the surface of the beetles or in the gut. These yeasts likely play an important role in the biology of the beetle. They may detoxify host compounds, alter the nutritional quality of the bark tissue or metabolize compounds in the gut of the beetle (Callaham and Shifrine, 1970; Leufvén *et al.*, 1988; Pignal *et al.*, 1988). Lu *et al.* (1957) found that *D. pseudotsugae* adults are attracted to some yeast associates. Yeasts have also been implicated in the interconversions of beetle pheromones (Leufvén *et al.*, 1988).

3.2.3 Ophiostomatales

Most members of *Ophiostoma*, the largest genus in this order, are associates of bark beetles (Harrington, 1988), and most bark beetles of conifers vector a few species of *Ophiostoma*

(Perry, 1991). Certain anamorphic genera, especially *Leptographium*, are closely tied with *Ophiostoma* and will be included in this section.

3.2.3.1. Biology

The association between blue stain fungi (the Ophiostomatales) and bark beetles has been known for a long time (Hartig, 1878; Münch, 1907) and reviewed several times (Francke-Grosmann, 1967; Graham, 1967; Whitney, 1982; Dowding, 1984; Beaver, 1989; Harrington, 1988). These relationships are varied, and few generalizations can be made safely. Chapters 6–8 will address these relationships with respect to beetle–pathogen–tree interactions.

Members of this order have small, unpigmented sexual spores (ascospores) in non-persistent asci and darkly pigmented fruiting bodies (ascocarps or perithecia). Most species have ascocarps with well-developed necks with an opening (ostiole) through which sticky ascospores emerge *en masse* (Fig. 3.1). The sticky ascospore drops are well adapted for insect dispersal and poorly suited for wind dissemination (Dowding, 1969). A few species, formerly accommodated in *Europhium*, form necks and ostioles rarely or not at all, but these species are similarly transmitted after the ascocarp wall ruptures to release the sticky spore mass. Interestingly, the ascospore drop of some of the Ophiostomatales will not disperse in water but will in conifer resin (Whitney and Blauel, 1972).

Fungus-feeding and predatory diptera, beetles, mites and other small animals are active in bark beetle galleries and may be important vectors of the Ophiostomatales (Leach *et al.*, 1934; Hetrick, 1949; Dowding, 1973, 1984; Bridges and Moser, 1983; Moser, 1985; Carpenter *et al.*, 1988; Levieux *et al.*, 1989; Chapter 7). Non-specific vectoring by such arthropods contributes to the generally varied mycoflora found in old beetle galleries

Fig. 3.1. *Ophiostoma* perithecia (black) in *Dendroctonus pseudotsugae* gallery in *Pseudotsuga menziesii* in western Oregon. Note fungus-feeding acarid mites (white) at center of photo. (Photo by T.D. Schowalter.)

(Francke-Grosmann, 1967; Solheim, 1986). Dowding (1984) has suggested that the relationship between bark beetles and sticky-spored fungi (Ophiostomatales) is in fact a relationship between the fungi and fungus-feeding arthropods, both of which are dependent upon the bark beetle. The dependence of bark beetles on the Ophiostomatales has not been well substantiated (Bridges *et al.*, 1985).

Ophiostoma species may be associated with boring insects other than bark beetles (Francke-Grosmann, 1967). *Ophiostoma huntii* and *L. wageneri* var. *ponderosum* are vectored primarily by bark beetles but have also been isolated from cerambycids or their galleries (Goheen and Cobb, 1978; Harrington, 1988). *Leptographium wageneri* var. *pseudotsugae* (Witcosky *et al.*, 1986) and *L. procerum* have been isolated from weevils as well as from bark beetles, and for *L. procerum*, weevils may be the most important vectors (Lewis and Alexander, 1986). Although the mutualistic associates of ambrosia beetles are generally thought to be unrelated to *Ophiostoma* (Batra, 1963), some *Ophiostoma* species may be contaminants or secondary associates of *Trypodendron* species and other ambrosia beetles and may cause blue staining around older beetle galleries (Bakshi, 1950; Francke-Grosmann, 1967; Hinds and Davidson, 1972; Carpenter *et al.*, 1988). A few *Ophiostoma* species have also been isolated from nitidulids and other beetles (Hinds, 1972; Juzwik and French, 1983).

Passively acquired spores of *Ophiostoma* species may adhere to the exoskeleton of bark beetles or may be eaten and pass through the digestive tract (Francke-Grosmann, 1967). In some of the bark beetle genera, special fungus carrying structures (mycangia) may be colonized by *Ophiostoma* species (Franke-Grosmann, 1963, 1967; Farris, 1969; Whitney and Farris, 1970; Barras and Perry, 1971; Whitney and Cobb, 1972; Beaver, 1989). Relatively few members of the Ophiostomatales have been shown to be vectored in this way, and they are not the only inhabitants of bark beetle mycangia. Mycangia and other beetle adaptations for vectoring fungi are addressed in Chapter 2.

Members of *Ophiostoma* typically inhabit the phloem and sapwood of beetle-attacked trees. They generally are more tolerant of conifer resin than are other wood-inhabiting fungi (Cobb *et al.*, 1968; de Groot, 1972), and they live primarily on readily available carbohydrates and other nutrients (Käärik, 1960; Barras and Hodges, 1969). Alteration of these constituents may favor the nutritional development of beetle larvae (Clark and Richmond, 1977), though this benefit may be facultative and not as important as with the aforementioned yeasts (Francke-Grosmann, 1967). With *O. ips*, the blue-stain fungus apparently has little effect on brood development of *Ips* bark beetles (Yearian *et al.*, 1972). On the other hand, *O. minus* is detrimental to the development of *D. frontalis* (Bridges and Perry, 1985); larvae and egg-laying adults avoid areas of the phloem colonized by *O. minus* (Franklin, 1970). *Ceratocystiopsis ranaculosus* and the other mycangial fungus of *D. frontalis* (the unidentified basidiomycete) enhance brood development, but perhaps more importantly, they inhibit the development of the detrimental *O. minus* (Bridges, 1983; Bridges and Perry, 1985). Brand *et al.* (1976) showed that a *Sporothrix* species (that could have been either *O. nigrocarpum* or *C. ranaculosus* (Harrington and Zambino, 1990)) isolated from *D. frontalis* galleries is capable of oxidizing *trans*-verbenol to verbenone, both important beetle pheromones.

Many, but not all, of the *Ophiostoma* spp. associated with tree-killing bark beetles are pathogenic to plants. By causing lesions that expand around the points of beetle attack and by colonizing the sapwood, they are thought to contribute to tree death, stopping host defense reactions and creating an environment conducive to beetle brood development (Berryman, 1972; Chapter 8). *Ophiostoma polonicum*, associated with *Ips typographus*, and *O. dryocoetidis*, associated with *Dryocoetes confusus*, are highly pathogenic, and mature trees have been killed when inoculated with these fungi in a manner simulating bark beetle attack (Molnar, 1965; Horntvedt *et al.*, 1983). *Ophiostoma clavigerum*, associated with *Dendroctonus ponderosae*, is another noteworthy pathogen (Reid *et al.*, 1967). An associate of *Scolytus ventralis*, *Trichosporium symbioticum* (perhaps related to the anamorphs of *Ophiostoma*) is capable of causing lesions and may aid the beetle in killing *Abies* (Wright, 1935; Raffa and Berryman, 1982). Many root-inhabiting *Leptographium* species (e.g. *L. procerum* and *L. terebrantis*) are also pathogenic, causing lesions in phloem and staining of the sapwood in the vicinity of bark beetle or weevil activity (Harrington and Cobb, 1983; Wingfield *et al.*, 1988a; Klepzig *et al.*, 1991).

The associates of *D. brevicomis* and *D. frontalis* are not as pathogenic as some of the fungi mentioned above. *Ophiostoma minus* causes larger lesions than the other associates when seedlings or larger trees are inoculated (Bramble and Holst, 1940; Owen *et al.*, 1987; Parmeter *et al.*, 1989), but it may not be pathogenic enough to kill the trees by itself. If *O. minus* is important in killing trees attacked by *D. frontalis*, it is surprising that in some epidemic populations of *D. frontalis*, few of the beetles carry this blue-stain fungus (Bridges *et al* ., 1985; Bridges and Moser, 1986).

3.2.3.2. Taxonomy

In his monograph on *Ceratocystis*, Hunt (1956) accepted Bakshi's (1951) broad concept of *Ceratocystis* Ellis & Halst. and placed all the modern-day Ophiostomatales in that genus. Griffin (1968), Olchowecki and Reid (1974), Upadhyay (1981) and Hutchison and Reid (1988) also have used the name *Ceratocystis* in a broad sense. However, others (von Arx, 1974; de Hoog, 1974; Weijman and de Hoog, 1975; de Hoog and Scheffer, 1984; Harrington, 1987, 1988) have argued that there are at least two unrelated groups within *Ceratocystis sensu lato*. The two groups can be separated on the basis of anamorphs (von Arx, 1974; Harrington, 1981; de Hoog and Scheffer, 1984), cell wall composition (Weijman and de Hoog, 1975), cycloheximide sensitivity (Harrington, 1981), and biology (Harrington, 1987, 1988). Similarity of ascocarps between the two groups is likely due to convergent evolution towards insect dissemination and not due to relatedness (Harrington, 1987). As recognized here, most of the bark beetle associates formerly considered in *Ceratocystis* are placed in the genus *Ophiostoma* H. & P. Sydow.

Ceratocystis now is restricted to species with *Chalara* (Corda) Rabenh. anamorphs. Eleven species, including the type species *Ceratocystis fimbriata*, were recognized by de Hoog (1974). Kile and Walker (1987) discussed the *Chalara* species described since Nag Raj and Kendrick's (1975) monograph. The biology of *Ceratocystis* and *Chalara* species is varied, including species important as rotters of fruit and vegetables, causes of stain-cankers or wilts of woody angiosperms, soilborne pathogens and saprophytes (Nag Raj and Kendrick, 1975). Insect vectors are known for many of the species, but only the blue-stain fungus *Ceratocystis coerulescens* (and the possible synonym *C. laricicola*) has been clearly associated with bark beetles (Dowding, 1984; Redfern *et al.*, 1987). Like *C. coerulescens*, *C. fagacearum* may be vectored by bark beetles, though nitidulid beetles appear to be more important vectors (Gibbs and French, 1980; Juzwik and French, 1983).

Members of *Ceratocystiopsis* have elongated ascospores (Upadhyay and Kendrick, 1975) but are otherwise similar to *Ophiostoma* (von Arx, 1981; de Hoog and Scheffer, 1984). However, species such as *C. proteae* (Wingfield *et al.*, 1988b) appear to have affinities to other orders, and the genus is becoming heterogeneous. I prefer to keep those *Ceratocystiopsis* species with *Ophiostoma* affinities in *Ophiostoma*, but at least one bark beetle-associated species cannot be transferred there readily. *Ceratocystiopsis ranaculosus* is one of the two common inhabitants of the mycangium of *D. frontalis* and is apparently the fungus previously referred to as *Ceratocystis minor* var. *barrasii* (Harrington and Zambino, 1990). The very small ascocarps of *C. ranaculosus* are shaped like an acorn and exude small, elongate ascospores that develop a bulbous swelling at one end (Bridges and Perry, 1987).

There are well over 100 described species in *Ophiostoma*, a unique and relatively homogeneous group of fungi that are primarily vectored by bark beetles or other subcortical insects. The genus *Europhium* was erected for *Ophiostoma*-like species without ascocarp necks, but this is a variable character and these species have been transferred to *Ophiostoma* (Harrington, 1987). Division of the large genus *Ophiostoma* may be warranted, but delineations based on anamorphs do not appear to divide *Ophiostoma* into natural groups (Harrington, 1988).

The conidia of *Ophiostoma* may accumulate in sticky drops at the apex of conidiophores (Fig. 3.2) and be dispersed by insects and mites (Dowding, 1969). The conidiophores vary substantially in complexity and in how the individual conidia are produced. Upadhyay and

Fig 3.2. *Leptographium* conidia in *Dendroctonus pseudotsugae* gallery in *Pseudotsuga menziesii* in western Oregon. (Photo by T.D. Schowalter.)

Kendrick (1975) recognized 14 anamorphic genera for conidial states that could be considered *Ophiostoma*, although this number could be greatly reduced (Harrington, 1988). The sole character distinguishing some of these anamorphic genera is the mode of proliferation of conidiogenous cells after the initial conidium is produced. However, use of this as the sole criterion appears inappropriate (de Hoog and Scheffer, 1984; Tsuneda and Hiratsuka, 1984; Wingfield, 1985; Harrington, 1988). More than one proliferation type may be found in a single isolate of some species and even on the same conidiogenous apparatus. Disregarding proliferation of conidiogenous cells, *Pesotum* is a synonym of *Graphium*, and *Verticicladiella* a synonym of *Leptographium*. Most of the species in *Phialocephala* are morphologically similar to *Leptographium* but are distinct in their conidiogenesis (their conidia form deep within the conidiogenous cell rather than on the surface), and their biology also differs (Harrington, 1988).

Another distinction among the anamophic genera is in the complexity of conidiophores. *Graphium* produces synnemata (pigmented hyphae fused in parallel to form a stalk), at the apex of which the hyphae branch and ultimately differentiate into conidiogenous cells. By contrast, *Leptographium* species are mononematous (a single, pigmented hypha makes up the stalk; Fig. 3.2). *Sporothrix* and *Hyalorhinocladiella* species, on the other hand, have unpigmented, micronematous (small) conidiophores. Upadhyay's (1981) genus for species with both mononematous and synnematous conidiophores, *Graphiocladiella*, is unnecessary; the anamorph of *O. clavigerum* could be considered a *Graphium* (Harrington, 1988).

Tsuneda and Hiratsuka (1984) demonstrated for *O. clavigerum* that repeated transfers on agar media result in a shift from complex to simpler forms of conidiophores and conidia. Deterioration progressed from synnematous to mononematous to micronematous conidiophores, and finally to yeast forms. Conidia also decreased in size, lost septations and changed shape. Hence, the taxonomic status of *O. clavigerum* and three other *Ophiostoma* species separated solely on conidial morphology needs to be investigated (Harrington, 1988).

3.3 ROOT PATHOGENS PREDISPOSING TREES TO BARK BEETLES

The various abiotic and biotic agents predisposing trees to bark beetles are addressed in Chapter 4. The best known of the biotic agents are the root disease fungi that kill trees directly but also weaken and render them susceptible to bark beetle attack.

Root pathogens of conifers can be divided into two broad groups; the root-rotters and the non-decay fungi. The former group are wood decay fungi and, in the case of pathogens on North American conifers, are members of the hymenomycetes, a class of Basidiomycotina that have a layer of basidia (spore bearing structures) exposed at maturity to release wind-disseminated basidiospores. Two genera of non-decay fungi are important on North American conifers: *Leptographium* and *Phytophthora* (Mastigomycotina: Oomycetes). Both genera of pathogens are capable of colonizing woody xylem, but they do not cause appreciable decay of wood.

3.3.1 Hymenomycetes

In North American forests, *Armillaria* species, *Heterobasidion annosum*, *Phellinus weirii*, *Phaeolus schweinitzii* and *Inonotus tomentosus* (and the closely related *I. circinatus*) are the most important root-rotters on conifers. Their biologies are similar in many ways.

3.3.1.1. Biology

Mortality caused by the root-rotting hymenomycetes may occur within a few years of infection, or infected trees may survive for decades or centuries. *Heterobasidion annosum* and *Armillaria* species on certain hosts can colonize and kill the cambium of roots in advance of wood decay, and their hosts may die in a matter of months or years (Hodges, 1969; Shigo and Tippett, 1981). In more resistant hosts, these and other pathogens may colonize only the oldest growth rings of the root and butt wood (Hodges, 1969; Rizzo and Harrington, 1988), that part of the xylem least able to effectively respond to the pathogen and prevent further encroachment (Gibbs, 1968). This "butt rot" type of infection results in cull and reductions in growth rate (Whitney and MacDonald, 1985), and the diseased trees are predisposed to bark beetles (Cobb *et al.*, 1974), pathogens, windthrow and windsnap (Rizzo and Harrington, 1988).

Underground spread of root-infecting hymenomycetes results in roughly circular "infection centers" in the forest. A clone (single genotype) of the pathogen grows outward, using the root system of the killed tree as a food base from which to attack adjacent susceptible tree species (Childs, 1963; Myren and Patton, 1971; Korhonen, 1978a,b; Anderson *et al.*, 1979; Shaw, 1980; Chase and Ullrich, 1983; Barrett and Blakeslee, 1989). Movement from colonized roots to new trees may be restricted to root contacts (Hodges, 1969), but some species form mycelial cords (undifferentiated hyphal strands) or rhizomorphs (with a differentiated rind) (Fig. 3.3) that facilitate spread through soil (Thompson, 1984; Rishbeth, 1985). With most of these hymenomycetes, the clone may persist for decades as a saprophyte in colonized stumps to infect the next generation of trees (Hodges, 1969; Hansen, 1979; Lewis and Hansen, 1989). Clones of different fungal species may overlap, and more

Fig. 3.3. Rhizomorphs of *Armillaria* sp. (Photo by G.M. Filip.)

Fig. 3.4. Mushrooms of *Armillaria ostoyae* on *Pinus ponderosa*. (USDA Forest Serv. Photo by B. Tkacz.)

than one species may be found in the roots of a given tree (Goheen and Filip, 1980; Filip and Goheen, 1982).

Initiation of new infection centers are relatively rare events for most of these root-rotters. Infection of freshly cut stumps by airborne basidiospores are important in the origination of infection centers of *H. annosum* (Hodges, 1969) and, to a lesser extent, *Armillaria* species (Rishbeth, 1988). *Heterobasidion annosum* (Hodges, 1969), *P. schweinitzii* (Barrett and Blakeslee, 1989) and *I. tomentosus* (Whitney, 1961; Myren and Patton, 1971) also can infect roots via basidiospores percolating through soil to start new infection centers.

3.3.1.2. Taxonomy

Basidiocarps of *Armillaria* are mushrooms (Fig. 3.4), placing it in the order Agaricales. Watling *et al.* (1982) considered *Armillaria* (Fr.:Fr.) Staude (family Tricholomataceae) in a narrow sense, and the later described genus name *Armillariella* (P. Karst.) P. Karst. as invalid; *A. mellea* was designated as the type species for both genera. Table 3.1 lists the North American species of *Armillaria.*

Aside from *Armillaria*, the most important genera of root-rotters on conifers belong in the order of non-gilled fungi, the order Aphyllophorales. Each produces a polypored basidiocarp in which the basidia line the inside of opened tubes. This configuration places these genera in the family Polyporaceae under the old Friesian system of classification. For a more modern, but still conservative treatment of families, see Gilbertson and Ryvarden (1986).

The genus *Heterobasidion* was erected with *H. annosum* as the type species. Of the three recognized species (Buchanan, 1988), the most important pathogen is *H. annosum*, also known as *Fomes annosus* and *Fomitopsis annosa*. The anamorph of *H. annosum* is *Spiniger meineckellus* (=*Oedocephalum lineatum*).

Table 3.1 North American species, intersterility groups and synonyms of *Armillaria*

Intersterility group[a]	Species and synonyms
I	*Armillaria ostoyae* (=*A. obscura*)
II	*Armillaria gemina*
III	*Armillaria calvescens*
IV	Interfertile with Group V
V	*Armillaria sinapina*, but partially interfertile with *Armillaria cepistipes*
VI	*Armillaria mellea*
VII	*Armillaria gallica* (=*A. lutea*, =*A. bulbosa*)
VIII	Interfertile with Group VI
IX	Unnamed
X	Unnamed, but partially interfertile with *A. cepistipes*
XI	Interfertile with *Armillaria cepistipes* (=species F, Morrison *et al.*, 1985
	Armillaria tabescens (=*Clitocybe tabescens*)

[a]Roman numerals of Anderson (1986).

Gilbertson (1974) transferred a species earlier known as *Fomitoporia weirii* or *Poria weirii* to *Phellinus*. Species of *Phellinus* and *Inonotus* are closely related members of the family Hymenochaetaceae. Most *Inonotus* species have softer, annual basidiocarps with a monomitic hyphal system (one type of hypha in the context), whereas *Phellinus* has tougher and generally perennial basidiocarps (Fig. 3.5) with a dimitic hyphal system (Gilbertson and Ryvarden, 1987). In spite of similarities of basidiocarps, *Phaeolus schweinitzii* does not appear to belong in the Hymenochaetaceae (Gilbertson and Ryvarden, 1987), although some would place it in the order Hymenochaetales (Larsen and Cobb-Poulle, 1990).

3.3.1.3. Host specialization

Host ranges of many root pathogens are relatively broad, but with three of the pathogens discussed above, variation in pathogenicity and host ranges recently has been reexamined. In most of these cases, intersterility and other genetic criteria have been used to support separations based on minor morphological differences.

Armillaria mellea, as it was previously considered, now encompasses a number of species with varying geographic distributions, host ranges and pathogenicity (Shaw and Kile, 1991). Although the modern species differ in basidiocarp morphology, fruiting bodies are seasonal and are not produced every year. Also, most of the species do not fruit *in vitro* with any regularity. Delineations of species and identifications are based largely on intersterility (Harrington *et al.*, 1992).

Nine intersterile groups are recognized in North America (Anderson, 1986); all but two (groups IX and X) are formally named as species (Table 3.1; Bérubé and Dessureault, 1989). *Armillaria tabescens* lacks an annulus but is otherwise similar to *A. mellea* and related species. There are fewer species of *Armillaria* in Europe than in North America, and some species are common to both continents. The relationships of groups V, X and XI and the European forms of *A. cepistipes* are in need of clarification (Morrison *et al.*, 1985; Bérubé and Dessureault, 1989).

Species of *Armillaria* have been reported on a wide range of angiosperms and gymnosperms (Raabe, 1962). Because of the recent recognition of species within *A. mellea, sensu lato*, the host and geographic ranges of individual *Armillaria* species is rather poorly known. *Armillaria mellea, sensu stricto*, is an important pathogen on various hardwood

Fig. 3.5. Fruiting bodies of *Phellinus weirii* (arrows) on bark of *Picea engelmannii*. (Photo by G.M. Filip.)

species and to a lesser extent on conifers in milder regions of Europe, Asia and North America (Guillaumin *et al.*, 1989). *Armillaria ostoyae* is more widespread than *A. mellea*; it occurs throughout the northern temperate forests of Europe, Asia and North America (Morrison *et al.*, 1985; Dumas, 1988; Guillaumin *et al.*, 1989; Blodgett and Worrall, 1992). Hardwoods, particularly species of *Betula* and *Acer*, appear to be more common hosts of *A. ostoyae* in North America than in Europe (personal observation), but on both continents, conifers are killed more commonly than hardwoods. This is probably the most important species on conifers in North America.

The pathogenicity of the other North American species of *Armillaria* is less clear. *Armillaria sinapina* appears to be weakly pathogenic on hardwoods and conifers, based on field observations (Morrison *et al.*, 1985; Dumas 1988). However, Mallett and Hiratsuka (1988) and Shaw and Loopstra (1988) have reported pathogenicity to conifer seedlings. Shaw and Loopstra (1988) also found group IX isolates to be pathogenic to *Picea* seedlings. *Armillaria tabescens* is known in the southeastern USA as a pathogen of both hardwoods and conifers (Rhoads, 1942; Barnard *et al.*, 1985).

Heterobasidion annosum is most important on conifers, but a number of hardwood hosts are also known (Webb and Alexander 1985). The type specimen is from *Betula*.

Host-specialized groups of *H. annosum* have been identified through seedling inoculations (Worrall *et al.*, 1983; Stenlid and Swedjemark, 1988; Harrington *et al.*, 1989); these groups are partially intersterile and differ slightly in morphology (Korhonen, 1978b). Genetic variation within the intersterility groups and the genetic basis for partial intersterility have been studied (Chase and Ullrich, 1988; Chase *et al.*, 1989; Otrosina *et al.*, 1989). These groups may prove to be distinct taxa. They differ from each other to the same extent that the recently described *H. araucariae* (Buchanan, 1988) differs from *H. annosum*, but they are as yet unnamed.

It would appear that the "P" (pine) intersterility group has a broader host range and wider distribution than the other intersterility groups (Korhonen, 1978b). It is responsible for most of the mortality of pines, although saplings and seedlings of pines may be killed by the "S" (spruce) intersterility group. The "P" group is widespread throughout the northern temperate region (Korhonen 1978b) and occurs in both eastern and western North America (Chase and Ullrich, 1988; Harrington *et al.*, 1989).

The "S" group occurs in Europe, where it is found mostly causing butt rot in *Picea abies*, and isolates interfertile with "S" testers from Europe have been reported from conifers in Asia (Korhonen, 1978b). A third group, the "F" group, has been recognized recently from Italy on *Abies alba* (Korhonen *et al.*, 1989). A western North American "fir" group that is interfertile with Korhonen's "S" group causes root and butt rot on *Abies* spp. and other conifers in western North America (Worrall *et al.*, 1983; Chase and Ullrich, 1988; Harrington *et al.*, 1989; Otrosina *et al.*, 1989). The relationships between the North American "fir" group and the "F" and "S" groups from Europe have not been fully explored, but there is some intercompatibility among the three populations (Korhonen *et al.*, 1989).

Two groups are recognized in *Phellinus weirii*: a cedar form (that includes the type specimen) and a fir form. The host range of *P. weirii* is relatively broad, but observations suggest that the fir pathogen rarely attacks *Thuja plicata*, and the cedar pathogen occurs primarily on *T. plicata*. These two forms differ slightly in size of setal hyphae, type of spore germination and characteristics in culture (Larsen and Lombard, 1989). Another difference is in the persistence of the basidiocarps; the cedar form has perennial basidiocarps while those of the fir form are annual (more typical of *Inonotus* species). These differences are not great, however, and there is some overlap, at least as much as is found among the intersterility groups of *H. annosum*. Sexuality in *Phellinus* is not well understood, but using a unique phenotypic reaction between homokaryons and putative heterokaryons, Angwin and Hansen (1989) presented evidence that the host-specialized groups of *P. weirii* are only partially interfertile.

Phellinus weirii occurs in the Pacific Northwest, British Columbia, Alaska and Asia. The two North American forms have overlapping distributions, particularly in Idaho and British Columbia (Angwin and Hansen, 1989). The distribution of the cedar form appears to be more northerly than that of the fir form.

3.3.2 Non-decay fungi

Root diseases caused by species of *Leptographium* were recently reviewed (Harrington and Cobb, 1988). As discussed under *Ophiostoma* anamorphs, *Leptographium* and *Verticicladiella* have been synonymized, and the former has nomenclatural priority. *Leptographium wageneri*, the best known of the species and the most important in predisposing conifers to bark beetles, is the only member of the genus to be discussed here.

Leptographium wageneri is vectored by root-feeding bark beetles, primarily species of *Hylastes* (Goheen and Cobb, 1978; Harrington *et al.*, 1985; Witcosky *et al.*, 1986), though root-feeding weevils also may be important (Witcosky *et al.*, 1986). Conidia are produced in sticky drops on top of conidiophores within beetle galleries for passive acquisition. New infection centers may be initiated when contaminated beetles (from broods developing in diseased trees) are attracted to disturbed stands, dig through soil in search of suitable roots for breeding or feeding, and bore into roots of living trees (Harrington et al., 1985). The disease in *Pseudotsuga menziesii* is strongly associated with stand disturbance (Hansen, 1978; Harrington *et al.*, 1983).

Once in the xylem of a living root, hyphae of *L. wageneri* colonize only the tracheids, moving systemically through the host xylem in a manner similar to that of the vascular wilt fungi in angiosperms (Smith, 1967; Hessburg and Hansen, 1987; Cobb, 1988). The fungus can grow from diseased roots for short distances through the soil to infect small-diameter roots of neighboring trees (Hicks *et al.*, 1980; Hessburg and Hansen, 1986). Thus, "infection centers" of *L. wageneri* can develop in a manner similar to root rots caused by hymenomycetes (Cobb *et al.*, 1982).

Leptographium wageneri now is recognized as three host-specialized varieties (Harrington and Cobb, 1986, 1987) with overlapping geographic distributions (Zambino and Harrington, 1989). The species originally was described from *Pinus monophylla*; and var. *wageneri* is specialized on the pinyons and found in the southwestern United States and north to southern Idaho. *Leptographium wageneri* var. *pseudotsugae* is a pathogen on *Pseudotsuga* throughout the western USA and British Columbia, Canada. Variety *ponderosum* occurs on the hard pines (and rarely *Tsuga*) in the Pacific Coast states, Idaho, Montana and British Columbia.

Host specialization of the three varieties has been demonstrated with inoculations of seedlings and mature trees (Harrington and Cobb, 1984). Morphologically, the three varieties differ only slightly (Harrington and Cobb, 1986, 1987). Isozyme analysis (Otrosina and Cobb, 1987; Zambino and Harrington, 1989) and vegetative (heterokaryon) compatibility testing (Zambino and Harrington, 1990) indicate limited variation in the species and confirm the delineation of three infraspecific taxa.

Ophiostoma wageneri (≡*Ceratocystis wageneri*) was described as the teleomorph before the varieties of *L. wageneri* were named (Goheen and Cobb, 1978), but it has been suggested (Harrington and Cobb, 1986) that it is the teleomorph of *L. wageneri* var. *ponderosum*. Because the anamorph–teleomorph connection has not been firmly established for each variety and the significance of sexual reproduction in this fungus has been questioned (Zambino and Harrington, 1989), the use of the teleomorph name is not encouraged (Harrington, 1988).

Phytophthora cinnamomi differs from *L. wageneri* and the hymenomycetes discussed above in that the fungus primarily kills only the small-diameter feeder roots of affected pines (Zentmeyer, 1980; Barnard *et al.*, 1985). Swimming zoospores released from sporangia are the asexual spores of this fungus, and wet conditions tend to favor this and other species of *Phytophthora* (Zentmeyer, 1980). Site conditions are very important for disease development. *Pinus echinata* on eroded soils of poor fertility are particularly vulnerable to *Phytophthora cinnamomi* (Belanger *et al.*, 1986). Trees under 20 years of age are rarely infected, and infected trees are often killed by bark beetles (Belanger *et al.*, 1986).

Fortunately, *P. cinnamomi* does not persist in soils of many of our coniferous forests (Zentmeyer, 1980), and in the United States, only the pine forests of the southeast are seriously threatened by this pathogen. Another species of *Phytophthora*, *P. lateralis*, is apparently an introduced pathogen that attacks native stands of *Chamaecyparis lawsoniana* in southwestern Oregon and adjacent California (Roth *et al.*, 1987). Infected trees often are attacked by bark beetles.

3.4 CONCLUSIONS

Much awaits to be discovered about the symbiosis between microorganisms and bark beetles. A wide variety of microbes are associated with bark beetles, but few have been intensively studied. Results from these studies have yielded few strong conclusions and fewer generalizations. Beneficial and detrimental effects on the beetles have been found. The pathogenicity of some *Ophiostoma* species to trees has been demonstrated, but the significance of such pathogenicity to bark beetle biology is still debated.

The taxonomy of the Ophiostomatales has been scrutinized over the last two decades, and generic limits of the teleomorphs and the associated anamorph genera now appear to be better defined. There is general agreement among taxonomists on the split between *Ophiostoma* and *Ceratocystis*; most of the important bark beetle associates now are placed in the former genus. The anamorph genus *Verticicladiella* has been synonymized with *Leptographium*, and this taxonomic change also has been generally accepted.

Among the fungi important in predisposing conifers to bark beetle attack, root pathogens would appear to be the most important. There are similarities in the biology of the root-rotting hymenomycetes, but root pathogens in *Phytophthora* and *Leptographium* are distinct.

Variation within many of these root pathogens has been re-evaluated recently. The species once known as *Armillaria mellea* is now recognized as a number of closely related, intersterile species with distinct morphology and biology. Host-specialized groups within *Heterobasidion annosum*, *Phellinus weirii* and *Leptographium wageneri* are now known, and in *L. wageneri* these groups have taxonomic status. With the developing array of molecular tools to characterize variation in fungi, further refinement of the taxonomy of these important pathogens is expected. These findings should be of value in recognizing host ranges of the pathogens and in management of the diseases they cause (Chapter 11).

REFERENCES

Anderson, J.B. (1986). Biological species of *Armillaria* in North America: redesignation of groups IV and VIII and enumeration of voucher strains for other groups. *Mycologia* **78**, 837–839.

Anderson, J.B., Ullrich, R.C., Roth, L.F., and Filip, G.M. (1979). Genetic identification of clones of *Armillaria mellea* in coniferous forests in Washington. *Phytopathology* **69**, 1109–1111.

Angwin, P.A. and Hansen, E.M. (1989). Population structure of *Phellinus weirii*. *In:* "Proceedings of the Seventh International Conference on Root and Butt Rots" (D. J. Morrison, ed.), pp. 371–380. Forestry Canada, Pacific Forestry Centre, Victoria, BC.

Bakshi, B.K. (1950). Fungi associated with ambrosia beetles in Great Britain. *Trans. Br. Mycol. Soc.* **33**, 111–120.

Bakshi, B.K. (1951). Studies on four species of *Ceratocystis* with a discussion of fungi causing sapstain in Britain. Commonw. Mycol. Inst. Mycol. Pap. 35, 16 pp.

Barnard, E.L., Blakeslee, G.M., English, J.T., Oak, S.W. and Anderson, R.L. (1985). Pathogenic fungi associated with sand pine root disease in Florida. *Plant Dis.* **69**, 196–199.

Barras, S.J. and Hodges, J.D. (1969). Carbohydrates of innerbark of *Pinus taeda* as affected by *Dendroctonus frontalis* and associated microorganisms. *Can. Entomol.* **101**, 489–493.

Barras, S.J. and Perry, T.J. (1971). Gland cells and fungi associated with prothoracic mycangium of *Dendroctonus adjunctus* (Coleoptera: Scolytidae). *Ann. Entomol. Soc. Am.* **64**, 123–126.

Barras, S.J. and Perry, T.J. (1972). Fungal symbionts in the prothoracic mycangium of *Dendroctonus frontalis* (Coleoptera: Scolytidae). *Z. Ang. Entomol.* **71**, 95–104.

Barrett, D.K. and Blakeslee, G.M. (1989). Apparent differences in the infection biology of *Phaeolus schweinitzii* as observed in pine stands in Florida and spruce stands in Britain. *In* "Proceedings of the Seventh International Conference on Root and Butt Rots" (D. J. Morrison, ed). pp. 603–611. Vernon and Victoria, BC, Canada.

Batra, L.R. (1963). Ecology of ambrosia fungi and their dissemination by beetles. *Trans. Kans. Acad. Sci.* **66**, 213–236.

Beaver, R.A. (1989). Insect-fungus relationships in the bark and ambrosia beetles. *In* "Insect-Fungus Interactions" (N. Wilding, N.M. Collins, P.M. Hammond and J.F. Webber, eds). pp.121–143. Academic Press, London.

Belanger, R.P., Hedden, R.L. and Tainter, F.H. (1986). Managing Piedmont forests to reduce losses from the littleleaf disease–southern pine beetle complex. USDA. For. Serv. Agric. Handbook. 649, 19 pp.

Berryman, A.A. (1972). Resistance of conifers to invasion by bark beetle-fungus associations. *BioScience* **22**, 598–602.

Bérubé, J.A. and Dessureault, M. (1989). Morphological studies of the *Armillaria mellea* complex: two new species, *A. gemina* and *A. clavescens*. *Mycologia* **81**, 216–225.

Blackwell, M.J., Moser, J.C. and Wisniewski, J. (1988). Ascospores of *Pyxidiophora* on mites associated with beetles in trees and wood. *Mycol. Res.* **92**, 397–403.

Blodgett, J.T. and Worrall, J.J. (1992). Distributions and hosts of *Armillaria* species in New York. *Plant Dis.* **76**, 166–169.

Bramble, W.C. and Holst, E.C. (1940). Fungi associated with *Dendroctonus frontalis* in killing shortleaf pines and their effect on conduction. *Phytopathology* **30**, 881–899.

Brand, J.M. and Barras, S.J. (1977). The major volatile constituents of a basidiomycete associated with the southern pine beetle. *Lloydia* **40**, 318–399.

Brand, J.M., Bracke, J.W., Markovetz, A.J., Wood, D.L. and Browne, L.E. (1975). Production of verbenol pheromone by a bacterium isolated from bark beetles. *Nature* **254**, 136–137.

Brand, J.M., Bracke, J.W., Britton, L.N.F Markovetz, A.J. and Barras, S.J. (1976). Bark beetle pheromones: production of verbenone by a mycangial fungus of *D. frontalis*. *J. Chem. Ecol.* **2**, 195–199.

Bridges, J.R. (1981). Nitrogen-fixing bacteria associated with bark beetles. *Microb. Ecol.* **7**, 131–137.

Bridges, J.R. (1983). Mycangial fungi of *Dendroctonus frontalis* (Coleoptera: Scolytidae) and their relationship to beetle population trends. *Environ. Entomol.* **12**, 858–861.

Bridges, J.R. and Moser, J.C. (1983). Role of phoretic mites in transmission of bluestain fungus, *Ceratocystis minor*. *Ecol. Entomol.* **8**, 9–12.

Bridges, J.R. and Moser, J.C. (1986). Relationship of phoretic mites (Acari: Tarsonemidae) to the bluestaining fungus, *Ceratocystis minor*, in trees infested by southern pine beetle (Coleoptera: Scolytidae). *Environ. Entomol.* **15**, 951–953.

Bridges, J.R. and Perry, T.J. (1985). Effects of mycangial fungi on gallery construction and distribution of bluestain in southern pine beetle-infested pine bolts. *J. Entomol. Sci.* **20**, 271–275.

Bridges, J.R. and Perry, T.J. (1987). *Ceratocystiopsis ranaculosus* sp. nov. associated with the southern pine beetle. *Mycologia* **79**, 630–633.

Bridges, J.R., Marler, J.E. and McSparrin, B.H. (1984). A quantitative study of the yeasts and bacteria associated with laboratory-reared *Dendroctonus frontalis* Zimm. (Coleoptera, Scolytidae). *Z. Ang. Entomol.* **97**, 261–267.

Bridges, J.R., Nettleton, W.A. and M.D. Connor. (1985). Southern pine beetle (Coleoptera: Scolytidae) infestations without the bluestain fungus, *Ceratocystis minor*. *J. Econ. Entomol.* **78**, 325–327.

Buchanan, P.K. (1988). A new species of *Heterobasidion* (Polyporaceae) from Australasia. *Mycotaxon* **32**, 325–337.

Callaham, R.Z. and Shifrine, M. (1970). The yeasts associated with bark beetles. *Forest Sci.* **16**, 146–154.

Carpenter, S.E., Harmon, M.E., Ingham, E.R., Kelsey, R.G., Lattin, J.D. and Schowalter, T.D. (1988). Early patterns of heterotroph activity in conifer logs. *Proc. Royal Soc. Edinburgh* **94B**, 33–43.

Chase, T.E. and Ullrich, R.C. (1983). Sexuality, distribution, and dispersal of *Heterobasidion annosum* in pine plantations of Vermont. *Mycologia* **75**, 825–831.

Chase, T.E. and Ullrich, R.C. (1988). *Heterobasidion annosum*, root- and butt-rot of trees. *Adv. Plant Pathol.* **6**, 501–510.

Chase, T.E., Ullrich, R.C., Otrosina, W.J., Cobb, F.W., Jr. and Taylor, J.W. (1989). Genetics of intersterility in *Heterobasidion annosum*. *In* "Proceedings of the Seventh International Conference on Root and Butt Rots" (D. J. Morrison, ed.) pp. 11–19. Forestry Canada, Pacific Forestry Centre., Victoria, BC.

Childs, T.W. (1963). *Poria weirii* root rot. *Phytopathology* **53**, 1124–1127.

Clark, E.W. and Richmond, J.A. (1977). Variations of free and triglyceride fatty acids in phloem of *Pinus taeda* infected by *Ceratocystis minor*. *Turrialba* **27**, 377–383.

Cobb, F.W., Jr. (1988). *Leptographium wageneri*, cause of black-stain root disease: a review of its discovery, occurrence and biology with emphasis on pinyon and ponderosa pine. *In* "*Leptographium* Root Diseases on Conifers" (T. C. Harrington and F. W. Cobb, Jr., eds) pp. 41–62. American Phytopathological Society Press, St. Paul, MN.

Cobb, F.W., Jr., Kristic, M., Zavarin, E. and Barber, H. W., Jr. (1968). Inhibitory effects of volatile oleoresin components on *Fomes annosus* and four *Ceratocystis* species. *Phytopathology* **58**, 1327–1335.

Cobb, F.W., Jr., Parmeter, J.R., Jr., Wood, D.L. and Stark, R.W. (1974). Root pathogens as agents predisposing ponderosa pine and white fir to bark beetles. *In* "Proceedings of the Fourth International Conference on *Fomes annosus*" (E.G. Kuhlman, ed.) pp. 8–15. USDA Forest Serv., Washington, DC.

Cobb, F.W., Jr., Slaughter, G.W, Rowney, D.L. and DeMars, C.J. (1982). Rate of spread of *Ceratocystis wageneri* in ponderosa pine stands in the central Sierra Nevada. *Phytopathology* **72**, 1359–1362.

Conn, J.E., Borden, J.H., Hunt, D.W.A., Holman, J., Whitney, H.S., Spanier, O.J., Pierce, H. D., Jr. and Oehlschlager, A.C. (1984). Pheromone production by axenically reared *Dendroctonus ponderosae* and *Ips paraconfusus* (Coleoptera: Scolytidae). *J.Chem. Ecol.* **10**, 281–290.

de Groot, R.C. (1972). Growth of wood-inhabiting fungi in saturated atmospheres of monoterpenoids. *Mycologia* **64**, 863–870.

de Hoog, G.S. (1974). The genera *Blastobotrys, Sporothrix, Calcarisporium* and *Calcarisporiella* gen. nov. *Stud. Mycol.* **7**. 1–84.

de Hoog, G.S. and Scheffer, R.J. (1984). *Ceratocystis* versus *Ophiostoma*: a reappraisal. *Mycologia* **76**, 292–299.

Dowding, P. (1969). The dispersal and survival of spores of fungi causing bluestain in pine. *Trans. Br. Mycol. Soc.* **52**, 125–137.

Dowding, P. (1970). Colonization of freshly bared pine sapwood surfaces by staining fungi. *Trans. Br. Mycol. Soc.* **53**, 399–412.

Dowding, P. (1973). Effects of felling time and insecticide treatment on the interrelationships of fungi and arthropods in pine logs. *Oikos* **24**, 422–429.

Dowding, P. (1984). The evolution of insect-fungus relationships in the primary invasion of forest timber. *In* "Invertebrate–Microbial Interactions" (J.M. Anderson, A.D.M. Raynor and D.W.H. Walton, eds), pp. 133–153. Cambridge University Press, New York, 349 pp.

Dumas, M.T. (1988). Biological species of *Armillaria* in the mixedwood forest of northern Ontario. *Can. J. For. Res.* **18**, 872–874.

Farris, S.H. (1969). Occurrence of mycangia in the bark beetle *Dryocoetes confusus* (Coleoptera: Scolytidae). *Can. Entomol.* **101**, 527–532.

Filip, G.M. and Goheen, D.J. (1982) Tree mortality caused by root pathogen complex in Deschutes National Forest, Oregon. *Plant Dis.* **66**, 240–248.

Fox, J.W., Wood, D.L. and Koehler, C.S. (1990). Distribution and abundance of engraver beetles (Scolytidae: *Ips* species) on Monterey pines infected with pitch canker. *Can. Entomol.* **122**, 1157–1166.

Fox, J.W., Wood, D.L., Koehler, C.S. and O'Keefe, S.T. (1991). Engraver beetles (Scolytidae: *Ips* species) as vectors of the pitch canker fungus, *Fusarium subglutinans*. *Can. Entomol.* **123**, 1355–1367.

Francke-Grosmann, H. (1963). Some new aspects in forest entomology. *Annu. Rev. Entomol.* **8**, 415–438.

Francke-Grosmann, H. (1967). Ectosymbiosis in wood-inhabiting insects. *In* "Symbiosis," Vol II (S.M. Henry, ed.) pp. 141–205. Academic Press, New York.

Franklin, R.T. (1970). Observations on the blue stain–southern pine beetle relationship. *J. Georgia Entomol.* **5**, 53–57.

Gibbs, J.N. (1968). Resin and the resistance of conifers to *Fomes annosus*. *Ann. Bot.* **32**, 649–665.

Gibbs, J.N. and French, D.W. (1980). The transmission of oak wilt. USDA Forest Serv. Res. Pap. NC-185. USDA Forest Serv. North Central Forest Exp. Stn., East Lansing, MI, 17 pp.

Gilbertson, R.L. (1974). "Fungi That Decay Ponderosa Pine." University of Arizona Press, Tucson, 197 pp.

Gilbertson, R.L. and Ryvarden, L. (1986 and 1987). "North American Polypores," Vols 1 and 2. Fungiflora, Oslo, 885 pp.

Goheen, D.J. and Cobb, F.W., Jr. (1978). Occurrence of *Verticicladiella wagenerii* and its perfect state, *Ceratocystis wageneri* sp. nov., in insect galleries. *Phytopathology* **68**, 1192–1195.

Goheen, D.J. and Filip, G.M. (1980). Root pathogen complexes in Pacific Northwest forests. *Plant Dis.* **64**, 793–794.

Graham, K. (1967). Fungal–insect mutualism. *Annu. Rev. Entomol.* **12**, 105–126.

Griffin, H.D. (1968). The genus *Ceratocystis* in Ontario. *Can. J. Bot.* **46**, 689–718.

Guillaumin, J.J., Mohammed, C. and Berthelay, S. (1989). *Armillaria* species in the northern temperate hemisphere. *In* "Proceedings of the Seventh International Conference on Root and Butt Rots" (D.J. Morrison, ed.) pp. 27–43. Forestry Canada, Pacific Forestry Centre, Victoria, BC.

Hansen, E.M. (1978). Incidence of *Verticicladiella wagenerii* and *Phellinus weirii* in Douglas-fir adjacent to and away from roads in western Oregon. *Plant Dis. Rep.* **62**, 179–181.

Hansen, E.M. (1979). Survival of *Phellinus weirii* in Douglas-fir stumps after logging. *Can. J. For. Res.* **9**, 484–488.

Harrington, T.C. (1980). Release of airborne basidiospores from the pouch fungus, *Cryptoporus volvatus*. *Mycologia* **72**, 926–936.

Harrington, T.C. (1981). Cycloheximide sensitivity as a taxonomic character in *Ceratocystis*. *Mycologia* **73**, 1123–1129.

Harrington, T.C. (1987). New combinations in *Ophiostoma* of *Ceratocystis* species with *Leptographium* anamorphs. *Mycotaxon* **28**, 39–43.

Harrington, T.C. (1988). *Leptographium* species, their distributions, hosts and insect vectors. *In* "*Leptographium* Root Diseases on Conifers" (T.C. Harrington and F.W. Cobb, Jr., eds), pp. 1–39. American Phytopathological Society Press, St. Paul, MN.

Harrington, T.C. and Cobb, F.W., Jr. (1983). Pathogenicity of *Leptographium* and *Verticicladiella* spp. isolated from roots of western North American conifers. *Phytopathology* **73**, 596–599.

Harrington, T.C. and Cobb, F.W., Jr. (1984). Host specialization of three morphological variants of *Verticicladiella wageneri*. *Phytopathology* **74**, 286–290.

Harrington, T.C. and Cobb, F.W., Jr. (1986). Varieties of *Verticicladiella wageneri*. *Mycologia* **78**, 562–567.

Harrington, T.C. and Cobb, F.W., Jr. (1987). *Leptographium wageneri* var. *pseudotsugae*, var. nov., cause of black stain root disease on Douglas-fir. *Mycotaxon* **30**, 501–507.

Harrington, T.C. and Cobb, F.W., Jr., eds (1988). "*Leptographium* Root Diseases on Conifers." American Phytopathological Society Press, St. Paul, MN, 149 pp.

Harrington, T.C. and Zambino, P.J. (1990). *Ceratocystiopsis ranaculosus*, not *Ceratocystis minor* var. *barrasii*, is the mycangial fungus of the southern pine beetle. *Mycotaxon* **38**, 103–115.

Harrington, T.C., Furniss, M.M. and Shaw, C.G. (1981). Dissemination of hymenomycetes by *Dendroctonus pseudotsugae* (Coleoptera: Scolytidae). *Phytopathology* **71**, 551–554.

Harrington, T.C., Reinhart, C., Thornburgh, D.A. and Cobb, F.W., Jr. (1983). Association of black-stain root disease with precommercial thinning of Douglas-fir. *Forest Sci.* **29**, 12–14.

Harrington, T.C., Cobb, F.W., Jr. and Lownsbery, J.W. (1985). Activity of *Hylastes nigrinus*, a vector of *Verticicladiella wageneri*, in thinned stands of Douglas-fir. *Can. J. For. Res.* **15**, 519–523.

Harrington, T.C., Worrall, J.J. and Rizzo, D.M. (1989). Compatibility among host-specialized isolates of *Heterobasidion annosum* from western North America. *Phytopathology* **79**, 290–296.

Harrington, T.C., Worrall, J.J. and Baker, F.A. (1992). *Armillaria*. *In* "Methods for Research on Soilborne Phytopathogenic Fungi" (L.L. Singleton, J.D. Mihail and C. Rush, eds). American Phytopathological Society Press, St. Paul, MN, pp. 81–85.

Hartig, R. (1878). "Die Zersetzungserscheinungen de Holzes, der Nadelbaume und der Eiche in forstlicher, botanischer und chemischer Richtung." Berlin, 151 pp.

Helton, W.A., Johnson, J.B. and Dilbeck, R.D. (1988). Arthropods as carriers of fungal wood-rotting pathogens of pome and stone fruit orchards. *Plant. Dis.* **72**, 1077.

Hessburg, P.F. and Hansen, E.M. (1986). Mechanisms of intertree transmission of *Verticicladiella wageneri* in young Douglas-fir. *Can. J. For. Res.* **16**, 1250–1254.

Hessburg, P.F. and Hansen, E.M. (1987). Pathological anatomy of black stain root disease of Douglas-fir. *Can. J. Bot.* **65**, 962–971.

Hetrick, L.A. (1949). Some overlooked relationships of southern pine beetle. *J. Econ. Entomol.* **42**, 466–469.

Hicks, B.R., Cobb, F.W., Jr. and Gersper, P.L. (1980). Isolation of *Ceratocystis wageneri* from forest soil with a selective medium. *Phytopathology* **70**, 880–883.

Himes, W.E. and Skelly, J.M. (1972). An association of the black terpentine beetle, *Dendroctonus terebrans*, and *Fomes annosus* in loblolly pine. *Phytopathology* **62**, 270 (Abstr.)

Hinds, T.E. (1972). Insect transmission of *Ceratocystis* species associated with aspen cankers. *Phytopathology* **62**, 221–225.

Hinds, T.E. and Davidson, R.W. (1972). *Ceratocystis* species associated with the aspen ambrosia beetle. *Mycologia* **64**, 405–409.

Hodges, C.S. (1969). Modes of infection and spread of *Fomes annosus*. *Annu. Rev. Phytopath.* **7**, 247–266.

Horntvedt, R., Christiansen, E., Solheim, H. and Wang, S. (1983). Artificial inoculation with *Ips typographus* associated blue-stain fungi can kill healthy Norway spruce trees. *Med. Norsk Inst. Skogforskn.* **38**, 1–20.

Hunt, D.W.A. and Borden, J.H. (1989). Terpene alcohol pheromone production by *Dendroctonus ponderosae* and *Ips paraconfusus* (Coleoptera: Scolytidae) in the absence of readily culturable microorganisms. *J. Chem. Ecol.* **15**, 1433–1463.

Hunt, J. (1956). Taxonomy of the genus *Ceratocystis*. *Lloydia* **19**, 1–58.

Hunt, R.S. and Cobb, F.W. (1982). Potential arthropod vectors and competing fungi of *Fomes annosus* in pine stumps. *Can. J. Plant Pathol.* **4**, 247–253.

Hutchison, L.J. and Reid, J. (1988). Taxonomy of some potential wood-staining fungi from New Zealand. 1. Ophiostomataceae. *New Zeal. J. Bot.* **26**, 63–81.

Juzwik, J. and French, D.W. (1983). *Ceratocystis fagacearum* and *C. piceae* on the surface of free-flying and fungus-mat-inhabiting nitidulids. *Phytopathology* **73**, 1164–1168.

Käärik, A. (1960). Growth and sporulation of *Ophiostoma* and some other blueing fungi on synthetic media. *Symb. Bot.* **16**, 1–159.

Kile, G.A. and Walker, J. (1987). *Chalara australis* sp. nov (Hyphomycetes), a vascular pathogen of *Nothofagus cunninghamii* (Fagaceae) in Australia and its relationship to other *Chalara* species. *Aust. J. Bot.* **35**, 1–32.

Klepzig, K.D., Raffa, K.F. and Smalley, E.B. (1991). Association of insect-fungal complexes with red pine decline in the Lake States. *Forest Sci.* **37**, 1119–1139.

Korhonen, K. (1978a). Intersterility groups of *Heterobasidion annosum*. *Commun. Inst. Forest. Fenn.* **94(6)**, 1–25.

Korhonen, K. (1978b). Infertility and clonal size in the *Armillariella mellea* complex. *Karstenia* **18**, 31–42.

Korhonen, K., Capretti, P., Moriondo, F. and Mugnai, L. (1989). A new breeding group of *Heterobasidion annosum* found in Europe. *In* "Proceedings of the Seventh International Conference on Root and Butt Rots" (D.J. Morrison, ed.), pp. 20–26. Forestry Canada, Pacific Forestry Centre, Victoria, BC.

Larsen, M.J. and Cobb-Poulle, L.A. (1990). "*Phellinus* (Hymenochaetaceae). A Survey of World Taxa." Fungiflora, Oslo, Norway.

Larsen, M.J. and Lombard, F.F. (1989). Taxonomy and nomenclature of *Phellinus weirii* in North America. *In* "Proceedings of the Seventh International Conference on Root and Butt Rots" (D.J. Morrison, ed.), pp. 573–578. Forestry Canada, Pacific Forestry Centre, Victoria, BC.

Leach, J.G., Orr, L.W. and Christensen, C. (1934). The inter-relationships of bark beetles and blue-staining fungi in felled Norway pine timber. *J. Agric. Res.* **49**, 315–341.

Leufvén, A. and Nehls, L. (1986). Quantification of different yeasts associated with the bark beetle, *Ips typographus*, during its attack on a spruce tree. *Microb. Ecol.* **12**, 237–243.

Leufvén, A., Bergstrom, G. and Falsen, E. (1988). Oxygenated monoterpenes produced by yeasts, isolated from *Ips typographus* (Coleoptera: Scolytidae) and grown in phloem medium. *J. Chem. Ecol.* **14**, 353–362.

Levieux, J., Lieuteir, F., Moser, J.C. and Perry, T.J. (1989). Transportation of phytopathogenic fungi by the bark beetle *Ips sexdentatus* Boerner and associated mites. *J. Appl. Entomol.* **108**, 1–11.

Lewis, K.J. and Alexander, S.A. (1986). Insects associated with the transmission of *Verticicladiella procera*. *Can. J. For. Res.* **16**, 1330–1333.

Lewis, K.J. and Hansen, E.M. (1989). Survival of *Inonotus tomentosus* and the infection of young stands. *In* "Proceedings of the Seventh International Conference on Root and Butt Rots" (D.J. Morrison, ed.), pp. 238–251. Forestry Canada, Pacific Forestry Centre, Victoria, BC.

Lu, K.C., Allen, D.G. and Bollen, W.B. (1957). Association of yeasts with the Douglas-fir beetle. *Forest Sci.* **3**, 336–343.

Mallet, K.I. and Hiratsuka, Y. (1988). Inoculation studies of lodgepole pine with Alberta isolates of the *Armillaria mellea* complex. *Can. J. For. Res.* **18**, 292–296.

Molnar, A.C. (1965). Pathogenic fungi associated with a bark beetle on alpine fir. *Can. J. Bot.* **43**, 563–570.

Moore, G.E. (1971). Mortality factors caused by pathogenic bacteria and fungi of the southern pine beetle in North Carolina. *J. Invert. Pathol.* **17**, 28–37.

Morrison, D.J., Chu, D. and Johnson, A.L.S. (1985). Species of *Armillaria* in British Columbia. *Can. J. Plant Pathol.* **7**, 242–246.

Moser, J.C. (1985). Use of sporothecae by phoretic tarsonemus mites to transport ascospores of coniferous bluestain fungi. *Trans. Br. Mycol. Soc.* **84**, 750–753.

Münch, E. (1907). Die Blaufaule des Nadelholzes II. *Naurw. Z. Land Forst.* **62**, 297–323.

Myren, D.T. and Patton, R.F. (1971). Establishment and spread of *Polyporus tomentosus* in pine and spruce plantations in Wisconsin. *Can. J. Bot.* **49**, 1033–1040.

Nag Raj, T.R. and Kendrick, B. (1975). "A Monograph of *Chalara* and Allied Genera." Wilfrid Laurier University Press, Waterloo, Ontario, 200 pp.

Olchowecki, A. and Reid, J. (1974). Taxonomy of the genus *Ceratocystis* in Manitoba. *Can. J. Bot.* **52**, 1675–1711.

Otrosina, W.J. and Cobb, F.W., Jr. (1987). Analysis of allozymes of three distinct variants of *Verticicladiella wageneri* isolated from conifers in western North America. *Phytopathology* **77**, 1360–1363.

Otrosina, W.J., Chase, T.E., Cobb, F.W., Jr. and Taylor, J.W. (1989). Isozyme structure of *Heterobasidion annosum* isolates relating to intersterility genotype. *In* "Proceedings of the Seventh International Conference on Root and Butt Rots" (D.J. Morrison, ed.), pp. 406–416. Forestry Canada, Pacific Forestry Centre, Victoria, BC.

Owen, D.R., Lindahl, K.Q., Jr., Wood, D.L. and Parmeter, J.R., Jr. (1987). Pathogenicity of fungi isolated from *Dendroctonus valens, D. brevicomis*, and *D. ponderosae* to pine seedlings. *Phytopathology* **77**, 631–636.

Parmeter, J.R., Jr., Slaughter, G.W., Chen, Mo-Mei, Wood, D.L. and Stubbs, H.A. (1989). Single and mixed inoculations of ponderosa pine with fungal associates of *Dendroctonus* spp. *Phytopathology* **79**, 768–772.

Perry, T.J. (1991). A Synopsis of the Taxonomic Revisions in the Genus *Ceratocystis* Including a Review of Bluestaining Species Associated with *Dendroctonus* Bark Beetles. USDA Forest Serv. Gen. Tech. Rpt. SO-86. USDA Forest Serv. Southern Forest Exp. Stn., New Orleans, LA, 16 pp.

Pettey, T.M. and Shaw, C.G. (1986). Isolation of *Fomitopsis pinicola* from in-flight bark beetles (Coleoptera: Scolytidae). *Can. J. Bot.* **64**, 1507–1509.

Pignal, M.C., Chararas, C., and Bourgeay-Causse, M. (1988). Yeasts from *Ips sexdentatus* (Scolytidae). Enzymatic activity and vitamin excretion. *Mycopathologia* **103**, 43–48.

Raabe, R.D. (1962). Host list of the root rot fungus, *Armillaria mellea*. *Hilgardia* **33**, 25–88.

Raffa, K.F. and Berryman, A.A. (1982). Accumulation of monoterpenes and associated volatiles following inoculation of grand fir with a fungus transmitted by the fir engraver beetle, *Scolytus ventralis* (Coleoptera: Scolytidae). *Can. Entomol.* **114**, 797–810.

Redfern, D.B., Stoakley, J.T., Steele, H. and Minter, D.W. (1987). Dieback and death of larch caused by *Ceratocystis laricicola* sp. nov. following attack by *Ips cembrae*. *Plant Pathol.* **36**, 467–480.

Reid, R.W., Whitney, H.S. and Watson, J.A. (1967). Reactions of lodgepole pine to attack by *Dendroctonus ponderosae* Hopkins and blue stain fungi. *Can. J. Bot.* **45**, 1115–1126.

Rhoads, A.S. (1942). Notes on *Clitocybe* rot of bananas and other plants in Florida. *Phytopathology* **32**, 487–496.

Rishbeth, J. (1985). Infection cycle of *Armillaria* and host response. *Eur. J. For. Pathol.* **15**, 332–341.

Rishbeth, J. (1988). Stump infection by *Armillaria* in first-rotation conifers. *Eur. J. For. Pathol.* **18**, 401–408.

Rizzo, D.M. and Harrington, T.C. (1988). Root and butt rot fungi on balsam fir and red spruce in the White Mountains, New Hampshire. *Plant Dis.* **72**, 329–331.

Roth, L.F., Harvey, R.D., Jr. and Kliejunas, J.T. (1987) Port-Orford-cedar root disease. USDA Forest Serv. FPM-PR-294-87, USDA Forest Serv. Forest Pest Manage., Region 6, Portland, OR, 11 pp.

Shaw, C.G., III. (1980). Characteristics of *Armillaria mellea* on pine root systems in expanding centers of root rot. *Northw. Sci.* **54**, 137–145.

Shaw, C.G., III and Kile, G.A., eds (1991). *Armillaria* root disease. USDA Agric. Handbook 691, USDA Forest Serv., Washington, DC, 233 pp.

Shaw, C.G., III and Loopstra, E.M. (1988). Identification and pathogenicity of some Alaskan isolates of *Armillaria. Phytopathology* **78**, 971–974.

Shifrine, M. and Phaff, H.J. (1956). The association of yeasts with certain bark beetles. *Mycologia* **48**, 41–55.

Shigo, A.L. and Tippett, J.T. (1981). Compartmentalization of decayed wood associated with *Armillaria mellea* in several tree species. USDA Forest Serv. Res. Pap. NE-488. USDA Forest Serv. Northeastern Forest Exp. Stn., Hamden, CT, 20 pp.

Smith, R.S., Jr. (1967). *Verticicladiella* root disease of pines. *Phytopathology* **57**, 935–938.

Solheim, H. (1986). Species of Ophiostomataceae isolated from *Picea abies* infested by the bark beetle *Ips typographus. Nord. J. Bot.* **6**, 199–207.

Stenlid, J. and Swedjemark, G. (1988). Differential growth of S- and P-isolates of *Heterobasidion annosum* in *Picea abies* and *Pinus sylvestris. Trans. Br. Mycol. Soc.* **90**, 209–213.

Thompson, W. (1984). Distribution, development, and functioning of mycelial cord systems of decomposer basidiomycetes of the deciduous woodland floor. *In* "The Ecology and Physiology of the Fungal Mycelium" (D.H. Jennings and A.D.M. Raynor, eds), pp. 185–214. Cambridge University Press, Cambridge.

Tsuneda, A. and Hiratsuka, Y. (1984). Sympodial and annellidic conidiation in *Ceratocystis clavigera. Can. J. Bot.* **62**, 2618–2624.

Upadhyay, H.P. (1981). "A Monograph of *Ceratocystis* and *Ceratocystiopsis*." University of Georgia Press, Athens, GA, 176 pp.

Upadhyay, H.P. and Kendrick, W.B. (1975). Prodromus for a revision of *Ceratocystis* (Microascales, Ascomycetes) and its conidial states. *Mycologia* **67**, 798–805.

Verrall, A.F. (1941). Dissemination of fungi that stain logs and lumber. *J. Agric. Res.* **63**, 549–558.

von Arx, J.A. (1974). "The Genera of Fungi Sporulating in Pure Culture," 2nd edn. J. Cramer, Vaduz, 315 pp.

von Arx, J.A. (1981). On *Monilia sitophila* and some families of Ascomycetes. *Sydowia* **34**, 13–29.

Watling, R., Kile, G.A. and Gregory, N.M. (1982). The genus *Armillaria* – nomenclature, typification, the identity of *Armillaria mellea* and species differentiation. *Trans. Br. Mycol. Soc.* **78**, 271–285.

Webb, R.S. and Alexander, S.A. (1985). An updated host index for *Heterobasidion annosum*. Virginia Polytechnic Institute Information Series No. 85-2. Virginia Polytechnic Institute, Blacksburg, VA, 27 pp.

Weijman, A.C.M. and de Hoog, G.S. (1975). On the subdivision of the genus *Ceratocystis. Antonie Leeuwenhoek J. Microbiol. Serol.* **41**, 353–360.

Whitney, H.S. (1971). Association of *Dendroctonus ponderosae* (Coleoptera: Scolytidae) with blue stain fungi and yeasts during brood development in lodgepole pine. *Can. Entomol.* **103**, 1495–1503.

Whitney, H.S. (1982). Relationships between bark beetles and symbiotic organisms. *In* "Bark Beetles in North American Conifers. A System for the Study of Evolutionary Biology" (J.B. Mitton and K. Sturgeon, eds), pp. 183–211. University of Texas Press, Austin.

Whitney, H.S. and Blauel, R.A. (1972). Ascospore dispersion in *Ceratocystis* spp. and *Europhium clavigerum* in conifier resin. *Mycologia* **64**, 410–414.

Whitney, H.S. and Cobb, F.W., Jr. (1972). Non-staining fungi associated with the bark beetle *Dendroctonus brevicomis* (Coleoptera: Scolytidae) on *Pinus ponderosa. Can. J. Bot.* **50**, 1943–1945.

Whitney, H.S. and Farris, S.H. (1970). Maxillary mycangium in the mountain pine beetle. *Science* **167**, 54–55.

Whitney, H.S., Ritchie, D.C., Borden, J.H. and Stock, A.J. (1984). The fungus *Beauveria bassiana* (Deuteromycotina: Hyphomycetaceae) in the western balsam bark beetle, *Dryocoetes confusus* (Coleoptera: Scolytidae). *Can. Entomol.* **116**, 1419–1424.

Whitney, H.S., Bandoni, R.J. and Oberwinkler, F. (1987). *Entomocorticium dendroctoni* gen. et sp.

nov. (Basidiomycotina), a possible nutritional symbiote of the mountain pine beetle in lodgepole pine in British Columbia. *Can. J. Bot.* **65**, 95–102.

Whitney, R.D. (1961). Root wounds and associated root rots of white spruce. *For. Chron.* **37**, 401–411.

Whitney, R.D. and MacDonald, G.B. (1985). Effects of root rot on the growth of balsam fir. *Can. J. For. Res.* **15**, 890–895.

Wingfield, M.J. (1985). Reclassification of *Verticicladiella* based on conidial development. *Trans. Br. Mycol. Soc.* **85**, 81–93.

Wingfield, M.J., Capretti, P. and MacKenzie, M. (1988a). *Leptographium* spp. as root pathogens of conifers. An international perspective. *In* "*Leptographium* Root Diseases on Conifers" (T.C. Harrington and F.W. Cobb, Jr., eds), pp. 113–128. American Phytopathological Society Press, St. Paul, MN.

Wingfield, M.J., van Wyk, P.S. and Marasas, W.F.O. (1988b). *Ceratocystiopsis proteae* sp. nov. with a new anamorph genus. *Mycologia* **80**, 23–30.

Witcosky, J.J., Schowalter, T.D. and Hansen, E.M. (1986). *Hylastes nigrinus* (Coleoptera: Scolytidae), *Pissodes fasciatus*, and *Steremnius carinatus* (Coleoptera: Curculionidae) as vectors of black stain root disease of Douglas-fir. *Environ. Entomol.* **15**, 1090–1095.

Worrall, J.J., Parmeter, J.R., Jr. and Cobb, F.W., Jr. (1983). Host specialization of *Heterobasidion annosum*. *Phytopathology* **73**, 304–307.

Wright, E. (1935). *Trichosporium symbioticum*, n. sp., a wood-staining fungus associated with *Scolytus ventralis*. *J. Agric. Res.* **50**, 525–539.

Yearian, W.C., Gouger, R.J. and Wilkinson, R.C. (1972). Effects of the bluestain fungus, *Ceratocystis ips*, on development of *Ips* bark beetles in pine bolts. *Ann. Entomol. Soc. Am.* **65**, 481–487.

Zambino, P.J. and Harrington, T.C. (1989). Isozyme variation within and among host-specialized varieties of *Leptographium wageneri*. *Mycologia* **8**, 122–133.

Zambino, P.J. and Harrington, T.C. (1990). Heterokaryosis and vegetative compatibility in *Leptographium wageneri*. *Phytopathology* **80**, 1460–1469.

Zentmeyer, G.A. (1980). "*Phytophthora cinnamomi* and the Diseases it Causes." Monogr. 10. American Phytopathological Society Press, St. Paul, MN, 96 pp.

Zink, P., and Fengel, D. (1988). Studies on the colouring matter of blue-stain fungi. Part 1. General characterization and the associated compounds. *Holzforschung* **42**, 217–220.

PART III
Factors Influencing Interactions

–4–
Abiotic and Biotic Predisposition

T. D. PAINE[1] and F. A. BAKER[2]
[1]*Department of Entomology, University of California, Riverside, CA, USA*
[2]*Department of Forest Resources, Utah State University, Logan, Utah, USA*

4.1 INTRODUCTION

Healthy trees typically limit insect and pathogen activity through physical and chemical defenses (see Chapter 5). Although adapted insects and pathogens can exploit healthy trees, their abundance usually increases on unhealthy trees with impaired defenses. Forest health or the condition of any particular stand within a forest may be reflected in the levels of bark beetle or pathogen activity. That is, insects and diseases may be both indicators of overall forest or stand stress and the proximal agents of forest or stand decline.

Manion (1981) introduced the concept of three groups of factors involved in tree or forest declines: predisposing factors, incitants, and contributing factors. Predisposing factors weaken an intolerant plant. Adverse climate, moisture regime and soil fertility, and air pollution pose long term stresses and predispose plants to other mortality factors. Incitants are short duration, relatively acute events (disturbances) that have a drastic effect. Incitants include storm damage, early or late frost, drought, flooding, salt spray, etc. Contributing factors, including most pests, such as bark beetles and pathogenic fungi, further the decline of a weakened host. Although these organisms may be associated with the ultimate death of the tree, they often are considered to be the cause of tree demise. In fact, hazard rating systems for many bark beetles are based on the physical appearance of individual trees. Apparently unhealthy trees are considered to be susceptible to bark beetle colonization

BEETLE–PATHOGEN INTERACTIONS IN CONIFER FORESTS
ISBN 0-12-628970-0

(Dunning, 1928; Person, 1928; Keen, 1936, 1943; Cobb *et al.*, 1968a,b, 1974; Stark *et al.*, 1968).

We expand this concept by including as predisposing factors those aggressive or virulent organisms, such as root pathogens, that can weaken vigorously growing trees. Thus, decline in forest health can be attributed to both abiotic and biotic factors that contribute to the susceptibility of trees to insects and pathogens.

It is often difficult to classify environmental factors as predisposing, incitant, or contributing factors because of interactions. For example, limiting amounts of available water have been associated with moisture stress and increased activity of bark beetles, and some root pathogens. However, the amount of water available to a particular tree is a function of microsite conditions, such as soil texture, daily temperatures, and the abundance of competing vegetation. Also, root rotting fungi can reduce the volume of fine roots and thereby impair the ability of the tree to acquire sufficient water and nutrients. The interaction of these abiotic and biotic factors, rather than their individual effects, can be critical. In the absence of competitors and root pathogens, the tree might have sufficient water available to prevent stress.

Further complicating our understanding of stress is the ability of trees to store considerable amounts of water, mineral nutrients and carbohydrates. Storage of critical resources may allow trees to survive relatively short periods of stress or, if the stress is chronic or severe, to decline over a period of years (Waring, 1987; Waring *et al.*, 1987). Scientists are only beginning to recognize in trees the physiological processes that are affected by stress (see Chapter 5).

4.2 ABIOTIC PREDISPOSING FACTORS

A variety of abiotic variables can predispose trees to bark beetles and pathogens. Studies relating bark beetle and pathogen activity to abiotic factors are summarized in Table 4.1.

4.2.1 Moisture Stress

One of the best recognized factors contributing to tree stress and predisposition to bark beetles and pathogens is moisture availability (Hicks, 1980; Mattson and Haack, 1987a, b; Chapter 5). Moisture availability is a major criterion included in hazard rating systems for evaluating stand susceptibility to bark beetles or pathogens (Lorio, 1968; Hicks *et al.*, 1978, 1981; Hertel, 1981; Lorio *et al.*, 1990). Both insufficient and excess water can stress trees and increase their susceptibility to invading organisms.

Water deficits can affect a broad range of physiological processes, many of which have been associated with tree resistance to insect or pathogen colonization (Chapter 5). Moisture deficit and shade have been found to cause fine root mortality in *Pseudotsuga menziesii* (Marshall, 1986). However, moisture stress apparently did not cause root mortality until roots exhausted their reserves of starch and sugar. Vité (1961) demonstrated that oleoresin exudation pressure was reduced in *Pinus ponderosa* stressed by water deficit. Lorio and Hodges (1968a) and Hodges and Lorio (1975) artificially induced water deficits in *Pinus taeda* and observed reduced oleoresin exudation pressure as well as changes in monoterpenes and resin acids comprising the oleoresin.

Bark beetles and pathogens often show greater colonization success and survival rates in moisture stressed trees. In a study of artificially stressed *Pinus taeda*, *Dendroctonus*

Table 4.1. Abiotic factors predisposing conifers to bark beetles and pathogens

Stress factor	Species affected	References
Drought	*Dendroctonus brevicomis*	Dunning (1928)
		Keen (1936, 1943)
	Dendroctonus frontalis	Lorio (1968)
		Hicks (1980)
	Scolytus ventralis	Ferrell (1978)
Excess moisture	*Dendroctonus frontalis*	Lorio (1968)
		Hicks *et al.* (1978)
	Phytophthora cinnamomi	Lorio and Hodges (1968b)
Soil/light	*Dendroctonus frontalis*	Bellandger *et al.* (1977)
	Dendroctonus ponderosae	Waring and Pitman (1985)
	Phytophthora cinnamomi	Fraedrich and Tainter (1985)
	Phellinus weirii	Entry *et al.* (1986)
	Armillaria spp.	Entry *et al.* (1986)
Physical injury	*Dendroctonus frontalis*	Coulson *et al.* (1983)
		Blanche *et al.* (1985)
	Armillaria spp.	McDonald *et al.* (1987)
Storm damage/fire	*Dendroctonus brevicomis*	Johnson (1966)
	Leptographium wageneri	Hansen (1978)
		Hadfield *et al.* (1986)
	Phaeolus schweinitzii	Geiszler *et al.* (1980)
Air pollution	*Dentroctonus brevicomis*	Stark *et al.* (1968)
	Dendroctonus ponderosae	Stark *et al.* (1968)
	Ips spp.	Sierpinski (1985)
	Heterobasidion annosum	James and Cobb (1982)

frontalis showed higher colonization and survival rates in water stressed trees compared to untreated trees (Lorio and Hodges, 1977). Ferrell (1978) demonstrated that water stressed *Abies concolor* were more susceptible to colonization by *Scolytus ventralis* than were well-hydrated trees. Otrosina and Ferrell (1989) reported that both greenhouse and field inoculations of *Abies* with *Trichosporium symbioticum*, a fungal associate of *S. ventralis*, caused longer cambial lesions in moisture stressed trees than in less stressed trees. Heavy infestations of *Phoradendron bolleanum* in the upper crown of *A. concolor* also induced water stress and increased the colonization success of *S. ventralis* (Ferrell, 1974).

Water-logged soil conditions also have been associated with increased tree susceptibility to bark beetles and pathogens (Lorio, 1968; Hicks *et al.*, 1978; Blanche *et al.* 1983). *Pinus taeda* growing on wet sites showed reduced oleoresin exudation pressure and increased infection of roots by *Phytophthora cinnamomi* (Lorio and Hodges, 1968b). Lorio *et al.* (1972) demonstrated that reduced fine root biomass under wet conditions contributed to increased tree susceptibility to bark beetles. Similarly, Paine and Stephen (1987) found that *Pinus taeda* growing on a poorly drained site had weaker induced defensive responses when inoculated with fungal associates of *D. frontalis* than did trees on a well drained site. Paine *et al.* (1987) also found differences in the quantitative composition of monterpenes in xylem resin of trees growing on the two sites.

4.2.2 Soil properties

Soil texture and fertility affect tree health and resistance to bark beetles and pathogens. *Phytophthora cinnamomi* is widely distributed and present in both diseased and disease-free stands of *Pinus echinata* and *P. taeda*, but *Phytophthora cinnamomi* develops only in stands where soils have poor internal drainage (Fraedrich and Tainter, 1989). After age 30, tree mortality increases and stand growth slows. Fraedrich and Tainter (1989) reported that essentially anaerobic conditions are necessary in the rhizosphere before root tips become predisposed to infection by *P. cinnamomi*. The difference in susceptibility between *Pinus echinata* and *P. taeda* could not be explained solely by their tolerance to anaerobic conditions. *Dendroctonus frontalis* preferentially attacks trees infected by *Phytophthora cinnamomi*, causing additional loss of stand productivity (Oak, 1985).

Soil composition and nutrient availability influence tree condition. High clay content at some sites has been correlated with increased infection of roots by *P. cinnamomi* and increased numbers of *D. frontalis* infestations (Bellanger *et al.*, 1977). Clay content may increase tree susceptibility to invading organisms through interaction with soil moisture (e.g. water penetration and holding capacity) or nutrients (e.g. chelation of cations), or through mechanical restriction of root production.

Soil fertility, particularly nitrogen levels, may directly affect tree vigor, growth rate, and susceptibility to invading organisms (Waring and Pitman, 1985). Interaction between availability of nutrients and light affected the susceptibility of *Tsuga mertensiana* seedlings to *Phellinus weirii*, and the susceptibility of *Pinus monticola* seedlings to *Armillaria* spp. (Entry *et al.*, 1986). Shaded seedlings and seedlings from which nitrogen was withheld developed more extensive foliar symptoms when inoculated with *P. weirii* than did untreated plants. Application of nitrogen reduced symptom development. These authors suggested that increased nitrogen availability, and perhaps increased light, improves tree resistance and limits infection in stands regenerating after pathogen-induced tree mortality.

Evidence from infested stands appears to support this hypothesis. Matson and Boone (1984) found that nitrogen availability was low in old-growth stands, but was 2- to 4-fold higher in areas with high levels of tree mortality, caused by *Phellinus weirii*, and subsequent regeneration. Nitrogen availability returned to old-growth levels in 85-year-old stands, that were infected again by the pathogen (Matson and Waring, 1984). However, these studies dealt with *P. weirii* on minor hosts growing on extreme sites. The disease does not behave similarly in lower elevation stands of *Pseudotsuga menziesii*.

4.2.3 Storm and fire damage

Storm events can injure trees and increase their susceptibility to bark beetles and pathogens. Lightning strikes to standing trees (Fig. 4.1) impair defense mechanisms and have been shown to increase tree susceptibility to colonization by bark beetles (Blanche *et al.*, 1983; Coulson *et al.*, 1983, 1985; Chapter 8). Increased frequency of attack by bark beetles has been observed on lightning struck trees (Johnson, 1966), and the number of bark beetle infestations has been associated with lightning strikes (Hodges and Pickard, 1971; Lorio and Bennett, 1974). Blanche *et al.* (1985) showed that resin flow nearly stopped during the three weeks following a lightning strike; resin chemistry also changed significantly (see Chapter 8). Lightning also may induce changes in the carbohydrate and nitrogen content of the inner bark (Smith, 1968; Hodges and Pickard, 1971). Beetles colonizing lightning struck trees may have greater reproductive success because of reduced host resistance and increased nutritional suitability (Johnson, 1966; Hodges and Pickard, 1971).

Wind can break branches and topple trees (Fig. 4.2), providing suitable resources for bark beetles and pathogenic fungi. Wind also may impair defense mechanisms against bark beetles and root pathogens in a manner that is often overlooked. Harrington (1986)

Fig. 4.1. Lightning injury (long scar and split bark on the tree to the right) to *Pinus taeda* in the southwestern US. (Photo by T.D. Schowalter.)

observed crown dieback in overstory *Picea rubens* and *Abies balsamea* in open stands on shallow, rocky soils subjected to extreme winds in the Appalachian Mountains in New Hampshire. Associated with these crown symptoms were loss of fine roots and reduced tree growth. Harrington (1986) attributed this condition to excessive root movement caused by tree sway induced by strong winds. Subsequently, Rizzo and Harrington (1988) showed that fine roots moved as much as 60 mm during tree movements. This movement resulted

Fig. 4.2. Windthrow in a *Pinus strobus* stand in Minnesota. (Photo by T.D. Schowalter.)

in breakage or abrasion of fine roots, creating wounds through which pathogens entered and causing xylem tissues to become non-conducting. Root loss, along with branch breakage, contributed to crown dieback and growth decline.

Fire damage to trees can promote bark beetle or pathogen entry and survival. Even non-catastrophic ground fires cause heat stress and damage to lower boles and subsurface roots, especially to trees near substantial litter accumulation or fallen logs (Geiszler *et al.*, 1980).

Geiszler *et al.* (1980) demonstrated that fire injury to roots and lower bole predisposed *Pinus contorta* to infection by *Phaeolus schweinitzii*, which subsequently increased tree susceptibility to *Dendroctonus ponderosae*.

4.3 BIOTIC PREDISPOSING FACTORS

Pathogens and insects can initiate tree declines and facilitate colonization by other insects or pathogens. Studies relating bark beetle and pathogen activity to biotic predisposing factors are summarized in Table 4.2.

4.3.1 Fungal diseases

Fungal pathogens can predispose trees to attack by other pathogens or insects. There is a substantial literature on the association of bark beetles with trees infected by root disease fungi. In parts of southeastern North America, *Dendroctonus frontalis* infestations frequently are associated with centers of *Heterobasidion annosum* infection (Alexander *et al.*, 1981; Skelly *et al.*, 1981). In western North America, root diseases and attack by bark beetles are associated on many hosts (Chapter 9). Partridge and Miller (1972) observed that bark beetles occurred in a greater proportion of trees with root diseases than in trees without root disease. *Leptographium wageneri* and *H. annosum* predisposed *Pinus ponderosa* to *D. brevicomis* and *D. ponderosae* (Cobb *et al.*, 1974; Goheen, 1976; Goheen and Cobb, 1980). *Dendroctonus ponderosae* and *D. valens* were attracted more frequently to *P. ponderosa* severely infected with *L. wageneri* than to symptomless trees (Goheen, 1985).

Hertert *et al.* (1975) concluded that endemic populations of *S. ventralis* were maintained in root-diseased *Abies grandis*. Endemic populations of *D. ponderosae* often prefer trees with root disease, especially that caused by *Armillaria* spp. (Livingston *et al.*, 1983; Hinds *et al.*, 1984; Johnson, 1984; Lessard *et al.*, 1985; Tkacz and Schmitz, 1986). *Armillaria* spp.

Table 4.2. Biotic factors predisposing conifers to bark beetles and pathogens

Species affected	Predisposing factor	References
Dendroctonus frontalis	*Heterobasidion annosum*	Alexander *et al.* (1981) Skelly *et al.* (1981)
	Phytophthora cinnamomi	Oak (1985)
Dendroctonus brevicomis	*Leptographium wageneri*	Cobb *et al.* (1974) Goheen (1976)
	Dendroctonus valens	Owen (1985)
Dendroctonus ponderosae	*Heterobasidion annosum*	Goheen and Cobb (1980)
	Cronartium commandrae	Rasmussen (1987)
	Armillaria spp.	Livingston *et al.* (1983) Hinds *et al.* (1984) Johnson (1984) Lessard *et al.* (1985)
Scolytus ventralis	*Arceuthobium abietinum*	Filip (1984)
	Cytospora abietis	Filip (1984)
	Phoradendron bolleanum	Ferrell (1974)
Armillaria spp.	*Choristoneura fumiferana*	Raske and Sutton (1986)
	Choristoneura pinus	Mallett and Volney (1989)
	Cronartium ribicola	Kulhavey *et al.* (1984)

often are associated with stress conditions, occurring frequently in trees stressed by other root diseases (Hansen and Goheen, 1988). However, in drier, interior forests of western North America, *Armillaria* spp. are considered more aggressive pathogens of *Abies*, *Pinus*, and *Pseudotsuga* than in wet, coastal forests (Wargo and Shaw, 1985).

Stem rusts and canker diseases also predispose trees to other pathogens and insects. Rasmussen (1987) observed more attacks by *D. ponderosae* in *Pinus contorta* var. *latifolia* infected with *Cronartium comandrae* than in uninfected trees. Kulhavey *et al.* (1984) proposed a sequence in which initial attack of *Pinus monticola* branches and stems by *Cronartium ribicola* weakens the tree, allowing *Armillaria* to colonize the roots. Trees further weakened by *Armillaria* then were attacked by *D. ponderosae* and later by less aggressive bark beetles. *Fusarium subglutinans* is a widespread canker disease of *Pinus* spp. in the southern US. Fox *et al.* (1990) reported that *Pinus radiata* infected with F. *subglutinans* were more attractive and susceptible to *Ips* spp.

4.3.2 Mistletoes

Other plant pathogens can predispose trees to insect or fungal pathogens. In central Oregon, *Abies arandis* infected by *Arceuthobium abietinum* and *Cytospora abietis* are subsequently attacked by *S. ventralis* (Fig. 4.3; Filip, 1984). *Arceuthobium* spp . may weaken their hosts, allowing trees to be killed by bark beetles (McCambridge, 1980; McCambridge *et al* ., 1982) or by *Coloradia pandora* defoliation (Wagner and Mathiasen, 1985). In some host–parasite combinations, bark beetles do not preferentially attack and kill *Arceuthobium*-infected trees, or the preference is not strong (Johnson *et al.*, 1976; McGregor, 1978; Rasmussen, 1987). Ferrell (1974) suggested that *Abies concolor* infested with *Phoradendron bolleanum* were

Fig. 4.3. *Abies grandis* mortality caused by *Arceuthobium abietinum* and *Scolytus ventralis*. (Photo by G.M. Filip.)

under greater water stress and produced less oleoresin, which in turn increased their susceptibility to attack by S. *ventralis*, compared to uninfested trees.

Miller and Keen (1960) observed that *Pinus ponderosa* mortality due to bark beetles was more than ten times greater in trees infested with *Arceuthobium campylopodum* (Fig. 4.4) than in uninfected trees. These authors also noted that *D. valens* initiated colonization of susceptible trees, and that these trees were four times more likely to be colonized subsequently by *D. brevicomis* than the "average" tree in the stand. Owen (1985) found that trees colonized by *D. valens* had a higher likelihood of being killed by *Leptographium wageneri*, *D. brevicomis* and/or *D. ponderosae* than did trees without *D. valens* attacks. Owen *et al.* (1987) also found that *Leptographium terebrantis*, the fungus associated with *D. valens*, was highly pathogenic to *P. ponderosa* seedlings.

In many *Arceuthobium*–host combinations, infested trees can live for many years before succumbing. In others, *Arceuthobium* causes large "witches' brooms" and tree mortality occurs within a relatively-few years of infection. The rapid mortality may result from predisposition of trees by *Arceuthobium* to insects and pathogenic fungi. For example, *Armillaria ostoyae* is found commonly on *Pinus banksiana* declining from *Arceuthobium* infection, but is absent from the roots of adjacent, more vigorous trees (Baker, personal observation). Declining *Arceuthobium*-infected *Picea mariana* often have *Leptographium* spp. on their roots whereas adjacent vigorous infested trees do not (Baker, personal observation).

4.3.3 Defoliators and other insects

Defoliation may predispose trees to attack by pathogens and bark beetles. Defoliation by *Choristoneura occidentalis* (Fig. 4.5) increased attacks on *Pseudotsuga menziesii* by *D.*

Fig. 4.4. *Arceuthobium campylopodum,* female plant, on *Pinus ponderosa.* (USDA Forest Serv. Photo by J. Thompson.)

Fig. 4.5. Larva and pupa of *Choristoneura occidentalis* on *Abies concolor*. (USDA Forest Serv. photo.)

pseudotsugae in Montana (Van Sickle, 1987). Raske and Sutton (1986) reported an episode of decline and mortality of *Picea mariana* after severe defoliation by *C. fumiferana*. *Polygraphus rufipennis* and a species of *Armillaria* were associated with the killed, defoliated trees. They also examined the fine roots of trees with varying degrees of crown damage caused by *C. fumiferana*. Trees with crown dieback had fewer fine roots than did undamaged trees, and trees attacked by *P. rufipennis* had even fewer fine roots. These authors speculated that *C. fumiferana* defoliation caused a reduction in the quantity of fine roots, which, as long as 2–3 years after recovery, contributed to water stress and allowed the bark beetle and root pathogen to cause further mortality. Mallett and Volney (1989) found *Armillaria* spp. on the roots of all dead *Pinus banksiana* defoliated for 2 years by *C. pinus*. The fungus was present on 60% of trees with dead tops and on only 20% of healthy trees.

Berryman and Wright (1978) reported that defoliation by *Orgyia pseudotsugata* increased vulnerability of *Abies grandis* and *Pseudotsuga menziesii* to *S. ventralis* and *D. pseudotsugae*, respectively. Fewer beetles were required to kill defoliated trees, and more offspring were produced in defoliated trees.

Other insects also may predispose trees to bark beetle or pathogen colonization. For example, *Ips* spp. often mine near resinous material exuded from sites of *Synanthedon sequoiae* feeding (Fox *et al.*, 1990).

4.4 ANTHROPOGENIC PREDISPOSING FACTORS

4.4.1 Air pollution

Air pollution is a relatively recent form of abiotic stress imposed on plant communities (Smith, 1981; Alstad *et al.*, 1982; Hain, 1987; Riemer and Whittaker, 1989). Cobb *et al.*

(1968b) suggested that photochemical atmospheric pollution has direct effects on the epidemiology of bark beetles. In one of the earliest studies, *Dendroctonus brevicomis* and *D. ponderosae* colonized smog-injured *Pinus ponderosa* more frequently than uninjured trees in the San Bernardino Mountains of California (Stark *et al.*, 1968). The damaged trees had lower oleoresin exudation pressure, reduced rate and yield of resin flow, thinner and drier phloem (Cobb *et al.* 1968a), as well as lowered concentrations of soluble sugars and polysaccharides (Miller *et al.*, 1968). More recently, increased tree mortality caused by *Ips* spp. in areas of moderate pollution damage has been reported from Europe (Sierpinski, 1985).

James and Cobb (1982) found no difference in pathogenicity between isolates of *Heterobasidion annosum* from areas with high ozone concentrations and those from areas with little or no pollution. However, trees with foliar injury caused by ozone were more susceptible to *H. annosum* than were undamaged trees in both field and greenhouse inoculation trials (James *et al.*, 1980). Sanitation or salvage logging to remove ozone damaged trees may increase the incidence and severity of this pathogen, which is most damaging in partially cut stands subject to frequent disturbance such as logging (Chapters 3 and 11).

Air pollution may contribute in some part to the decline of *Picea rubens* in the Appalachian Mountains in the eastern United States, but there is little evidence to suggest that pollution is predisposing trees to bark beetles in this region (Johnson and McLaughlin, 1986). *Armillaria* is associated with, but apparently not responsible for, the decline and mortality of *P. rubens* in northeastern forests (Carey *et al.*, 1984).

4.4.2 Forest management

A number of construction or management activities have been shown to promote tree susceptibility to bark beetles and pathogens in forests. New infection centers of *Leptographium wageneri* often are associated with road construction, precommercial thinning, and tractor logging (Hansen, 1978; Hadfield *et al.*, 1986; Witcosky *et al.*, 1986). Accumulating evidence indicates that *Armillaria* root disease differs in severity among sites, depending on degree of site disturbance (McDonald *et al.*, 1987). Although wounding may permit entry of the pathogens, soil compaction and damage to fine roots also may stress trees and induce physiological changes that attract beetle vectors (Witcosky *et al.*, 1987).

Any stress or damage can predispose trees to subsequent infection by pathogens and/or attack by bark beetles. Obviously, not all injury or stress will necessarily result in tree death. The extent of tree stress or injury, the size and proximity of bark beetle or pathogen populations, and the ability of surrounding trees to inhibit insect or pathogen spread determine the probability of an infestation (Paine *et al.*, 1984). Silvicultural practices in a stand can enhance resistance of trees or, under certain conditions, can predispose the stand to pathogens or insects (Kuhlman *et al.*, 1976; Schowalter, 1986; Baker, 1988; Filip and Schmitt, 1990; Chapter 11).

Many of the hazard rating systems developed for pine–bark beetle interactions include some estimate of tree vigor (e.g. average radial growth, percent live crown, or tree height), site quality (e.g. soil depth, elevation, slope, or percent clay), and tree density (e.g. species basal area) (Amman, 1977; Bellanger *et al.*, 1977; Hicks *et al.*, 1980; Ku *et al.*, 1980; Lorio and Sommers, 1985). While such rating systems give resource managers a basis for predicting potential for infestations and for assessing the need for treatments to reduce forest susceptibility, all rating systems are based on an underlying assumption that measured variables reflect abiotic and biotic stresses. Recommendations for silvicultural treatments are based on relieving those stresses (Chapter 11). However, if silvicultural practices are improperly applied, the risk to the stand can be increased.

Thinning is a common recommendation for relieving stresses due to competition for limited resources. Selectively reducing the number of trees in a stand can increase the growth and resistance of the remaining trees (Nebeker *et al.*, 1983; Brown *et al.*, 1987; Filip *et al.*, 1992). However, thinning also can increase the availability and susceptibility of decaying stem and root material for some bark beetles and pathogens (e.g. Witcosky *et al.*, 1986). Nebeker *et al.* (1983) presented a conceptual model that suggested that the abundance of slash could lead to increased populations of *Ips* spp., cerambycid beetles and the insect-vectored nematode, *Bursaphelenchus xylophilus*. In addition, the process of thinning can wound remaining trees and injure roots, providing entry courts for pathogens and ultimately reducing tree resistance to other organisms (Nebeker and Hodges, 1985; Wood *et al.*, 1985; Filip *et al.*, 1992).

Mueller-Dombois (1987) discussed forest declines, the death of groups of neighboring trees, in response to natural phenomena. Declines usually reflect multiple factors acting in concert. To date, no forest decline has been shown to result solely from abiotic or biotic agents.

4.5 FUTURE TRENDS

Major research efforts have addressed interactions among bark beetles, pathogens, and their host trees throughout North America and elsewhere (Waters *et al.*, 1985). Much of this effort has focused on the expression of resistance mechanisms in *Pinus* spp. It has become evident that tree resistance is affected greatly by both abiotic and biotic conditions. The importance of the interactions of trees with site, stand, and other environmental conditions to bark beetle and pathogen epidemiologies has been recognized and incorporated into infestation growth and proliferation models (Sharpe *et al.*, 1985; Chapter 10). However, there are growing concerns that the environment is changing and that these changes may alter the interaction among trees, beetles and pathogens (Chapter 1).

4.5.1 Atmospheric pollution

Levels of atmospheric pollution can directly affect tree and forest health, as discussed above (Cobb *et al.*, 1968b; Hain, 1987; Riemer and Whittaker, 1989). Current studies are focused on separating the effects of different classes of pollutants (e.g. ozone, SO_X, NO_X, peroxyacetyl nitrate (PAN), acid deposition). However, the interactive effects among the pollutants may be critical to plants, particularly if pollution levels increase progressively across wide geographic areas. Urban areas are encroaching on forests at an increasing rate as cities expand and as recreational communities in rural areas grow. The long-term effects of these demographic changes and associated levels of atmospheric pollution are not yet clear.

4.5.2 Global warming

The effect of increased levels of atmospheric CO_2 and potential warming of the globe also are not clear (Graham *et al.*, 1990). The higher CO_2 content may foster increased plant growth (Carlson, 1982). In areas of moderate climates, the length of the growing season might increase. However, the increased amounts of available carbon may alter the carbon/nitrogen balance of the tree so that nitrogen becomes limiting for growth. Suitable latitudinal and elevational ranges of tree species are likely to change, imposing stresses on trees at the margins of their current ranges (Franklin *et al.*, 1992; Gear and Huntley, 1991).

Bark beetles and pathogens are likely to become more active as global warming stresses trees at the margins of their geographic ranges. Increasing temperatures also will affect beetle and fungus development, survival and reproduction. Beetle development rate depends on temperature (Miller and Keen, 1960). Temperatures must exceed minimum thresholds for larvae to develop and adults to become active (McCambridge, 1974; McClelland and Hain, 1979). However, survival of adults flying between trees can be reduced at high temperatures (Coulson *et al.*, 1979). Global warming may lengthen the season of activity for bark beetles and pathogens, putting trees at risk for a longer period of time.

There is a great deal of uncertainty about the potential changes in precipitation patterns that could result from increased atmospheric temperatures. If rainfall increases in forests, then growth and resistance may be enhanced, assuming that increased precipitation does not become excessive or predispose trees to infection by pathogenic fungi. If rainfall is insufficient to meet evapotranspiration demand, trees may become more stressed and susceptible to colonization by bark beetles and root pathogens, as currently indicated by unprecedented bark beetle activity throughout the western US after 5 years of drought.

Atmospheric changes and enhanced activity of insects and pathogens may alter plant communities. Some species might experience increased mortality and reduced reproductive success. Other species could develop a competitive advantage in the community and show population growth. Consequently, the geographic distributions of individual species or entire communities may expand or contract (Gear and Huntley, 1991). Insects and diseases may be the apparent proximal agents associated with these changes in plant communities, but the increased levels of pollutants, CO_2, and temperature may be the factors driving the changes.

4.6 CONCLUSIONS

Tree condition and forest health are major factors influencing bark beetle and pathogen epidemiologies. A number of abiotic and biotic factors reduce tree vigor and increase susceptibility to invasion by these organisms. Major abiotic factors include moisture limitation or flooding, soil properties that limit nutrient availability, and injury resulting from storms or fire. Biotic factors include prior infection by pathogenic fungi or mistletoes, and infestation by defoliators or other insects that weaken host trees. In addition, anthropogenic factors, such as atmospheric pollution and forest management practices, often stress trees and create susceptibility targets for bark beetles and pathogens. Global warming is likely to exacerbate tree stress, especially near the margins of species geographical ranges.

Management of bark beetle–pathogen interactions requires attention to predisposing factors. Identification of trees and stands in susceptible condition is necessary to remedy stressful conditions, where possible, or to apply other management strategies, as appropriate (see Chapters 9 and 11).

REFERENCES

Alexander, S.A., Skelly, J.M. and Webb, R.S. (1981). Effects of *Heterobasidion annosum* on radial growth in southern pine beetle infested loblolly pine. *Phytopathology* **71**, 479–481.

Alstad, D.N., Edmunds, G.F., Jr. and Weinstein, L.H. (1982). Effects of air pollutants on insect populations. *Annu. Rev. Entomol.* **27**,369–384.

Amman, G.D. (1977). The role of mountain pine beetle in lodgepole pine ecosystems, impact on succession. *In* "Arthropods in Forest Ecosystems" (W.J. Mattson, ed.), pp. 3–18. Springer-Verlag, New York.

Baker, F.A. (1988). The influence of forest management on pathogens. *Northw. Environ. J.* **4**, 229–246.

Bellanger, R.P., Hatchell, G.E. and Moore, G.E. (1977). Soil and stand characteristics related to southern pine beetle infestations: a progress report for Georgia and North Carolina. *In* "Proc. Sixth Southern Forest Soils Workshop," pp. 99–107. USDA Forest Serv., Southeast Area State & Private Forestry, Atlanta, GA.

Berryman, A.A. and Wright, L.C. (1978). Defoliation, tree condition, and bark beetles. *In* "The Douglas-fir Tussock Moth: a Synthesis" (M.H. Brookes, R.W. Stark and R.W. Campbell, eds), pp. 81–88. USDA Forest Serv. Tech. Bull. 1585. USDA Forest Serv., Washington, DC.

Blanche, C.A., Hodges, J.D., Nebeker, T.E. and Moehring, D.M. (1983). Southern pine beetle: the host dimension. Mississippi Agric. & Forestry Exp. Stn. Bull. 917, Mississippi State University, Mississippi State, MS, 29 pp.

Blanche, C.A., Hodges, J.D. and Nebeker, T.E. (1985). Changes in bark beetle susceptibility indicators in a lightning-struck loblolly pine. *Can. J. For. Res.* **15**, 397–399.

Brown, M.W., Nebeker, T.E. and Honea, C.R. (1987). Thinning increases loblolly pine vigor and resistance to bark beetles. *South. J. Appl. For.* **11**, 28–31.

Carlson, R.W. (1982). The influence of elevated atmospheric CO2 on the growth and photosynthesis of early successional species. *In* "Effects of Gaseous Air Pollution in Agriculture and Horticulture" (M.H. Unsworth and D.P Ormrod, eds), pp. 489–491. Butterworth Scientific, London.

Carey, A.C., Miller, E.A., Geballe, G.T., Wargo, P.M., Smith, W.H. and Siccama, T.G. (1984). *Armillaria mellea* and decline of red spruce. *Plant Dis.* **68**, 794–795.

Cobb, F.W., Jr., Wood, D.L., Stark, R.W. and Miller, P.R. (1968a). Effect of injury upon physical properties of oleoresin, moisture content, and phloem thickness. *Hilgardia* **39**, 127–134.

Cobb, F.W., Jr., Wood, D.L., Stark, R.W. and Parmeter, J.R., Jr. (1968b). Theory on the relationships between oxidant injury and bark beetle infestation. *Hilgardia* **39**, 141–152.

Cobb, F.W., Jr., Parmeter, J.R., Jr., Wood, D.L. and Stark, R.W. (1974). Root pathogens as agents predisposing ponderosa pine and white fir to bark beetles. *In* "Proc. Fourth International Conference on *Fomes annosus*," pp. 8–15. USDA Forest Serv., Washington, DC.

Coulson, R.N., Pulley, P.E., Pope, D.N. Fargo, W.S., Gagne, J.A. and Kelly, C.L. (1979). Estimation of survival and allocation of adult southern pine beetles between trees during development of an infestation. *In* "Dispersal of Forest Insects: Evaluation, Theory, and Management Implications" (A.A. Berryman and L. Safranyik, eds), pp. 194–212, Washington State University, Pullman, WA.

Coulson, R.N., Hennier, P.B., Flamm, R.O., Rykiel, E.J., Hu, L.C. and Payne, T.L. (1983). The role of lightning in the epidemiology of the southern pine beetle. *Z. Ang. Entomol.* **96**, 182–193.

Coulson, R.N., Saunders, M.C., Payne, T.L., Flamm, R.O., Wagner, T.L., Hennier, P.B. and Rykiel, E.J. (1985). A conceptual model of the role of lightning in the epidemiology of the southern pine beetle. *In* "The Role of the Host in Population Dynamics of Forest Insects" (L. Safranyik, ed.), pp. 136–146. Canadian Forestry Serv., Pacific Forestry Centre, Victoria, BC.

Dunning, D. (1928). A tree classification system for the selection forests of the Sierra Nevada. *J. Agric. Res.* **36**, 755–771.

Entry, J.A., Martin, N.E., Cromack, K., Jr. and Stafford, S.G. (1986). Light and nutrient limitation in *Pinus monticola*, seedling susceptibility to *Armillaria* infection. *For. Ecol. Manage.* **17**, 189–198.

Ferrell, G.T. (1974). Moisture stress and fir engraver (Coleoptera: Scolytidae) attack in white fir infected by true mistletoe. *Can. Entomol.* **106**, 315–318.

Ferrell, G.T. (1978). Moisture stress threshold of susceptibility to fir engraver beetles in pole-size white firs. *Forest Sci.* **24**, 85–92.

Filip, G.M. (1984). Dwarf mistletoe and cytospora canker decrease grand fir growth in central Oregon. *Forest Sci.* **30**, 1071–1079.

Filip, G.M. and Schmitt, C.L. (1990). R_x for *Abies*: Silvicultural options for diseased firs in Oregon and Washington. USDA Forest Serv. Gen. Tech. Rpt. PNW-GTR-252. USDA Forest Serv. Pacific Northwest Res. Stn., Portland, OR, 34 pp.

Filip, G.M., Wickman, B.E., Mason, R.R., Parks, C.A. and Hosman, K.P. (1992). Thinning and nitrogen fertilization in a grand fir stand infested with western spruce budworm. Part III. Tree wound dynamics. *Forest Sci.* **38**,.265–274.

Fox, J.W., Wood, D.L. and Koehler, C.S. (1990). Distribution and abundance of engraver beetles (Scolytidae: *Ips* species) on Monterey pines infected with pitch canker. *Can. Entomol.* **122**,1157–1166.

Fraedrich, S.W. and Tainter, F.H. (1989). Effect of dissolved oxygen concentration on the relative susceptibility of shortleaf and loblolly pine root tips to *Phytophthora cinnamomi. Phytopathology* **79**, 1114–1118.

Franklin, J.F., Swanson, F.J., Harmon, M.E., Perry, D.A., Spies, T.A., Dale, V.H., McKee, A., Ferrell, W.K., Means, J.E., Gregory, S.V., Lattin, J.D., Schowalter, T.D. and Larson, D. (1992). Effects of global climatic change on forests in Northwestern North America. *In* "Global Warming and Biological Diversity" (R.L. Peters and T.E. Lovejoy, eds). pp 244–257. Yale University Press, New Haven, CT.

Gear, A.J. and Huntley, B. (1991). Rapid changes in the range limits of Scots pine 4000 years ago. *Science* **251**,544–547.

Geiszler, D.R., Gara, R.I., Driver, C.H., Gallucci, V.F. and Martin, R.E. (1980). Fire, fungi, and beetle influences on a lodgepole pine ecosystem in south-central Oregon. *Oecologia* **46**, 239–243.

Goheen, D.J. (1976). *Verticicladiella wagenerii* in *Pinus ponderosa*: epidemiology and interrelationships with insects. Ph.D. Dissertation, University of California, Berkeley, CA, 118 pp.

Goheen, D.J. and Cobb, F.W., Jr. (1980). Infestation of *Ceratocystis wageneri*-infected ponderosa pine by bark beetles (Coleoptera: Scolytidae) in the central Sierra Nevada. *Can. Entomol.* **112**, 725-730.

Goheen, D.J., Cobb, F.W., Jr., Wood, D.L. and Rowney, D.L. (1985). Visitation frequencies of some insect species on *Ceratocystis wageneri*-infected and apparently healthy ponderosa pines. *Can. Entomol.* **117**, 1535–1543.

Graham, R.L., Turner, M.G. and Dale, V.H. (1990). How increasing CO2 and climate change affect forests. *BioScience* **40**,575–587.

Hadfield, J.S., Goheen, D.J., Filip, G.M., Schmitt, C.L. and Harvey, R.D. (1986). Root diseases in Oregon and Washington conifers. USDA Forest Serv. FPM Rpt. R6-250-86. USDA Forest Serv. Forest Pest Management, Region 6, Portland, OR, 27 pp.

Hain, F.P. (1987). Interactions of insects, trees and air pollutants. *Tree Physiol.* **3**, 93–102.

Hansen, E.M. (1978). Incidence of *Verticicladiella wageneri* and *Phellinus weirii* adjacent to and away from roads in western Oregon. *Plant Dis. Rep.* **62**, 179–181.

Hansen, E.M. and Goheen, D.J. (1988). Root disease complexes in the Pacific Northwest. *In* "Proc. Seventh International Conference on Root and Butt Rots" (D.J. Morrison, ed.), pp. 129–141. Forestry Canada, Pacific Forestry Centre, Victoria, BC.

Harrington, T.C. (1986). Growth decline of wind-exposed red spruce and balsam fir in the White Mountains. *Can. J. For. Res.* **16**, 232–238.

Hertel, G.D. (1981). Implementation experiences: an overview. *In* "Hazard-rating Systems in Forest Insect Pest Management" (R.L. Hedden, S.J. Barras and J.E. Coster, eds), pp. 13–21. USDA Forest Serv. Gen. Tech. Rpt. WO-27, USDA Forest Serv., Washington, DC.

Hertert, H.D., Miller, D.L. and Partridge, A.D. (1975). Interaction of bark beetles (Coleoptera: Scolytidae) and root-rot pathogens in grand fir in northern Idaho. *Can. Entomol.* **107**, 899–904.

Hicks, R.R., Jr. (1980). Climate, site, and stand factors. *In* "The Southern Pine Beetle" (R.C. Thatcher, J.L. Searcy, J.E. Coster and G.D. Hertel, eds), pp. 55–68. USDA Forest Serv. Tech. Bull. 1631, USDA Forest Serv., Washington, DC.

Hicks, R.R., Jr., Howard, J.E., Coster, J.E. and Watterston, K.G. (1978). The role of tree vigor in susceptibility of loblolly pine to southern pine beetle. *In* "Proc. Fifth North American Forest Biology Workshop" (C.A. Hollis and A.E. Squillace, eds), pp. 177–181. University of Florida Press, Gainesville, FL.

Hicks, R.R., Jr., Howard, J.E., Watterston, K.G. and Coster, J.E. (1980). Rating forest stand susceptibility to southern pine beetle in east Texas. *For. Ecol. Manage.* **2**, 269–283.

Hicks., R.R., Jr., Howard, J.E., Watterston, K.G. and Coster, J.E. (1981). Rating east Texas stands for southern pine beetle susceptibility. *South. J. Appl. For.* **5**, 7–10.

Hinds, T.E., Fuller, L.R., Lessar, E.D. and Johnson, D.W. (1984). Mountain pine beetle infestation and *Armillaria* root disease of ponderosa pine in the Black Hills of South Dakota. USDA Forest Serv. Tech. Rpt. R2-30, USDA Forest Serv., Forest Pest Manage., Region 2, Fort Collins, CO, 7 pp.

Hodges, J.D. and Lorio, P.L., Jr. (197S). Moisture stress and composition of xylem oleoresin in loblolly pine. *Forest Sci.* **21**, 283–290.

Hodges, J.D. and Pickard, L.S. (1971). Lightning in the ecology of the southern pine beetle, *Dendroctonus frontalis* (Coleoptera: Scolytidae). *Can. Entomol.* **103**, 44–51.

James, R.L. and Cobb, F.W., Jr. (1982). Variability in virulence of *Heterobasidion annosum* from ponderosa and Jeffrey pine in areas of high and low photochemical air pollution. *Plant Dis.* **66**, 835–837.

James, R.L., Cobb, F.W., Jr., Miller, P.R. and Parmeter, J.R., Jr. (1980). Effects of oxidant air pollution on susceptibility of pine roots to *Fomes annosus*. *Phytopathology* **70**, 560–563.

Johnson, A.H. and McLaughlin, S.B. (1986). The nature and timing of deterioration of red spruce in the northern Appalachian Mountains. *In* "Acid Deposition Long Term Trends," pp. 200–230. Nat. Acad. Press, Washington, DC.

Johnson, D.W. (1984). An assessment of root diseases in the Rocky Mountain region. USDA Forest Serv. Tech. Rpt. R2-29. USDA Forest Serv., Forest Pest Manage., Region 2, Fort Collins, CO.

Johnson, D.W., Yarger, L.C., Minnemeyer, C.D. and Pace, V.E. (1976). Dwarf mistletoe as a predisposing factor for mountain pine beetle attack of ponderosa pine in the Colorado Front Range. USDA Forest Serv. Tech. Rpt. R2-4. USDA Forest Serv., Forest Pest Manage., Region 2, Fort Collins, CO, 13 pp.

Johnson, P.C. (1966). Attractiveness of lightning-struck ponderosa pine trees to *Dendroctonus brevicomis* (Coleoptera: Scolytidae). *Ann. Entomol. Soc. Am.* **59**, 615.

Keen, F.P. (1936). Relative susceptibility of ponderosa pine to bark beetle attack. *J. Forestry* **34**, 919–927.

Keen, F.P. (1943). Ponderosa pine tree classes redefined. *J. Forestry* **41**, 249–253.

Ku, *T.T.*, Sweeney, J.M. and Shelburne, V.B. (1980). Site and stand conditions associated with southern pine beetle outbreaks in Arkansas – a hazard rating system. *South. J. Appl. For.* **4**, 103–106.

Kuhlman, E.G., Hodges, C.S., Jr. and Froelich, R.C. (1976). Minimizing losses to *Fomes annosus* in the southern United States. USDA Forest Serv. Res. Paper SE-151, USDA Forest Serv. Southeastern Forest Exp. Stn., Asheville, NC, 16 pp.

Kulhavey, D.L., Partridge, A.D. and Stark, R.W. (1984). Root diseases and blister rust associated with bark beetles (Coleoptera: Scolytidae) in western white pine in Idaho. *Environ. Entomol.* **13**, 813–817.

Lessard, G., Johnson, D.W., Hinds, T.E. and Hoskins, W.H. (1985). Association of *Armillaria* root disease with mountain pine beetle infestations on the Black Hills National Forest, South Dakota. USDA Forest Serv. Rpt. 85-4, USDA Forest Serv., Forest-Pest Manage., Region 2, Fort Collins, CO, 6 pp.

Livingston, W.J., Mangini, A.C., Kinzer, H.G. and Mielke, M.E. (1983). Association of root diseases and bark beetles (Coleoptera: Scolytidae) with *Pinus ponderosa* in New Mexico. *Plant Dis.* **67**, 674–676.

Lorio, P.L., Jr. (1968). Soil and stand conditions related to southern pine beetle activity in Hardin County, Texas. *J. Econ. Entomol.* **61**, 565–566.

Lorio, P.L., Jr. and Bennett, W.H. (1974). Recurring southern pine beetle infestations near Oakdale, Louisiana. USDA Forest Serv. Res. Paper SO-95, USDA Forest Serv., Southern Forest Exp. Stn., New Orleans, LA, 6 pp.

Lorio, P.L., Jr. and Hodges, J.D. (1968a). Oleoresin exudation pressure and relative water content of inner bark as indicators of moisture stress in loblolly pines. *Forest Sci.* **14**, 392–398.

Lorio, P.L., Jr. and Hodges, J.D. (1968b). Microsite effects on oleoresin exudation pressure of large loblolly pines. *Ecology* **49**, 1207–1210.

Lorio, P.L., Jr. and Hodges, J.D. (1977). Tree water status affects induced southern pine beetle attack and brood production. USDA Forest Serv. Res. Paper SO-135, USDA Forest Serv., Southern Forest Exp. Stn., New Orleans, LA, 7 pp.

Lorio, P.L., Jr. and Sommers, R.A. (1985). Potential use of soil maps to estimate southern pine beetle risk. *In* "Proc. Integrated Pest Management Symposium" (S.J. Branham and R.C. Thatcher, eds), pp. 239–245. USDA Forest Serv. Gen. Tech. Rpt. SO-56, USDA Forest Serv., Southern Forest Exp. Stn., New Orleans, LA.

Lorio, P.L., Jr., Howe, V.K. and Martin, C.N. (1972). Loblolly pine rooting varies with microrelief on wet sites. *Ecology* **53**, 1134–1140.

Lorio, P.L., Jr., Sommers, R.A., Blanche, C.A., Hodges, J.D. and Nebeker, T.E. (1990). Modeling pine resistance to bark beetles based on growth and differentiation balance principles. *In* "Process modeling of forest growth responses to environmental stress" (R.K. Dixon, R.S. Meldahl, G.A. Ruark, and W.G. Warren, eds), pp. 402–409. Timber Press, Portland, OR.

Mallett, K.I. and Volney, J.A. (1989). Correlation of *Armillaria* root rot with damage to trees defoliated by the jack pine budworm. *Phytopathology* **79**,1166.

Manion, P.D. (1981). "Tree Disease Concepts." Prentice-Hall, Englewood Cliffs, NJ, 399 pp.

Marshall, J.D. (1986). Drought and shade interact to cause fine-root mortality in Douglas-fir seedlings. *Plant Soil* **91**, 51–60.

Matson P.A. and Boone, R.D. (1984). Natural disturbance and nitrogen mineralization: wave-form dieback of mountain hemlock in the Oregon Cascades. *Ecology* **65**, 1511–1516.

Matson, P.A. and Waring, R.H. (1984). Effects of nutrient and light limitation on mountain hemlock: susceptibility to laminated root rot. *Ecology* **65**, 1517–1524.

Mattson, W.J. and Haack, R.A. (1987a). The role of drought in outbreaks of plant-eating insects. *BioScience* **37**, 110–118.

Mattson, W.J. and Haack, R.A. (1987b). The role of drought stress in provoking outbreaks of phytophagous insects. *In* "Insect Outbreaks" (P. Barbosa and J.C. Schultz, eds), pp. 365–407. Academic Press, San Diego.

McCambridge, W.F. (1974). Influence of low temperatures on attack, oviposition, and larval development of mountain pine beetle, *Dendroctonus ponderosae* (Coleoptera: Scolytidae). *Can. Entomol.* **106**, 979–984.

McCambridge, W.F. (1980). Some mountain pine beetle infestation characteristics in dwarf mistletoe infected and uninfected ponderosa pine. USDA Forest Serv. Res. Note RM-39, USDA Forest Serv., Rocky Mountain Res. Stn., Ft. Collins, CO, 2 pp.

McCambridge, W.F., Hawksworth, F.G., Edminster, C.B. and Laut, J.G. (1982). Ponderosa pine mortality resulting from a mountain pine beetle outbreak. USDA Forest Serv. Res. Paper RM-235, USDA Forest Serv., Rocky Mountain Res. Stn., Ft. Collins, CO, 7 pp.

McClelland, W.T. and Hain, F.P. (1979). Survival of declining *Dendroctonus frontalis* populations during a severe and a nonsevere winter. *Environ. Entomol.* **8**, 231–235.

McDonald, G.I., Martin, N.E. and Harvey, A.E. (1987). *Armillaria* in the Northern Rockies: pathogenicity and host susceptibility on pristine and disturbed sites. USDA Forest Serv. Res. Note INT-371, USDA Forest Serv., Intermountain Res. Stn., Odgen, UT, 5 pp.

McGregor, M.D. (1978). Management of mountain pine beetle in lodgepole pine stands in the Rocky Mountain area. *In* "Theory and Practice of Mountain Pine Beetle Management in Lodgepole Pine Forests: Symposium Proceedings" (A.A. Berryman, G.D. Amman and R.W. Stark, eds), pp. 129–139. University of Idaho, Moscow, ID.

Miller, J.M. and Keen, F.P. (1960). Biology and control of the western pine beetle, USDA Forest Serv. Misc. Publ. 800, USDA Forest Serv., Washington, DC, 381 pp.

Miller, P.R., Cobb, F.W., Jr., and Zavarin, E. (1968). Effect of injury upon oleoresin composition, phloem carbohydrates and phloem pH. *Hilgardia* **39**, 135–140.

Mueller-Dombois, D. (1987). Natural dieback in forests. *BioScience* **37**, 575–583.

Nebeker, T.E. and Hodges, J.D. (1985). Thinning and harvesting practices to minimize site and stand disturbance and susceptibility to bark beetle and disease attacks. *In* "Proc. Integrated Pest Management Symposium" (S.J. Branham and R.C. Thatcher, eds), pp. 263–271. USDA Forest Serv. Gen. Tech. Rpt. SO-56, USDA Forest Serv. Southern Forest Exp. Stn., New Orleans, LA.

Nebeker, T.E., Moehring, D.M., Hodges, J.D., Brown, M.W. and Blanche, C.A. (1983). Impact of thinning on host susceptibility. *In* "Proc. Second Biennial Southern Silvicultural Research Conference" (E.P. Jones, Jr., ed.), pp. 376–381. USDA Forest Serv. Gen. Tech. Rpt. SE-24, USDA Forest Serv. Southeastern Forest Exp. Stn., Asheville, NC.

Oak, S.W. (1985). Adaptation of littleleaf disease hazard rating for use in forest management in South Carolina national forests. *In* "Proc. Integrated Pest Management Symposium" (S.J. Branham and R.C. Thatcher, eds), pp. 246–251. USDA Forest Serv. Gen. Tech. Rpt. SO-56, USDA Forest Serv., Southern Forest Exp. Stn., New Orleans, LA.

Otrosina, W.J. and Ferrell, G.T. (1989). Effects of water stress on white fir wound response to *Trichosporium symbioticum*, fungal symbiont of the fir engraver. *Phytopathology* **79**, 1164.

Owen, D.R. (1985). The role of *Dendroctonus valens* and its vectored fungi in the mortality of ponderosa pine. Ph.D. Dissertation, University of California, Berkeley, CA.

Owen, D.R., Lindahl, K.Q., Jr., Wood, D.L. and Parmeter, J.R., Jr. (1987). Pathogenicity of fungi isolated from *Dendroctonus valens, D. brevicomis,* and *D. ponderosae* to ponderosa pine seedlings. *Phytopathology* **77**,631–636.

Paine, T.D. and Stephen, F.M. (1987). Influence of tree stress and site quality on the induced defense system of loblolly pine. *Can. J. For. Res.* **17**, 569–571.

Paine, T.D., Stephen, F.M. and Taha, H.A. (1984). Conceptual model of infestation probability based on bark beetle abundance and host tree susceptibility. *Environ. Entomol.* **13**, 619–624.

Paine, T.D., Blanche, C.A., Nebeker, T.E. and Stephen, F.M. (1987). Composition of loblolly pine resin defenses: comparison of monoterpenes from induced lesion and sapwood resin. *Can. J. For. Res.* **17**, 1202–1206.

Partridge, A.L. and Miller, D.L. (1972). Bark beetles and root rots related in Idaho conifers. *Plant Dis. Rep.* **56**, 498–500.

Person, H.L. (1928). Tree selection by the western pine beetle. *J. Forestry* **26**, 564–578.

Raske, A.G. and Sutton, W.J. (1986). Decline and mortality of black spruce caused by spruce budworm defoliation and secondary organisms. Canadian Forestry Serv. Info. Rpt. N-X-236, 29 pp.

Rasmussen, L.A. (1987). Mountain pine beetle selection of dwarf mistletoe and commandra blister rust infected lodgepole pine. USDA Forest Serv. Res. Note INT-367, USDA Forest Serv., Intermountain Res. Stn., Ogden, UT, 3 pp.

Riemer, J. and Whittaker, J.B. (1989). Air pollution and insect herbivores: observed interactions and possible mechanisms. *In* "Insect-Plant Interactions," Vol. 1 (E.A. Bernays, ed.), pp. 73–105. CRC Press, Boca Raton.

Rizzo, D.M. and Harrington, T.C. (1988). Root movement and root damage of red spruce and balsam fir on subalpine sites in the White Mountains, New Hampshire. *Can. J. For. Res.* **18**, 991–1001.

Schowalter, T.D. (1986). Ecological strategies of forest insects: the need for a community-level approach to reforestation. *New For.* **1**,57–66.

Sharpe, P.J.H., Wu, H., Cates, R.G. and Goeschl, J.D. (1985). Energetics of pine defense systems to bark beetle attack. *In* "Proc. Integrated Pest Management Symposium" (S.J. Branham and R.C. Thatcher, eds), pp. 206–223. USDA Forest Serv. Gen. Tech. Rpt. SO-56, USDA Forest Serv., Southern Forest Exp. Stn., New Orleans, LA.

Sierpinski, Z. (1985). Luftverunreinigungen und Forstschadlinge. *Z. Ang. Entomol.* **99**, 1.

Skelly, J.M., Alexander, S.A. and Webb, R.S. (1981). Association of annosus root rot with southern pine beetle attacks. *In* "Site, Stand, and Host Characteristics of Southern Pine Beetle Infestations" (J.E. Coster and J.L. Searcy, eds), pp. 50–67. USDA Forest Serv. Tech. Bull. 1612. USDA Forest Serv., Washington, DC.

Smith, C.C. (1968). Qualitative and quantitative changes in the inner bark of lightning-struck loblolly pines. M.S. Thesis, Northwest State College, Shreveport, LA, 39 pp.

Smith, W.H. (1981). "Air Pollution and Forests." Springer-Verlag, New York. 379 pp.

Stark, R.W., Miller, P.R., Cobb, F.W., Jr., Wood, D.L. and Parmeter, J.R., Jr. (1968). Incidence of bark beetle infestation in injured trees. *Hilgardia* **39**, 121–126.

Tkacz, B.M. and Schmitz, R.F. (1986). Association of an endemic mountain pine beetle population with lodgepole pine infected by Armillaria root disease in Utah. USDA Forest Serv. Res. Note INT-353. USDA Forest Serv. Intermountain Res. Stn., Odgen, UT, 7 pp.

Van Sickle, G.A. (1987). Host responses. *In* "Western Spruce Budworm" (M.H. Brookes, R.W. Campbell, J.J. Colbert, R.G. Mitchell and R.W. Stark, eds), pp. 57–70. USDA Forest Serv. Tech. Bull. 1694. USDA Forest Serv., Washington, DC.

Vité, J.P. (1961). The influence of water supply on oleoresin exudation pressure and resistance to bark beetle attack in *Pinus ponderosa. Contrib. Boyce Thompson Inst.* **21**, 37–66.

Wagner, M.R. and Mathiasen, R.L. (1985). Dwarf mistletoe–pandora moth interaction and its contribution to ponderosa pine mortality in Arizona. *Great Basin Nat.* **45**, 423–426.

Wargo, P.M. and Shaw, C.G. III. (1985). *Armillaria* root rot: the puzzle is being solved. *Plant Dis.* **69**, 826–832.

Waring, R.H. (1987). Characteristics of trees predisposed to die. *BioScience* **37**, 569–574.

Waring, R.H. and Pitman, G.B. (1985). Modifying lodgepole pine stands to change susceptibility to mountain pine beetle attacks. *Ecology* **66**, 889–897.

Waring, R.H., Cromack, K., Jr., Matson, P.A., Boone, R.D. and Stafford, S.G. (1987). Responses to pathogen-induced disturbance: decomposition, nutrient availability, and tree vigor. *J. Forestry* **60**, 219–227.

Waters, W.E., Stark, R.W. and Wood, D.L., eds (1985). "Integrated Pest Management in Pine-Bark Beetle Ecosystems." John Wiley & Sons, New York.

Witcosky, J.J., Schowalter, T.D. and Hansen, E.M. (1986). *Hylastes nigrinus* (Coleoptera: Scolytidae), *Pissodes fasciatus* and *Steremnius carinatus* (Coleoptera: Curculionidae) as vectors of black-stain root disease of Douglas-fir. *Environ. Entomol.* **15**, 1090–1095.

Witcosky, J.J., Schowalter, T.D. and Hansen, E.M. (1987). Host-derived attractants for the beetles *Hylastes nigrinus* (Coleoptera: Scolytidae) and *Steremnius carinatus* (Coleoptera: Curculionidae). *Environ. Entomol.* **16**, 1310–1313.

Wood, D.L., Stark, R.W., Waters, W.E., Bedard, W.D. and Cobb, F.W., Jr. (1985). Treatment tactics and strategies. *In* "Integrated Pest Management in Pine-Bark Beetle Ecosystems" (W.E. Waters, R.W. Stark and D.L. Wood, eds), pp. 121–139. John Wiley & Sons, New York.

–5–

Environmental Stress and Whole-tree Physiology

PETER L. LORIO, JR.
USDA Forest Services, Southern Forest Exp. Stn., Pineville, LA, USA

5.1 INTRODUCTION

Interactions among bark beetles, pathogens, and conifers constitute a triangle. Another triangle of interactions exists among the invading organisms (bark beetles and pathogens), the trees, and the environment. How important, variable or constant, simple or complex, is the role of trees in these triangles? Understanding the wide range of interactions that take place among trees, bark beetles, and pathogens, and between the organisms and their environments, requires consideration of tree responses to fluctuating environmental conditions and the effect of tree responses on tree-invading organisms (Loomis and Adams, 1983).

In recent years there has been considerable progress in research on interactions among bark beetles, microorganisms, and conifers. This chapter focuses on a few important aspects of tree physiology as it affects tree interactions with bark beetles and pathogens.

5.2 A BASIS FOR UNDERSTANDING STRESS EFFECTS

Biotic and abiotic stresses of various kinds (Chapter 4) are common during the growth and development of conifer forests and virtually all natural systems. Unfortunately, different

BEETLE–PATHOGEN INTERACTIONS IN CONIFER FORESTS
ISBN 0-12-628970-0

kinds of stress frequently are not distinguished, thus hindering our understanding of effects on tree physiological processes and tree interactions with other organisms (Sharpe *et al.*, 1985). That deficiency and the general perception that stress is inherently bad for trees and good for invading organisms are major obstacles to better understanding and management of bark beetle–pathogen–conifer interactions.

Stress due to water deficit occurs so frequently that, in the long run, it reduces plant growth and crop yield more than all other stresses combined (Kramer, 1983). For this reason, water supply has been the subject of much study concerning interactions among bark beetles, pathogens, and conifers and will be a focus of this chapter.

Our limited knowledge of whole-tree physiology, and particularly stress effects, is imperfectly integrated with the study of bark beetle, pathogen, and conifer interactions. Tree physiology generally has received little attention with regard to solving forestry problems (Kramer, 1986). This state of affairs indicates a need to provide a broad basis upon which to consider the effects of stress on conifers and how stresses, such as water deficits, may affect tree physiology and interactions of trees with other organisms.

Kramer (1986) attributes the limited use of physiology in forestry to a poor understanding of the role of plant physiology. He specifically focuses on the general lack of appreciation of the concept, developed many years ago, that the nature and limitations of physiological processes are controlled genetically, but that the environment determines actual productivity (Fig. 5.1). Clearly, heredity and environment influence plant growth and development through their combined effects on physiological processes. Klebs (1910) provided some of the earliest input to this concept, and his contributions are commonly recognized in both older and recent literature (Krauss and Kraybill, 1918; Loomis, 1932, 1953; Kramer and Kozlowski, 1979; Sachs, 1987). It is interesting and relevant to this chapter that Klebs (1910) worked with both microorganisms and higher plants.

Fig. 5.1. Diagram illustrating the role of physiology in forestry. Genetic potential and environment must operate through physiological processes in determining the quantity and quality of growth. Expression of resistance mechanisms of conifers to invasion by pathogens and bark beetles are likewise governed by these relationships. (From Kramer, 1986.)

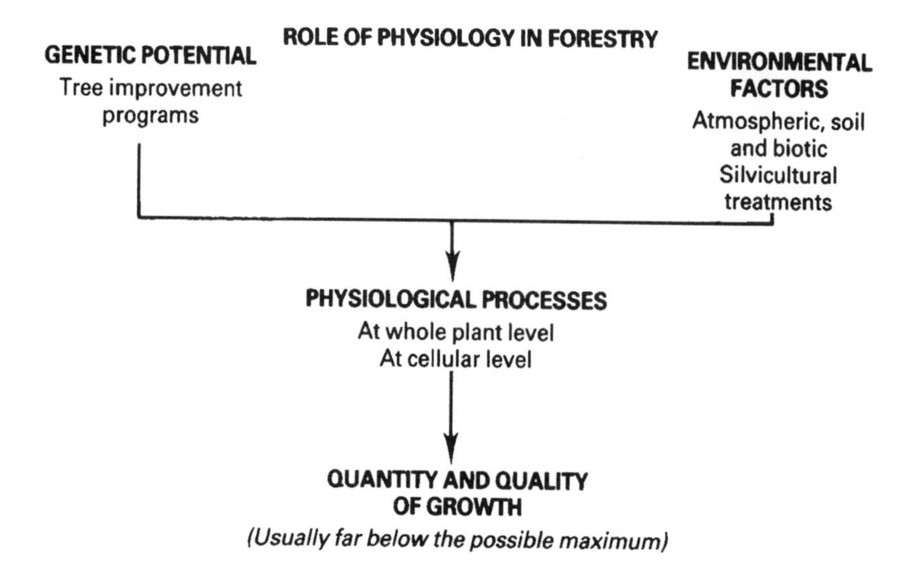

5.3 HOST TREE ONTOGENY AND PHENOLOGY

5.3.1 Ontogeny

Tree ontogeny (the development or course of development of an individual) may be considered in various contexts: a tree's life span from seedling to maturity; the development of cells and their differentiation and maturation; the formation of specialized tissues associated with plant development, and the morphological, physiological, and chemical changes that occur through the seasons of the year. Physiological changes normally associated with tree development, apart from severe environmental stresses, may have important effects on tree susceptibility to and suitability for herbivores and pathogens in general. For example, Coleman (1986) proposes that leaves are most subject to herbivore attack at the time of transition from being carbohydrate sinks (recipients) to sources (synthesizers).

Regarding bark beetle, pathogen, and conifer interactions, physiological changes associated with either maturation or aging are important. For example, when trees develop a capacity to flower and fruit (maturity) an additional demand is placed on photosynthates previously partitioned among vegetative growth and differentiation processes. Given limited energy and material resources within trees, the potential production of oleoresin, a primary factor in resistance of *Pinus* species to bark beetle attack (Hodges *et al.*, 1979), is affected by such developmental changes.

Aging eventually leads to slower rates of metabolism, slower wound healing, and reduced resistance to certain insects and fungi (Kozlowski, 1971). Typical hosts for *Dendroctonus frontalis* have reached maturity and have grown to large size. Such trees provide abundant food and habitat for colonizing beetles. Their long boles and large diameters provide large surface areas to accommodate attacking beetles and brood in numbers greatly exceeding that possible in small trees.

All of the various reproductive and vegetative growth and differentiation processes are energy dependent to some degree, and there are multiple interactions and competitions among them for available substrates. Carbohydrate relations during growth and development of *Pinus* strobili indicate that enlarging cones draw heavily on resources from 1-year-old needles early in the season and from current-year needles late in the season (Kramer and Kozlowski, 1979). Thus, it seems that cones supply little of their own carbohydrate needs for growth and development, that they can suppress both apical and cambial growth, and potentially reduce the synthesis of secondary metabolites (defensive chemicals) such as oleoresins. Some understanding of the relationships between reproductive and vegetative growth and development, and how they may be affected by a variety of environmental conditions over time, is essential to testing effects of environmental stresses. For example, variation in the timing, intensity, and duration of water deficits obviously will affect vegetative and reproductive processes differently (Hsiao *et al.*, 1976b).

5.3.2 Phenology

Tree phenology concerns relationships between climate and periodic biological phenomena, such as flowering, bud burst, foliage production, and leaf fall. Stanley (1958) illustrated the general reproductive growth cycle of the genus *Pinus* (Fig. 5.2), but indicated that exact time of pollination varies with species, latitude, site, and elevation. Environmental factors may cause other phenomena to occur earlier or later in a season, as well as shorten or lengthen their duration. Relationships among flowering, cone growth and development,

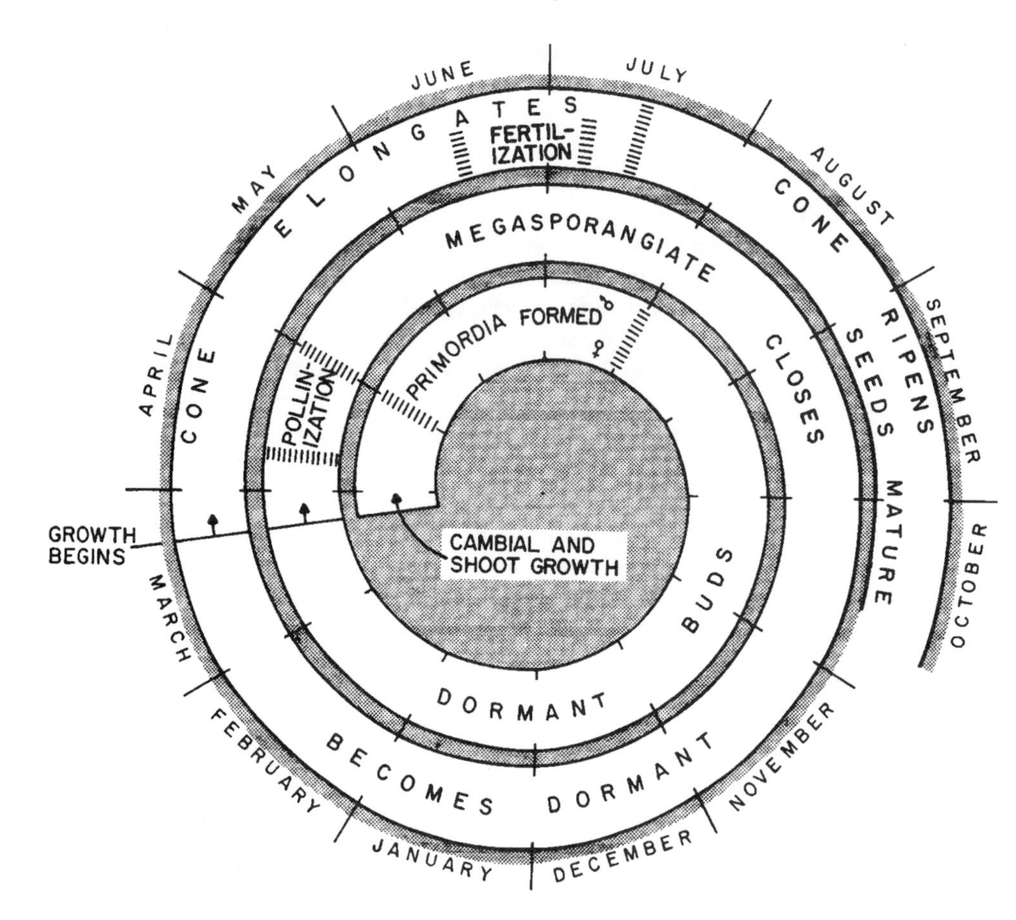

Fig. 5.2. Reproductive cycle of pine. (From Stanley, 1958.)

and other seasonal vegetative growth processes are of interest because competition among such processes for available photosynthates can affect tree resistance to herbivores or pathogens.

The transition from earlywood to latewood in *Pinus* species, the development of the current year's vertical resin ducts, and the seasonal course of resin flow are of particular interest in relation to bark beetle attack (Lorio, 1986). During earlywood formation, partitioning of carbohydrate to oleoresin synthesis is limited by strong demands for reproductive and vegetative growth processes. As rows of earlywood tracheids are produced, the target of beetle attack (the xylem/phloem interface) is displaced progressively from the principal oleoresin reservoir (the vertical resin ducts of the preceding and prior years). With the transition to latewood and the development of the current year's vertical resin ducts, the potential flow of resin at beetle attack sites in the the cambial region increases, as does the inferred resistance to beetle attack. Lorio *et al.* (1990) explored these relationships in testing a conceptual model of seasonal changes in the resistance of *P. taeda* to *D. frontalis* (Figs 5.3 and 5.4). The timing and duration of developmental processes, and associated changes in tree resistance to beetle attack, can be altered considerably during a growing season in response to environmental influences.

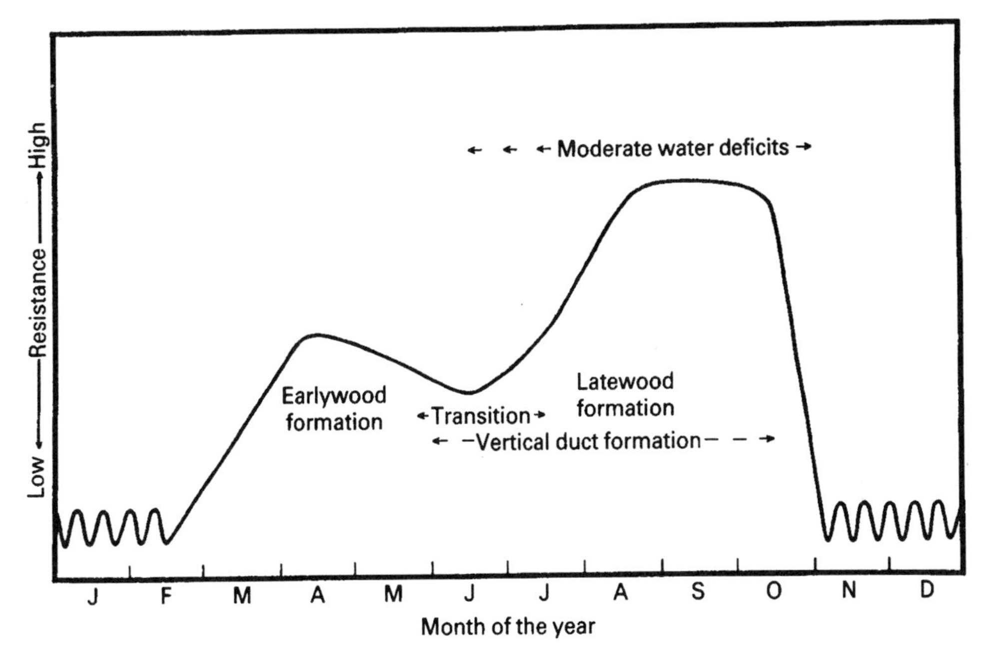

Fig. 5.3. A conceptual model of seasonal changes in pine resistance to southern pine beetle attack for years which have soil water balance patterns similar to that of the long-term average. Resistance to the earliest attacking beetles is considered to be highly dependent on the potential flow of oleoresin at the wound site. (From Lorio *et al.*, 1990.)

5.4 PHYSIOLOGICAL CHANGES ASSOCIATED WITH DEVELOPMENT

Growth response alone may be an unsafe measure of vigor (Kraus and Kraybill, 1918). Tree growth and resistance to bark beetle and fungal invasion are correlated to some degree (Waring, 1983), and resistance may reside in the amount of current and stored photosynthate available for production of defensive compounds (Waring and Pitman, 1985; Christiansen *et al.*, 1987; Dunn *et al.*, 1987). Factors that greatly influence both growth and the production of defensive chemicals include water deficits and excessive crowding of trees. Crowding leads to shortened crowns, reduced photosynthates to support growth and other processes, and thereby to reduced synthesis of defensive compounds. Severe water deficits can reduce cambial growth, but perhaps more importantly they reduce shoot and needle growth, cause premature needle fall, and reduce leaf surface area, with long-lasting effects on photosynthate supply to support growth, maintenance, and production of defensive chemicals. Apart from effects of crowding, water deficits, and other factors, important physiological changes take place in the course of ontogenetic and seasonal growth and development.

5.4.1 Partitioning of assimilate

Seasonal aspects of the competition between growth and the production of defensive chemicals in conifers have not been given much attention. In other systems, however, the integrative control of plant growth and development has been given a great deal of attention (Brouwer, 1983; Chapin *et al.*, 1987; Loomis *et al.*, 1990). Partitioning of carbon during a growing season is partially under genetic control (Fig. 5.1), but phenology and an array of

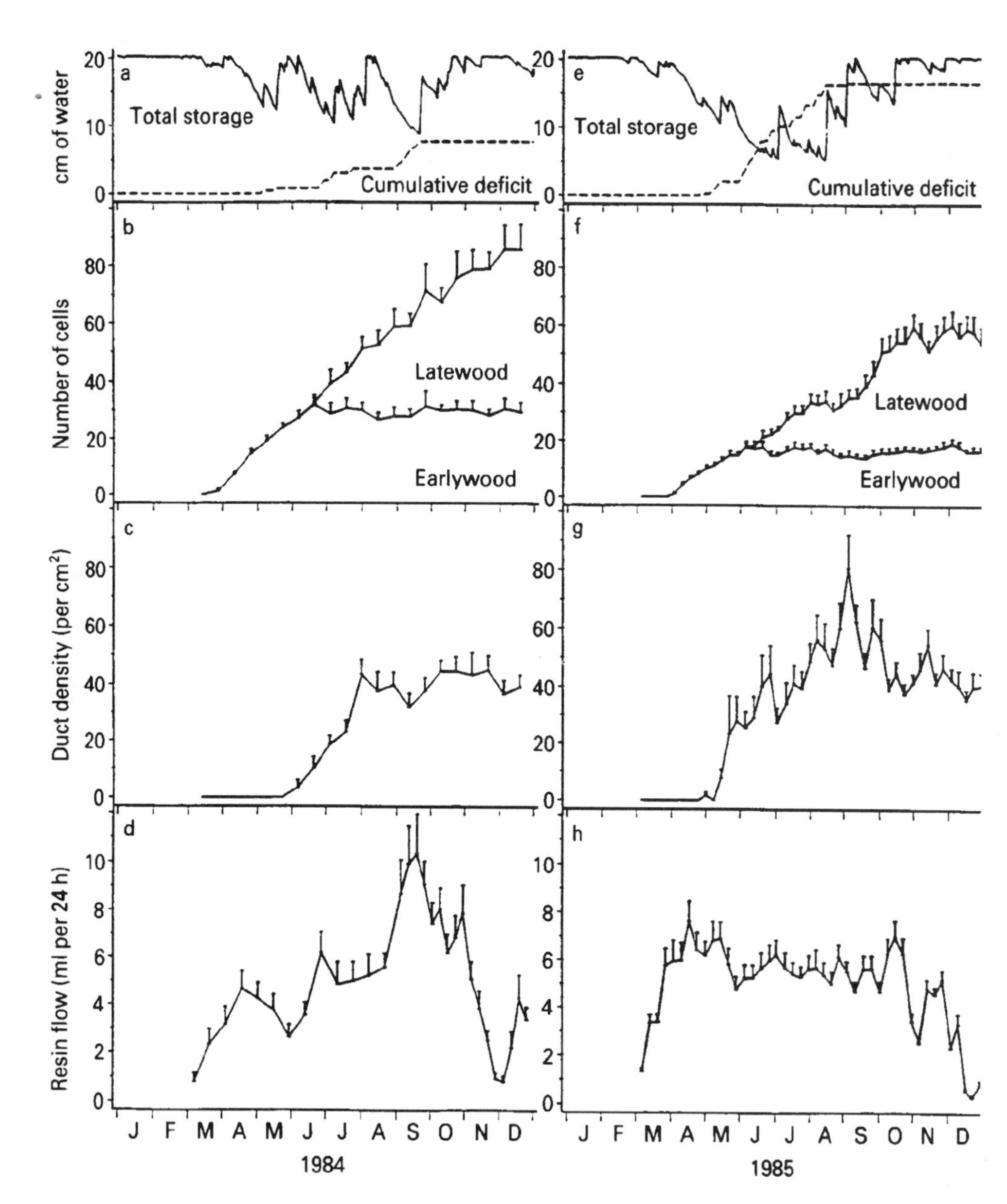

Fig. 5.4. Graphs of water regimes, the course of xylem growth and development, vertical resin duct formation, and resin yield in 1984 (a, b, c, d) and 1985 (e, f, g, h). Daily water storage and cumulative water deficits (a, e). Tree bole cambial growth with total tracheid counts, time of transition to latewood, and amount of earlywood and latewood formed (b, f). Vertical resin duct densities in the current year's growth ring (c, g). Oleoresin yield over 24-hour periods (d, h). Vertical bars are standard errors, $n = 10$ for 1984, $n = 16$ for 1985. Results for 1984 approximate the conceptual model shown in Figure 5.3, but those for 1985, a year abnormally dry in spring and early summer relative to long-term average conditions, do not. (From Lorio *et al.*, 1990.)

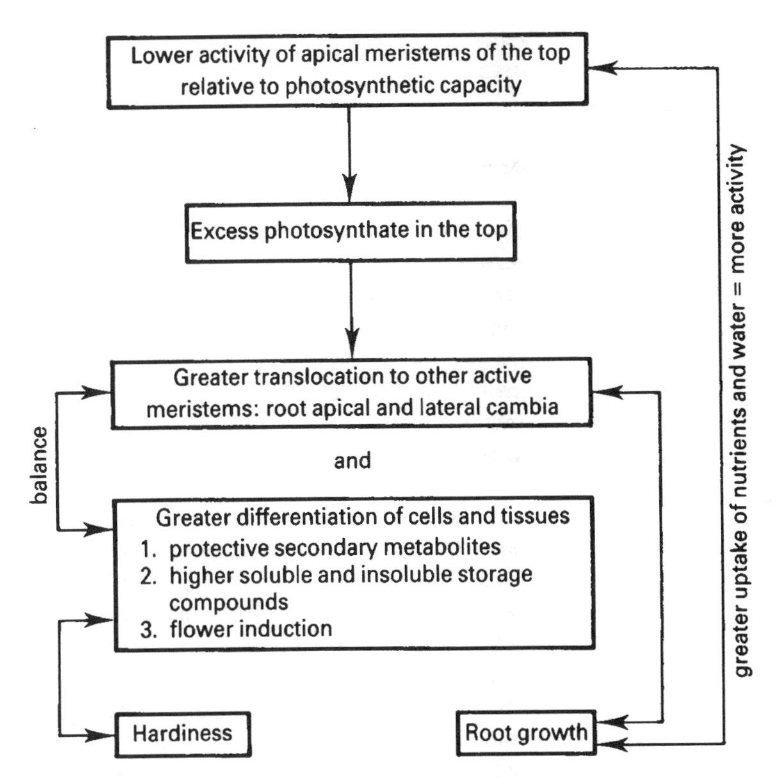

Fig. 5.5. Carbon partitioning among growth and developmental processes in response to greater or lesser nutrient and water supplies, illustrating the competition between growth and differentiation processes as well as among systems within a tree. (After Gordon and Smith, 1987.)

environmental conditions can greatly alter partitioning patterns, especially in the short term. Substrate supply is considered to be a major factor.

Supplies of carbon and nitrogen assimilates are normally limited, and various sinks compete for the limited supplies. Gordon and Smith (1987) illustrate how changes in nutrient and water supply may affect activity of apical meristems relative to photosynthesis and an array of growth and developmental processes (Fig. 5.5). If one process removes a substrate from circulation, another cannot use it. This competition for limited resources results in developmental changes (Trewavas, 1985). Loomis *et al.* (1990) refer to this coordination of plant growth and development as a "nutritional theory" of coordination, that is based on the work of Klebs (1910), Kraus and Kraybill (1918), and Loomis (1932). Loomis (1932) proposed the concept of growth-differentiation balance as a convenient and simplified scheme for predicting or explaining plant behavior.

5.4.2 Relevance to bark beetle–pathogen–tree interactions

Several studies illustrate the great seasonal variation in physiological responses to inoculation with fungi vectored by bark beetles (Reid and Shrimpton, 1971; Paine, 1984; Stephen and Paine, 1985; Cook *et al.*, 1986; Owen *et al.*, 1987; Raffa and Smalley, 1988). Much of the variation may be associated with differences in carbohydrate partitioning due to ontogeny and environment. For the study of interactions among bark beetles, fungi, and trees, it is especially important to recognize that significant chemical changes occur seasonally in conifer phloem and cambial tissue. For example, starch contents increased dramatically, but lipids decreased, from winter to spring in *Abies balsamea* (Little, 1970). Similar changes took place in *Pinus strobus,* and both starch and lipids declined during rapid cambial growth prior to mid-June (Parkerson and Whitmore, 1972). The work of Chung and Barnes (1980a,b) illustrates other dynamic changes in the seasonal production, consumption, and balance of photosynthates in *P. taeda.* Their results agree with those of Gordon and Larson (1968) who found close correlation between time of maturation of the current year's needles and the initiation of latewood formation in 4-year old *P. resinosa.* Using Mooney and Chu's (1974) assignments of biochemical constituents to different functional categories (i.e. metabolism, storage, structure, and protection), Chung and Barnes (1980a) found that seasonal pattern of photosynthate partitioning into protection components (phenolics) in shoots and needles was the inverse of that for structural components. Termination of shoot and needle elongation seemed to be correlated with a rapid increase in photosynthate partitioned to defensive constituents, similar to the findings of Mooney and Chu (1974) with *Heteromeles arbutifolia.*

Other studies, such as Bernard-Dagan's (1988) assessment of the seasonal variation of energy sources in organs and tissues of *Pinus pinaster* are relevant to research on tree interactions with invading organisms. She clearly illustrated the dramatic change from lipids to starch that takes place in secretory (epithelial) cells in needles and buds from winter to early spring. Monoterpene and diterpene compounds, major components of oleoresin, are formed during a very short time (a few days) when the epithelial cells are active. As tissues age the activity of the secretory cells decreases and eventually stops. Presumably, this same phenomenon would occur in epithelial cells associated with bole cambial growth and xylem formation. However, vertical resin ducts are usually absent in earlywood, except in juvenile xylem formed in or near the crown (Figs 5.3 and 5.4). Vertical resin ducts generally appear during and after the transition to latewood.

Many changes occur in the production and distribution of photosynthates in *Pinus* through the year. Bernard-Dagan's (1988) description of the seasonal pattern for *P. pinaster* may not differ greatly from what occurs in North American conifers. These basic physiological changes may vary somewhat with latitude and among species, but knowledge of the general patterns is extremely important for understanding bark beetle, pathogen, and tree interactions (Cheniclet *et al.*, 1988).

Until recently, bark beetle researchers have paid little attention to Kozlowski's (1969) advice concerning study of tree physiological changes associated with ontogeny. Physiological ecologists, however, have made considerable effort to study resource partitioning patterns in plants with respect to growth, reproduction, and defense against herbivores (Bazzaz *et al.*, 1987).

The seasonal ontogeny of southern *Pinus* species, in terms of vegetative and reproductive growth (Fig. 5.2), has great potential significance regarding the resistance of even the healthiest trees to bark beetle attack. For example, high priority demands by apical and cambial growth, and even higher demands by reproductive growth, limit the potential partitioning of photosynthates to secondary metabolic processes and the production of defensive compounds at the same time (Veihmeyer and Hendrickson, 1961; Mooney and Chu, 1974; Chung and Barnes, 1980a; Lorio and Sommers, 1986). However, environmental factors, such as moderate water deficits during the spring and early summer growth flush can alter relationships among ontogenetic processes and result in proportionately less of the currently available energy being used in growth and more in oleoresin synthesis (Lorio *et al.*, 1990).

The ontogeny and phenology of trees are associated with physiological changes of great significance to bark beetle, pathogen, and tree interactions. Further, one should recognize the importance of seasonal changes apart from environmental stress such as water deficits, as well as the need to evaluate the effects of water deficits in terms of their timing, severity, and duration.

5.5 WATER DEFICITS, TREE DEVELOPMENT AND PHYSIOLOGY

Because of its seemingly overall importance to tree growth and yield (Zahner, 1968), water deficit, rather than the variety of other biotic and abiotic stresses (Chapter 4), is the focus of this chapter. Tree responses to water deficits, as to other stresses, are related to the tree's stage of development.

5.5.1 Plant factors affecting responses

A plant's response to water stress depends on its metabolic activity, morphology, stage of growth, and yield potential (Gardner *et al.*, 1985). Water deficits during vegetative growth phases have effects entirely unlike effects during flowering and fruiting, and fruit and seed development and maturation (Hsiao *et al.*, 1976b). While cell growth is most sensitive to water deficit, xylem conductance reduction may not be affected until water potential is below -0.7 to -3.0 MPa (7 to 30 bars) (Hsiao *et al.*, 1976a; Tyree and Dixon, 1986; Jones, 1989). However, assessing the effects of water deficits on plants is not a simple problem.

Bradford and Hsiao (1982) provided a thorough discussion of inherent difficulties in understanding the direct effects of water deficits on physiological processes within plants, and on the long-term behavior of plants. They point out that because of feedback loops among various metabolic processes, it is difficult to distinguish between cause and effect following the onset of a perturbation. Further, in nature, water stress usually develops gradually, over days to weeks, providing ample time for coordinated responses of internal processes, and resulting in adaptation and enhancement of survival under stress.

5.5.2 General patterns of response

Some patterns of gross plant responses to water stress are generally applicable to a large number of plant species (Bradford and Hsiao, 1982). It appears that overall changes in response to stress almost always lead to adjustments in water use and supply that increase the probability of survival and propagation of the species. Bradford and Hsiao (1982) reviewed a number of gross responses and adaptations that occur with many species: reduced foliage growth during early water deficit, resulting in reduced transpiration and a rationing of water supply to meet the demands of a longer ontogenetic span; increases in roots relative to shoots; osmotic adjustment throughout the plant, resulting in maintained turgor and physiological functions at reduced water potential; changes in leaf shape and orientation, leaf abscission, and closing of stomates, resulting in reduced light interception, reduced transpiration, and conserved water. Studies by Linder *et al.* (1987) with *Pinus radiata*, and by Cregg *et al.* (1988) and Hennessey *et al.* (1992) with *P. taeda*, illustrate these important relationships.

Bradford and Hsiao (1982) concluded that insufficient attention has been given to plant responses to water stress that precede stomatal closure and changes in photosynthesis. In

Pinus, responses to such deficits no doubt include greater partitioning of photosynthates to oleoresin synthesis, an important response relative to resistance to bark beetle attack (Lorio and Sommers, 1986; Lorio *et al.*, 1990) .

5.5.3 Flowering and reproduction

The process of flower initiation in plants is complex and only partially understood, but Sachs (1987) notes that moisture and osmotic stress and reduced nitrogen nutrition can reduce growth, accelerate flowering, and result in increased carbohydrate concentration in shoots. In *Pinus*, which initiate strobili primordia in mid- to late-summer in the southern US, research has demonstrated important relationships between rainfall and flower initiation and cone production (Wenger, 1957; Bengtson, 1969; Shoulders, 1967, 1968, 1973; Dewers and Moehring, 1970; Grano, 1973).

Wenger (1957) found that *P. taeda* seed crops varied directly with May-to-July rainfall of the second preceding year, and suggested that irrigation in seed orchards might be beneficial. Early results of irrigation trials with *P. elliottii* var. *elliottii* were ambiguous (Bengtson, 1969). However, Dewers and Moehring (1970) found that irrigation of *P. taeda* in a 13-year-old seed orchard April through June, followed by drought treatment July through September, resulted in significantly higher conelet production than any other treatment. Results indicated that a period of soil water deficit in *P. taeda* stands during or just prior to initiation of ovulate primordia is conducive to cell differentiation into reproductive tissues. Late season irrigation may suppress conelet production. These results support Wenger's (1957) findings for natural stands of *P. taeda*, and Shoulder's (1967) findings for *P. palustris* and illustrate that timing of water deficit greatly influences the nature of tree responses.

Later studies by Shoulders (1973) with *P. elliottii* var. *elliottii* in Louisiana, and Grano (1973) with *P. taeda* in southern Arkansas, showed that flowering and seed yields were enhanced by plentiful spring rain followed by moderate water deficits during the summer. These results are consistent with results expected from consideration of plant growth and differentiation balance principles (Loomis, 1932, 1953), and clearly indicate the need for a broad approach to the study of conifer responses to stresses such as water deficits, and the potential effects of environmental stresses on carbon partitioning to growth, reproduction, and defensive compounds.

5.6 MECHANISMS OF TREE DEFENSE

5.6.1 Theories

There is an extensive literature dealing with theories of defensive mechanisms and relationships to energy supply in plants: apparency (Rhoades and Cates, 1976; Feeny, 1976), defensive chemistry (Rhoades, 1979; Schultz and Baldwin, 1982), carbon/nutrient balance (Bryant *et al.*, 1983), resource availability (Coley *et al.*, 1985; Bazzaz *et al.*, 1987), multiple modes (Mattson *et al.*, 1988; Berryman, 1988), to name a few. Other literature more specific to bark beetles includes Wood (1972), Cates and Alexander (1982), Raffa and Berryman (1983), Hodges *et al.* (1985), Matson and Hain (1985), Hain *et al.* (1985), Sharpe and Wu (1985), Sharpe *et al.* (1985), Lorio and Hodges (1985) and Lorio (1986).

The defensive mechanisms that function in conifers in response to invasions by bark beetles and associated pathogens seem to center on some aspects of their oleoresin systems.

In those conifers that possess a primary resin system, such as Pinus species, the potential rate and duration of flow from wounds is considered an important defensive mechanism (Chapter 8). However, induced responses to wounding by bark beetles or pathogens also may be important in Pinus and in other conifers not possessing preformed resin systems (Berryman, 1972; Chapter 8).

5.6.2 Oleoresin in conifers

Stark (1965) extensively reviewed the subject of oleoresin as a resistance mechanism to attack by forest insects. He considered the subject fundamentally important and of immediate practical applicability. The resin problem has been researched considerably since Stark's review, with much progress. See Chapter 8 for a discussion of the oleoresin system in *Pinus* species in relation to colonization by bark beetles. This section focuses on some aspects of host tree water status and the potential for resin to flow from bark wounds.

Oleoresin exudation pressure (OEP), measured with glass capillary manometers (Bourdeau and Schopmeyer, 1958), Bourdon gauges (Vité, 1961), or other devices such as potentiometer-type pressure transducers (Helseth and Brown, 1970), generally is perceived to be directly related to resin flow. However, little experimental evidence is available to support this idea.

Schopmeyer *et al.* (1954) reasoned that flow through radial resin ducts could be expressed by a modification of Poiseuille's equation for the flow of liquid through capillaries:

$$Y = [KN(ab)^2P]/\eta$$

where Y is the yield of resin over a fixed time, K is a constant, N is the number of radial resin ducts per unit area of wound surface, $(ab)^2$ is the average of squared products of the major and minor semiaxes (a and b) of the duct cross section, P is the exudation pressure at the point of discharge, and η is resin viscosity.

Bourdeau and Schopmeyer (1958) found a statistically significant relationship between resin yield per inch of wound width over two 2-week periods of commercially tapped *Pinus elliottii* var. *elliottii* and the function

$$[N(ab)^2P]/\eta$$

for variables as defined above. When P was removed, the relationship was not significant. They concluded that number and size of resin ducts, pressure, and viscosity were related to resin yield from wounds in *P. elliottii* var. *elliottii* stems and that pressure, viscosity, and pressure:viscosity ratios might be useful variables in progeny testing because they are under strong genetic control.

Since the work of Schopmeyer *et al.* (1954) and Bourdeau and Schopmeyer (1958), few studies have been conducted in which both OEP and resin flow were measured along with other variables. In contrast to the results of Bourdeau and Schopmeyer (1958), Barrett and Bengtson (1964) found no correlation between OEP and resin yield from seven seed sources of *P. elliottii* var. *elliottii*. They did find that the ratio of OEP to viscosity was related to flow, as did Bourdeau and Schopmeyer (1958), but viscosity alone was more closely related to flow than was the OEP:viscosity ratio. Hodges and Lorio (1971) found OEP of *P. taeda* to be positively correlated with the xylem water potential of twigs and the relative water content of needles (relative turgidity of Weatherley, 1950), but not correlated (r = 0.03) with oleoresin flow measured as described by Mason (1969). Subsequently, Hodges *et al.* (1981) reported no correlation between resin flow and the size or number of resin ducts for any of four *Pinus* species studied.

Vité (1961) used the work of Bourdeau and Schopmeyer (1958) as a basis for development of a technique for measuring OEP in *Pinus ponderosa* in an intensive study of relationships among OEP, water supply, phloem moisture, and severity of bark beetle attack. Vité did not measure oleoresin flow, but he considered OEP to be an expression of the water balance of a tree (and its oleoresin flow) and a dependable indicator of *P. ponderosa* resistance or susceptibility to bark beetle infestation. Vité and Wood (1961) found that a high percentage of successfully attacked trees had low OEP and suggested that OEP be used for risk rating individual trees and stands of second growth *P. ponderosa*. However, Wood (1962) reported that *Ips confusus* failed to colonize *P. ponderosa* with extremely low OEP (less than 1 bar) over a span of 26 days or longer. Stark (1965), Raffa and Berryman (1982), Matson and Hain (1985), and others have concluded that OEP is an unreliable predictor of tree susceptibility to bark beetles, as discussed below.

Lorio and Hodges (1968b) showed that OEP of *P. taeda* was very responsive to changes in atmospheric evaporative demand and soil water status, as did Vité (1961) for *P. ponderosa* and Bourdeau and Schopmeyer (1958) for *P. elliottii* var. *elliottii*, but they also did not measure resin flow. Cobb *et al.* (1968) measured both OEP and resin flow in *P. ponderosa* affected by photochemical air pollution and reported that OEP was reduced gradually as disease severity increased. However, resin flow rate remained relatively constant through intermediate stages of disease development. Total resin flow apparently followed the same pattern, but the limited capacity of the collection vials may have contributed to the similar yields for healthy and moderately diseased trees. Although OEP seems to be well related to tree water status, neither OEP nor tree water status appear to be directly related to resin flow. Several studies indicate a more complex or subtle relationship may exist (Harper and Wyman, 1936; Ostrom *et al.*, 1958; Mason, 1971).

Harper and Wyman (1936) studied the daily pattern of resin yield from *P. elliottii* var. *elliottii* and observed a decided decline in rate of resin yield whenever tree boles expanded between 6 a.m. and 4 p.m. A small increase in flow occurred when stems contracted. Further, they showed that resin yield peaked between 8 a.m. and 10 a.m. during the second day after wounding, when temperature was increasing at its fastest rate during the day, and rate of stem shrinkage was greatest. The time of their observed maximum resin yield for the day coincides with the time that OEP typically is dropping at a fast rate (Bourdeau and Schopmeyer, 1958; Vité, 1961; Lorio and Hodges, 1968a,b; Helseth and Brown, 1970; Hodges and Lorio, 1971).

Ostrom *et al.* (1958) found that acid applied to wounds made to the face of the xylem of *P. taeda*, *P. elliottii* var. *elliottii*, and *P. palustris* destroyed the cambium to about 2 cm above a wound, enlarged openings of ducts near the cambium by collapsing or destroying epithelial cells and ray parenchyma, and served chiefly to facilitate outflow of resin from ducts (essentially reducing OEP to zero in the affected area). Mason (1971) found that resin flow in *P. taeda* was affected more by overstocking in an unthinned stand than by temporary soil moisture stress. Induced drought alone did not affect the mean OEF for the season. He concluded that resin flow must be controlled mostly by factors only indirectly influenced by tree water status.

Perhaps the resin duct systems in conifers are not sufficiently similar to hydraulic systems with rigid-walled capillaries for a modification of Poiseuille's equation to predict flow adequately. It appears that resin flow from wounds made to the face of the xylem is enhanced by a reduction in turgor of the epithelial cells (Harper and Wyman, 1936; Ostrom *et al.*, 1958). With full turgor, as resin flow begins near the wound site, resin ducts may be rapidly closed as epithelial cells distend in response to reduced back pressure from the resin column. With less than full turgor, which commonly occurs in mid-morning, epithelial cells may distend less rapidly, perhaps incompletely closing the ducts, thus allowing faster and more prolonged resin flow than with full turgor.

Measurement of OEP may at times be useful to estimate the relative water status of trees. However, caution should be used in inferring any relationship with resin flow. It seems clear that considerably more detailed and elaborate studies will be required to describe adequately the diurnal and seasonal relationship between OEP and resin flow.

5.7 FACTORS AFFECTING EXPRESSION OF RESISTANCE

As suggested by Raffa and Berryman (1983), trees vary in their quantitative resistance to invading organisms due to genetic, environmental, and seasonal factors. This section focuses on some aspects of the growth and development of trees that could affect the potential expression of defense against bark beetles and pathogens. If the resin systems, primary or secondary, are important to defense, it seems likely that the partitioning of photosynthates would be very important to the expression of resistance to invasion. Accordingly, trees with large and efficient crowns that produce sufficient photosynthates to meet the demands for both vegetative and reproductive growth, as well as synthesis of differentiation products, such as allelochemicals, will likely be more resistant to invasion than trees with small, inefficient crowns. Several factors, including forest or stand structure, carbohydrate status of trees, and seasonal environmental factors, influence expression of resistance. The effects of these factors may vary with the developmental stage of the tree.

5.7.1 Forest and stand conditions

Site quality, species composition, stand density, tree ages and sizes, disease incidence, and other insect activity are among the many factors that can affect tree growth and development, and interactions with bark beetles (Waring and Schlesinger, 1985). Growth, especially bole radial growth, often has been viewed as indicative of tree susceptibility or resistance to beetle attack (Miller and Keen, 1960; Hicks, 1980; Waring, 1983). Relationships among growth, environment, and invading organisms are complex and confounded, but there is little doubt that crowded stand conditions result in reduced light absorption, leaf area, photosynthesis, growth, and potential production of allelochemicals, compared to uncrowded stands (Waring, 1983; Christiansen *et al.*, 1987; Lorio, 1980). Furthermore, merely the close proximity of neighboring trees in such stands greatly facilitates the movement of bark beetles from tree to tree (Gara and Coster, 1968). Close spacing may play a more important role in facilitating *D. frontalis* infestation growth during endemic periods than during epidemics (Johnson and Coster, 1978). Further, in stands with closed canopies, beetle aggregation to sources of pheromones may be more effective than in open stands where high insolation and convection disrupt pheromone plumes (Fares *et al.*, 1980).

Thinning often is recommended to reverse the growth reduction and susceptibility to bark beetle attack observed at high stand densities (Chapter 11). Few direct tests of thinning effects on stand susceptibility to bark beetles have been reported, but Brown *et al.* (1987) reported increased resistance of *P. taeda* to attack by *D. frontalis* following thinning of stands to 23 m^2ha^{-1} or less. Nebeker *et al.* (1985) and Wood *et al.* (1985) reviewed thinning practices for southern and western *Pinus* species and made recommendations for pest management, including consideration of activities that could aggravate pathogen problems. Amman and Schmitz (1988) and Amman *et al.* (1988) discussed several management strategies, including the use of partial cuts, to minimize losses of *P. contorta* var. *latifolia* to *D. ponderosae*. Several studies conducted in the western US demonstrated that thinning reduced tree mortality to bark beetles (Sartwell and Stevens, 1975; Mitchell *et al.*, 1983; McGregor *et al.*, 1987). As with the southern *Pinus* species, consideration must be given to potential problems resulting from stand disturbance (root diseases, windthrow, dwarf mistletoe, etc.) (Witcosky *et al.*, 1986).

Although the greatest potential benefit would seem to come from well-planned, early, and scheduled thinnings over the life of a stand, thinnings of mature stands can result in improved physiological condition of the remaining trees (more photosynthate production, leading to more allelochemics production). In some instances beetle attack is reduced so quickly after thinning that drastic changes in the microclimate may be the primary factor inhibiting beetle activity (Amman *et al.*, 1988; Bartos and Amman, 1989)

5.7.2 Carbohydrate status of host trees

The carbohydrate status of host trees determines tree ability to produce or mobilize compounds critical for defense against bark beetle attack (Christiansen *et al.*, 1987) or defense against infections by pathogenic fungi (Shigo, 1985). The little-explored literature on the formation of lightwood (oleoresin-saturated woody tissue) seems particularly relevant to this issue (Schwarz, 1983; Stubbs *et al.*, 1984; Chapter 8).

Pinus responses to paraquat, a bipyridilium herbicide, applied in appropriate concentrations to living stem tissue beneath the outer bark, results in extensive resin soaking of the xylem. Years of intensive research on this process in *Pinus* species demonstrated clearly that the current and stored photosynthate status of trees at the time of treatment is important to the response obtained (Brown *et al.*, 1976, 1979; Wolter and Zinkel, 1984). In general, spring applications (when carbohydrate status normally would be high) produces the best average oleoresin yields (Stubbs *et al.*, 1984). Physiological responses to paraquat application include ethylene synthesis and heightened respiration, both of which are associated with cell-wounding response and stress (Wolter and Zinkel, 1984). Fungal colonization of phloem and xylem in association with bark beetle attacks appear to elicit similar responses (Berryman, 1972; Raffa and Smalley, 1988; Chapter 8). The nature and extent of response probably is mediated by the same factors that influence response to paraquat applications.

5.7.3 Environment and tree development

Drought may play an important role in the development of insect outbreaks (White, 1974; Mattson and Haack, 1987; Chapter 4). Martinat (1987), however, questions the role of "climatic release" as a mechanism to explain widespread periodic outbreaks, partially because of methodological problems. Further, Connor (1988) and McQuate and Connor (1990a,b) report results of studies with *Corythucha arcuata* on *Quercus alba*, and *Epilachna varivestis* on *Glycine max* that are inconsistent with, and contrary to, White's (1974) hypothesis that water deficit leads to improved insect performance and abundance. Recently, Turchin *et al.* (1991) concluded that outbreaks of *Dendroctonus frontalis* are not driven by stochastic fluctuations of weather, but by some unknown population process acting in a delayed density-dependent manner. While severe water deficits may greatly reduce tree resistance to bark beetle attack (Vité, 1961; Lorio and Hodges, 1968a, 1977), it does not follow that bark beetle outbreaks will necessarily occur. Further, it is important to realize that general relationships vary among species of insects, pathogens, and trees.

In the south and southeast, severe droughts usually are associated with high temperatures during the mid- to late-summer, conditions that are typically unfavorable for adult beetle survival and brood development (Gagne *et al*., 1980). Moreover, drought seldom, if ever, develops so rapidly that plants do not have an opportunity to respond physiologically in a manner that will enhance their survival and reproduction. A developing drought should limit growth before it limits photosynthesis, resulting in more carbohydrate for processes other than growth. However, carbohydrate use for respiration may increase temporarily (Brix, 1962). Perhaps, if drought severely depletes carbohydrate levels by reducing effective leaf area, the response of photosynthesis to improved water conditions may be lower than necessary to meet needs for growth and development as drought is alleviated.

Inoculation studies with pathogens or fungal associates of bark beetles could benefit from consideration of environmental and developmental factors that can affect results. For example, Chou (1982), working with *P. radiata* and *Diplodia pinea*, a fungus that causes shoot dieback, found that susceptibility fluctuates with season, being high in spring–summer but low in autumn–winter. He concluded that seasonal predisposition is important in planning inoculation trials for any purpose. Studies with a variety of organisms and tree species support Chou's conclusion. Paine (1984) found significantly different rates and lengths of lesions formed in response to inoculation of the mycangial fungi of *Dendroctonus brevicomis* in *P. ponderosa* from spring through autumn. Owen *et al*. (1987) found *P. ponderosa* seedlings to be more susceptible to a number of fungi isolated from *Dendroctonus* spp. during shoot elongation than prior to shoot elongation. Myers (1986), working with several *Pinus* species and *Bursaphelenchus xylophilus*, found greater mortality from spring inoculations, when cambial activity was greater, than from late summer and autumn inoculations.

5.8 CONCLUSIONS

The greatest challenge in the study of interactions among bark beetles, pathogens, and host conifers is to integrate the roles of all three elements effectively. The physiology of host trees has been the least effectively integrated element in past research, partially because of a generally poor understanding of the role of plant physiology in forestry (Kramer, 1986). Excessive emphasis has been placed on presumed effects of stresses, such as water deficits, without consideration of host-tree ontogeny and phenology, and the importance of timing, intensity, and duration of stresses. Greater attention must be paid to the role of tree physiology, phenology and ontogeny in the triangle of interactions among invading organisms, trees, and environment.

Ultimately, adequate knowledge of tree, bark beetle, and pathogen interactions is needed to permit mechanistic modeling of "what if?" situations, an extremely difficult problem considering the myriad of organisms and conditions involved, and the complexities of forest systems. Progress could be enhanced greatly with more effective cooperation and development of integrated studies, as in approaches recommended by Landsberg (1986) and Linder *et al*. (1987), and by framing hypotheses in explanatory, rather than statistical, models, an approach strongly endorsed by Loomis and Adams (1983).

REFERENCES

Amman, G.D. and Schmitz, R.F. (1988). Mountain pine beetle–lodgepole pine interactions and strategies for reducing tree losses. *Ambio* **17**, 62–68.

Amman, G.D., McGregor, M.D., Schmitz, R.F. and Oakes, R.D. (1988). Susceptibility of lodgepole pine to infestation by mountain pine beetles following partial cutting of stands. *Can. J. For. Res.* **18**, 688–695.

Barrett, J.P. and Bengtson, G.W. (1964). Oleoresin yields for slash pines from seven seed sources. *Forest Sci.* **10**, 159–164.

Bartos, D.L. and Amman, G.D. (1989). Microclimate: an alternative to tree vigor as a basis for mountain pine beetle infestations. USDA Forest Serv. Intermountain Res. Stn. Res. Paper INT-400. USDA Forest Serv. Intermountain Res. Stn., Ogden, UT, 10 pp.

Bazzaz, F.A., Chiariello, N.R., Coley, P.D. and Pitelka, L.F. (1987). Allocating resources to reproduction and defense. *BioScience* **37**, 58–67.

Bengtson, G.W. (1969). Growth and flowering of clones of slash pine under intensive culture: early results. USDA Forest Serv. Southeastern For. Exp. Stn. Res. Paper SE-46. USDA Forest Serv. Southeastern Forest Exp. Stn., Asheville, NC, 9 pp.

Bernard-Dagan, C. (1988). Seasonal variations in energy sources and biosynthesis of terpenes in maritime pine. *In* "Mechanisms of Woody Plant Defenses against Insects" (W.J. Mattson, J. Levieux and C. Bernard-Dagan, eds), pp. 93–116. Springer-Verlag, New York.

Berryman, A.A. (1972). Resistance of conifers to invasion by bark beetle–fungus associations. *BioScience* **22**, 598–602.

Berryman, A.A. (1988). Towards a unified theory of plant defense. *In* "Mechanisms of Woody Plant Defenses against Insects" (W.J. Mattson, J. Levieux and C. Bernard-Dagan, eds), pp. 39–56. Springer-Verlag, New York.

Bourdeau, P.F. and Schopmeyer, C.S. (1958). Oleoresin exudation pressure in slash pine: its measurement, heritabity, and relation to oleoresin yield. *In* "Physiology of Forest Trees" (K.V. Thimann, ed), pp. 313–319. Ronald Press, New York.

Bradford, K.J. and Hsiao, T.C. (1982). Physiological responses to moderate water stress. *In* "Physiological Ecology II. Encyclopedia of Plant Physiology, New Series," Vol. 12B (O.L. Lange, P.S. Nobel, C.B. Osmond and H. Ziegler, eds), pp. 263–324. Springer-Verlag, Berlin.

Brix, H. (1962). The effect of plant water stress on the rates of photosynthesis and respiration in tomato plants and loblolly pine seedlings. *Physiol. Plant.* **15**, 10–20.

Brouwer, R. (1983). Functional equilibrium: sense or nonsense? *Neth. J. Agric. Sci.* **31**, 335–348.

Brown, C.L., Clason, T.R. and Michael, J.L. (1976). Paraquat induced changes in reserve carbohydrates, fatty acids and oleoresin content of young slash pines. *In* "Lightwood Research Coordinating Council Proceedings" (M.H. Esser, ed.), pp. 8–19. USDA Forest Serv. Southeastern For. Exp. Stn., Asheville, NC.

Brown, C.L., Sommer, H.E. and Birchem, R. (1979). Additional observations on utilization of reserve carbohydrates in paraquat induced resinosis. *In* "Proceedings, Sixth Annual Lightwood Research Conference," pp. 4–11. USDA Forest Serv. Southeastern For. Exp. Stn., Asheville, NC.

Brown, M.W., Nebeker, T.E. and Honea, C.R. (1987). Thinning increases loblolly pine vigor and resistance to bark beetles. *South. J. Appl. For.* **11**, 28–31.

Bryant, J.P., Chapin, F.S. III and Klein, D.R. (1983). Carbon/nutrient balance of boreal plants in relation to vertebrate herbivory. *Oikos* **40**, 357–368.

Cates, R.G. and Alexander, H. (1982). Host resistance and susceptibility. *In* "Bark Beetles in North American Conifers: A System for the Study of Evolutionary Biology" (J.B. Mitton and K.B. Sturgeon, eds), pp. 212–263. University of Texas Press, Austin, TX.

Chapin, F.S. III, Bloom, A.J., Field, C.B. and Waring, R.H. (1987). Plant responses to multiple environmental factors. *BioScience* **37**, 49–57.

Cheniclet, C., Bernard-Dagan, C. and Pauly, G. (1988). Terpene biosynthesis under pathological conditions. *In* "Mechanisms of Woody Plant Defenses against Insects" (W.J. Mattson, J. Levieux and C. Bernard-Dagan, eds), pp. 117–130. Springer-Verlag, New York.

Chou, C.K.S. (1982). Susceptibility of *Pinus radiata* seedlings to infection by *Diplodia pinea* as affected by pre-inoculation conditions. *New Zealand J. For. Sci.* **12**, 438–441.

Christiansen, E., Waring, R.H. and Berryman, A.A. (1987). Resistance of conifers to bark beetle attack: searching for general relationships. *For. Ecol. Manage.* **22**, 89–106.

Chung, H.H. and Barnes, R.L. (1980a). Photosynthate allocation in *Pinus taeda*. II. Seasonal aspects of photosynthate allocation to different biochemical fractions in shoots. *Can. J. For. Res.* **10**, 338–347.
Chung, H.H. and Barnes, R.L. (1980b). Photosynthate allocation in *Pinus taeda*. III. Photosynthate economy: its production, consumption and balance in shoots during the growing season. *Can. J. For. Res.* **10**, 348–356.
Cobb, F.W., Jr., Wood, D.L., Stark, R.W. and Miller, P.R. (1968). Photochemical oxidant injury and bark beetle (Coleoptera: Scolytidae) infestation in ponderosa pine. II. Effect of injury upon physical properties of oleoresin, moisture content, and phloem thickness. *Hilgardia* **39**, 127–134.
Coleman, J.S. (1986). Leaf development and leaf stress: increased susceptibility associated with sink–source transition. *Tree Physiol.* **2**, 289–299.
Coley, P.D., Bryant, J.P. and Chapin, F.S. III. (1985). Resource availability and plant antiherbivore defense. *Science* **230**, 895–899.
Connor, E.F. (1988). Plant water deficits and insect responses: the preference of *Corythucha arcuata* (Heteroptera: Tingidae) for the foliage of white oak, *Quercus alba*. *Ecol. Entomol.* **13**, 375–381.
Cook, S.P., Hain, F.P. and Nappen, P.B. (1986). Seasonality of the hypersensitive response by loblolly and shortleaf pine to inoculation with a fungal associate of the southern pine beetle (Coleoptera: Scolytidae) *J. Entomol. Sci.* **20**, 283–285.
Cregg, B.M., Dougherty, P.M. and Hennessey, T.C. (1988). Growth and wood quality of young loblolly pine trees in relation to stand density and climatic factors. *Can. J. For. Res.* **18**, 851–858.
Dewers, R.S. and Moehring, D.M. (1970). Effect of soil water stress on initiation of ovulate primordia in loblolly pine. *Forest Sci.* **16**, 219–221.
Dunn, J.P., Kimmerer, T.W. and Potter, D.A. (1987). Winter starch reserves of white oak as a predictor of attack by the two-lined chestnut borer, *Agrilus bilineatus* (Weber) (Coleoptera: Buprestidae). *Oecologia* **74**, 352–355.
Fares, Y., Sharpe, P.J.H. and Magnuson, C.H. (1980). Pheromone dispersion in forests. *J. Theoret. Biol.* **85**, 335–359.
Feeny, P.P. (1976). Plant apparency and chemical defense. *Rec. Adv. Phytochem.* **10**, 1–40.
Gagne, J.A., Coulson, R.N., Foltz, J.L., Wagner, T.L. and Edson, L.J. (1980). Attack and survival of *Dendroctonus frontalis* in relation to weather during three years in east Texas. *Environ. Entomol.* **9**, 222–229.
Gara, R.I. and Coster, J.E. (1968). Studies on the attack behavior of the southern pine beetle. III. Sequence of tree infestation within stands. *Contrib. Boyce Thompson Inst.* **24**, 77–85.
Gardner, F.P., Pearce, R.B. and Mitchell, R.L. (1985). "Physiology of Crop Plants." Iowa State University Press, Ames, IA, 327 pp.
Gordon, J.C. and Larson, P.R. (1968). Seasonal course of photosynthesis, respiration, and distribution of ^{14}C in young *Pinus resinosa* trees as related to wood formation. *Plant Physiol.* **43**, 284–288.
Gordon, J.C. and Smith, W.H. (1987). Tree roots and microbes: a high priority for high technology. *In* "Future Developments in Soil Science Research" (L. L. Boersma *et al.*, eds), pp. 423–431. Soil Science Society of America, Madison, WI.
Grano, C.X. (1973). Loblolly pine fecundity in South Arkansas. USDA Forest Service Research Note SO-159 USDA Forest Serv. Southern For. Exp. Stn., New Orleans, LA, 7 pp.
Hain, F.P., Cook, S.P., Matson, P.A. and Wilson, K.G. (1985). Factors contributing to southern pine beetle host resistance. *In* "Proceedings, Integrated Pest Management Research Symposium." USDA Forest Service Gen. Tech. Rpt. SO-56. pp. 154–160. USDA Forest Serv. Southern For. Exp. Stn., New Orleans, LA.
Harper, V.L. and Wyman, L. (1936). Variations in naval-stores yields associated with weather and specific days between chippings. USDA Technical Bulletin 510, Washington, DC, 35 pp.
Helseth, F.A. and Brown, C.L. (1970). A system for continuously monitoring oleoresin exudation pressure in slash pine. *Forest Sci.* **16**, 346–349.
Hennessey, T.C., Dougherty, P.M. and Cregg, B.M. (1992). Annual variation in needlefall patterns of a loblolly pine stand in relation to climate and stand density. *For. Ecol. Manage.* **51**, 329–338.
Hicks, R.R., Jr. (1980). Climatic, site, and stand factors. *In* "The Southern Pine Beetle" (R.C. Thatcher, J.L. Searcy, J.E. Coster and G.D. Hertel, eds), pp. 55–68. USDA Forest Service, and Science and Education Administration, Technical Bulletin 1631.
Hodges, J.D. and Lorio, P.L. Jr. (1971). Comparison of field techniques for measuring moisture stress in large loblolly pines. *Forest Sci.* **17**, 220–223.

Hodges, J.D., Elam, W.W., Watson, W.F. and Nebeker, T.E. (1979). Oleoresin characteristics and susceptibility of four southern pines to southern pine beetle (Coleoptera: Scolytidae) attacks. *Can. Entomol.* **111**, 889–896.
Hodges, J.D., Elam, W.W. and Bluhm, D.R. (1981). Influence of resin duct size and number on oleoresin flow in southern pines. USDA Forest Service Research Note SO-266, USDA Forest Serv. Southern For. Exp. Stn., New Orleans, LA, 3 pp.
Hodges, J.D., Nebeker, T.E., DeAngelis, J.D., Karr, B.L. and Blanche, C.A. (1985). Host resistance and mortality: a hypothesis based on the southern pine beetle–microorganism–host interactions. *Bull. Entomol. Soc. Am.* Spring, 31–35.
Hsiao, T.C., Acevedo, E., Fereres, E. and Henderson, D.W. (1976a). Water stress, growth, and osmotic adjustment. *Philosoph. Trans. Royal Soc. London. (B)* **273**, 479–500.
Hsiao, T.C., Fereres, E., Acevedo, E. and Henderson, D.W. (1976b). Water stress and dynamics of growth and yield of crop plants. *In* "Water and Plant Life. Problems and Modern Approaches" (O. L. Lange *et al.*, eds), pp. 281–305. Springer-Verlag, Berlin.
Johnson, P.C. and Coster, J.E. (1978). Probability of attack by southern pine beetle in relation to distance from an attractive host tree. *Forest Sci.* **24**, 574–580.
Jones, H.G. (1989). Water stress and stem conductivity. *In* "Environmental Stress in Plants" (J.H. Cherry, ed.), pp. 17–24. Springer-Verlag, Berlin.
Klebs, G. (1910). Alterations in the development and forms of plants as a result of environment. *Proc. Royal Soc. London* **82B**, 547–558.
Kozlowski, T.T. (1969). Tree physiology and forest pests. *J. Forestry* **67**, 118–123.
Kozlowski, T.T. (1971). "Growth and Development of Trees," Vol. I. Academic Press, New York, 443 pp.
Kramer, P.J. (1983). "Water Relations of Plants." Academic Press, New York, 489 pp.
Kramer, P.J. (1986). The role of physiology in forestry. *Tree Physiol.* **2**, 1–16.
Kramer, P.J. and Kozlowski, T.T. (1979). "Physiology of Woody Plants." Academic Press, New York, 811 pp.
Kraus, E.J. and Kraybill, H.R. (1918). Vegetation and reproduction with special reference to the tomato. Oregon Agric. College Exp. Stn. Bull. 149. Corvallis, OR, 89 pp.
Landsberg, J.J. (1986). Experimental approaches to the study of the effects of nutrients and water on carbon assimilation by trees. *Tree Physiol.* **2**, 427–444.
Linder, S., Benson, M.L., Myers, B.J. and Raison, R.J. (1987). Canopy dynamics and growth of *Pinus radiata*. I. Effects of irrigation and fertilization during a drought. *Can. J. For. Res.* **17**, 1157–1165.
Little, C.H.A. (1970). Derivation of the springtime starch increase in balsam fir (*Abies balsamea*). *Can. J. Bot.* **48**, 1995–1999.
Loomis, R.S. and Adams, S.S. (1983). Integrative analyses of host–pathogen relations. *Annu. Rev. Phytopathol.* **21**, 341–362.
Loomis, R.S., Luo, Y. and Kooman, P. (1990). Integration of activity in the higher plant. *In* "Theoretical Production Ecology: Reflections and Perspectives" (R. Rabinge, J. Goudriaan, H. van Keulen, F.W.T. Penning de Vries and H.H. van Laar, eds), pp. 105–124. Pudoc, Wageningen, The Netherlands.
Loomis, W.E. (1932). Growth–differentiation balance vs carbohydrate–nitrogen ratio. *Proc. Am. Soc. Hort. Sci.* **29**, 240–245.
Loomis, W.E. (1953). Growth correlation. *In* "Growth and Differentiation in Plants" (W. E. Loomis, ed.), pp. 197–217. Iowa State College Press, Ames, IA.
Lorio, P.L., Jr. (1980). Loblolly pine stocking levels affect potential for southern pine beetle infestation. *Southern J. Appl. For.* **4**, 162–165.
Lorio, P.L., Jr. (1986). Growth–differentiation balance: a basis for understanding southern pine beetle–tree interactions. *Forest Ecol. Manage.* **14**, 259–273.
Lorio, P.L., Jr. and Hodges, J.D. (1968a). Microsite effects on oleoresin exudation pressure of large loblolly pines. *Ecology* **49**, 1207–1210.
Lorio, P.L., Jr. and Hodges, J.D. (1968b). Oleoresin exudation pressure and relative water content of inner bark as indicators of moisture stress in loblolly pines. *Forest Sci.* **14**, 392–398.
Lorio, P.L., Jr. and Hodges, J.D. (1977). Tree water status affects induced southern pine beetle attack and brood production. USDA Forest Serv. Res. Paper SO-135, USDA Forest Serv. Southern For. Exp. Stn., 7 pp.
Lorio, P.L., Jr. and Hodges, J.D. (1985). Theories of interactions among bark beetles, associated

microorganisms, and host trees. *In* "Proceedings of the 3rd Biennial Southern Silvicultural Research Conference" (E. Shoulders, ed.), pp. 485–492. USDA Gen. Tech. Rpt. SO-54, USDA Forest Service Southern For. Exp. Stn., New Orleans, LA.

Lorio, P.L., Jr. and Sommers, R.A. (1986). Evidence of competition for photosynthates between growth processes and oleoresin synthesis in *Pinus taeda* L. *Tree Physiol.* **2**, 301–306.

Lorio, P.L., Jr., Sommers, R.A., Blanche, C.A., Hodges, J.D. and Nebeker, T.E. (1990). Modeling pine resistance to bark beetles based on growth and differentiation balance principles. *In* "Process Modeling of Forest Growth Responses to Environmental Stress" (R.K. Dixon, R.S. Meldahl, G.A. Ruark, and W.G Warren, eds), pp. 402–409. Timber Press, Portland, OR.

Martinat, P.J. (1987). The role of climatic variation and weather in forest insect outbreaks. *In* "Insect Outbreaks" (P. Barbosa and J.C. Schultz, eds), pp. 241–268. Academic Press, San Diego, CA

Mason, R.R. (1969). A simple technique for measuring oleoresin exudation flow in pines. *Forest Sci.* **15**, 56–67 .

Mason, R.R. (1971). Soil moisture and stand density affect oleoresin exudation flow in a loblolly pine plantation. *Forest Sci.* **17**, 170–177.

Matson, P.A. and Hain, F.P. (1985). Host conifer defense strategies: a hypothesis. *In* "The Role of the Host in the Population Dynamics of Forest Insects" (L. Safranyik, ed.), pp. 33–42. Canadian Forestry Service and USDA Forest Service, Victoria, BC.

Mattson, W.J. and Haack, R.A. (1987). The role of drought in outbreaks of plant-eating insects. *BioScience* **37**, 110–118 .

Mattson, W.J., Lawrence, R.K., Haack, R.A., Herms, D.A. and Charles, P.J. (1988). Defensive strategies of woody plants against different insect-feeding guilds in relation to plant ecological strategies and intimacy of association with insects. *In* "Mechanisms of Woody Plant Defenses against Insects" (W.J. Mattson, J. Levieux, and C. Bernard-Dagan, eds), pp. 3–38. Springer-Verlag, New York.

McGregor, M.D., Amman, G.D., Schmitz, R.F. and Oakes, R.D. (1987). Partial cutting lodgepole pine stands to reduce losses to the mountain pine beetle. *Can. J. For. Res.* **17**, 1234–1239.

McQuate, G.T. and Connor, E.F. (1990a). Insect responses to plant water deficits. I. Effect of water deficits in soybean plants on the feeding preference of Mexican bean beetle larvae. *Ecol. Entomol.* **15**, 419–431.

McQuate, G.T. and Connor, E.F. (1990b). Insect responses to plant water deficits. II. Effect of water deficits in soybean plants on the growth and survival of Mexican bean beetle larvae. *Ecol. Entomol.* **15**, 433–445.

Miller, J.M. and Keen, F.P. (1960). Biology and control of the western pine beetle. A summary of the first fifty years of research. USDA Forest Service Miscellaneous Publ. 800, USDA Forest Serv., Washington, DC, 381 pp.

Mitchell, R.G., Waring, R.H. and Pitman, G.B. (1983). Thinning lodgepole pine increases tree vigor and resistance to mountain pine beetle. *Forest Sci.* **29**, 204–211.

Mooney, H.A. and Chu, C. (1974). Seasonal carbon allocation in *Heteromeles arbutifolia*, a California evergreen shrub. *Oecologia* **14**, 295–306.

Myers, R.F. (1986). Cambium destruction in conifers caused by pinewood nematodes. *J. Nematol.* **18**, 398–402.

Nebeker, T.E., Hodges, J.D., Karr, B.K. and Moehring, D.M. (1985). Thinning practices in southern pines — with pest management recommendations. USDA Forest Serv. Tech. Bull. 1703. USDA Forest Serv., Washington, DC, 36 pp.

Ostrom, C.E., True, R.P. and Schopmeyer, C.S. (1958). Role of chemical treatment in stimulating resin flow. *Forest Sci.* **4**, 296–306.

Owen, D.R., Lindahl, K.Q., Jr., Wood, D.L. and Parmeter, J.R. (1987). Pathogenicity of fungi isolated from *Dendroctonus valens, D. brevicomis,* and *D. ponderosae* to ponderosa pine seedlings. *Phytopathology* **77**, 631–636.

Paine, T.D. (1984). Seasonal response of ponderosa pine to inoculation of the mycangial fungi from the western pine beetle. *Can. J. Bot.* **62**, 551–555.

Parkerson, R.H. and Whitmore, F.W. (1972). A correlation of stem sugars, starch, and lipid with wood formation in eastern white pine. *Forest Sci.* **18**, 178–183.

Raffa, K.F. and Berryman, A.A. (1982). Physiological differences between lodgepole pines resistant and susceptible to the mountain pine beetle and associated microorganisms. *Environ. Entomol.* **11**, 486–492.

Raffa, K.F. and Berryman, A.A. (1983). The role of host plant resistance in the colonization behavior and ecology of bark beetles (Coleoptera: Scolytidae). *Ecol. Monogr.* **53**, 27–49.

Raffa, K.F. and Smalley, E.B. (1988). Seasonal and long-term responses of host trees to microbial associates of the pine engraver, *Ips pini*. *Can. J. For. Res.* **18**, 1624–1634.

Reid, R.W. and Shrimpton, D.M. (1971). Resistant response of lodgepole pine to inoculation with *Europhium clavigerum* in different months and at different heights on stem. *Can. J. Bot.* **49**, 349–351.

Rhoades, D.F. (1979). Evolution of plant chemical defense against herbivores. *In* "Herbivores: their Interaction with Secondary Metabolites" (G.A. Rosenthal and D.H. Janzen, eds), pp. 3–54. Academic Press, New York.

Rhoades, D.F. and Cates, R.G. (1976). Toward a general theory of plant antiherbivore chemistry. *Rec. Adv. Phytochem.* **10**, 168–213.

Sachs, R.M. (1987). Roles of photosynthesis and assimilate partitioning in flower initiation. *In* "Manipulation of Flowering" (J.G. Atherton, ed.), pp. 317–340. Nottingham University, Buttersworth, London.

Sartwell, C. and Stevens, R.E. (1975). Mountain pine beetle in ponderosa pine: prospects for silvicultural control in second-growth stands. *J. Forestry* **73**, 136–140.

Schopmeyer, C.S., Mergen, F. and Evans, T.C. (1954). Applicability of Poiseuille's law to exudation of oleoresin from wounds on slash pine. *Plant Physiol.* **29**, 82–87.

Schultz, J.C. and Baldwin, I.T. (1982). Oak leaf quality declines in response to defoliation by gypsy moth larvae. *Science* **217**, 149–151.

Schwarz, O. J. (1983). Paraquat-induced lightwood formation in pine. *In* "Plant Growth Regulating Chemicals," Volume II (L.G. Nickell, ed.), pp. 77–97. CRC Press, Boca Raton, FL.

Sharpe, P.J.H. and Wu, H. (1985). A preliminary model of host susceptibility to bark beetle attack. *In* "The Role of the Host in the Population Dynamics of Forest Insects" (L. Safranyik, ed.), pp. 108–127 . Canadian Forestry Service and USDA Forest Service, Victoria, BC.

Sharpe, P.J.H., Wu, H., Cates, R.G. and Goeschl, J.D. (1985). Energetics of pine defense systems to bark beetle attack. *In* "Proceedings, Integrated Pest Management Research Symposium" (S.J. Branham and R.C. Thatcher, eds), pp. 206–223. USDA Forest Service Gen. Tech. Rep. SO-56, USDA Forest Serv. Southern Forest Experiment Station, New Orleans, LA.

Shigo, A.L. (1985). Stress and death of trees. *In* "Spruce-fir Management and Spruce Budworm" (D. Schmitt, ed.), pp. 31–38. USDA Forest Serv. Gen. Tech. Rpt NE-99, USDA Forest Service, Northeastern Forest Experiment Station, Broomall, PA.

Shoulders, E. (1967). Fertilizer application, inherent fruitfulness, and rainfall affect flowering of longleaf pine. *Forest Sci.* **14**, 376–383 .

Shoulders, E. (1968). Fertilization increases longleaf and slash pine flower and cone crops in Louisiana. *J. Forestry* **66**, 193–197.

Shoulders, E. (1973). Rainfall influences female flowering of slash pine. USDA Forest Service Research Note SO-150, USDA Forest Serv. Southern For. Exp. Stn, 7 pp.

Stanley, R.G. (1958). Methods and concepts applied to a study of flowering in pine. *In* "The Physiology of Forest Trees" (K.V. Thimann, ed.), pp. 583–599. Ronald Press, New York.

Stark, R.W. (1965). Recent trends in forest entomology. *Annu. Rev. Entomol.* **10**, 303–324.

Stephen, F.M. and Paine, T.D. (1985). Seasonal pattern of host tree resistance to fungal associates of the southern pine beetle. *Zeit. Ang. Entomol.* **99**, 113–122.

Stubbs, J., Roberts, D.R. and Outcalt, K.W. (1984). Chemical stimulation of lightwood in southern pines. USDA Forest Service Gen. Tech. Rpt. SE-25, USDA Forest Serv. Southeastern For. Exp. Stn., Asheville, NC, 51 pp.

Trewavas, A. (1985). A pivotal role for nitrate and leaf growth in plant development. *In* "Control of Leaf Growth" (N.K. Baker, W.J. Davies, and C. Ong, eds), pp. 77–91. Cambridge University Press, Cambridge.

Turchin, P., Lorio, P.L., Jr., Taylor, A.D. and Billings, R.F. (1991). Why do populations of southern pine beetles (Coleoptera: Scolytidae) fluctuate? *Environ. Entomol.* **20**, 401–409.

Tyree, M.T. and Dixon, M.A. (1986). Water stress induced cavitation and embolism in some woody plants. *Physiol. Plant.* **66**, 397–405.

Veihmeyer, F.J. and Hendrickson, A.H. (1961). Responses of a plant to soil-moisture changes as shown by guayule. *Hilgardia* **30**, 621–637.

Vité, J.P. (1961). The influence of water supply on oleoresin exudation pressure and resistance to bark beetle attack in *Pinus ponderosa*. *Contrib. Boyce Thompson Inst.* **21**, 37–66.

Vité, J.P. and Wood, D.L. (1961). A study on the applicability of the measurement of oleoresin exudation pressure in determining susceptibility of second growth ponderosa pine to bark beetle infestation. *Contrib. Boyce Thompson Inst.* **21**, 67–78.

Waring, R.H. (1983). Estimating forest growth and efficiency in relation to tree canopy area. *Adv. Ecol. Res.* **13**, 327–354.

Waring, R.H. and Pitman, G.B. (1985). Modifying lodgepole pine stands to change susceptibility to mountain pine beetle attack. *Ecology* **66**, 889–897.

Waring, R.H. and Schlesinger, W.H. (1985). "Forest Ecosystems: Concepts and Management." Academic Press, Orlando, FL, 340 pp.

Weatherley, P.E. (1950). Studies in the water relations of the cotton plant. I. The field measurement of water deficits in leaves. *New Phytol.* **49**, 81–97.

Wenger, K.F. (1957). Annual variation in the seed crops of loblolly pine. *J. Forestry* **55**, 567–569.

White, T.C.R. (1974). A hypothesis to explain outbreaks of looper caterpillars, with special reference to populations of *Selidosema suavis* in a plantation of *Pinus radiata* in New Zealand. *Oecologia* **16**, 279–301.

Witcosky, J.J., Schowalter, T.D. and Hansen, E.M. (1986). The influence of time of thinning on the colonization of Douglas-fir by three species of root-colonizing insects. *Can. J. For. Res.* **16**, 745–749.

Wolter, K.E. and Zinkel, D.F. (1984); Observations on the physiological mechanisms and chemical constituents of induced oleoresin synthesis in *Pinus resinosa. Can. J. For. Res.* **14**, 452–458.

Wood, D.L. (1962). Experiments on the interrelationship between oleoresin exudation pressure in *Pinus ponderosa* and attack by *Ips confusus* (Lec.) (Coleoptera: Scolytidae). *Can. Entomol.* **94**, 473–477.

Wood, D.L. (1972). Selection and colonization of ponderosa pine by bark beetles. *In* "Insect/Plant Relationships" (H.F. van Emden, ed.), pp. 101–117. Blackwell Scientific Publications, Oxford.

Wood, D.L., Stark, R.W., Waters, W.E., Bedard, W.D. and Cobb, F.W., Jr. (1985). Treatment tactics and strategies. *In* "Integrated Pest Management in Pine–Bark Beetle Ecosystems" (W.E. Waters, R.W. Stark and D.L. Wood, eds), pp. 121–139. John Wiley & Sons, New York.

Zahner, R. (1968). Water deficits and growth of trees. *In* "Water Deficits and Plant Growth" (T.T. Kozlowski, ed.), pp. 191–254. Academic Press, New York.

–6–

Strategies and Mechanisms of Host Colonization by Bark Beetles

KENNETH F. RAFFA,[1] THOMAS W. PHILLIPS[1,2] and SCOTT M. SALOM[3]

[1] *Department of Entomology, University of Wisconsin, Madison, WI, USA*

[2] *US Department of Agriculture, Agricultural Research Service, Madison, WI, USA*

[3] *Department of Entomology, Virginia Polytechnic Institute and State University, Blacksburg, VA, USA*

6.1 INTRODUCTION

Bark beetles have evolved complex strategies and mechanisms for discovering and exploiting the unique subcortical resources of suitable host trees. By contrast, pathogens typically have little control over their dispersal and contact with hosts. Hyphal growth and spore dissemination to new hosts are dictated by root proximity and air currents, respectively. Thus, pathogens vectored by bark beetles benefit from the increased efficiency of transport to and inoculation into suitable hosts. This chapter focuses on the strategies and mechanisms employed by bark beetle for utilizing the subcortical resources of conifers.

6.2 THE SUBCORTICAL ENVIRONMENT

This section focuses on aspects of the subcortical environment that affect colonization success of bark beetles. See Chapter 5 for underlying physiological processes contributing to the subcortical environment.

BEETLE–PATHOGEN INTERACTIONS IN CONIFER FORESTS
ISBN 0-12-628970-0

6.2.1 Characteristics of the subcortical environment

The evolutionary success of scolytids, and the difficulties they pose to forest managers, are largely attributable to features of their subcortical environment. This habitat provides good protection from environmental extremes, potential natural enemies that are not specifically adapted to finding subcortical insects and, to a certain extent, specialized predatory and parasitic species. However, exploitation of this habitat has required specialized adaptations.

The subcortical substrate of woody plants is comprised of outer xylem (wood) and inner phloem (inner bark) tissues that arise throughout the life of the plant from a medial group of lateral meristem cells called the vascular cambium (Kramer and Kozlowski, 1979). The young outer portion of the xylem (sapwood) serves to conduct sap, strengthen the stem, and in some cases act as a storage reservoir for food. Sap consists of a dilute solution of mineral salts and organic substances, including nitrogenous compounds, carbohydrates, enzymes, and growth regulators. Aged xylem tissue, made up of dead cells in the center of the tree (heartwood), provides mechanical support. Xylem tissue is comprised of longitudinal tracheids and epithelial cells as well as transverse ray parenchyma and epithelial cells. Resin canals, intercellular spaces that store resin secreted from epithelial cells, are a normal feature in *Pinus*, *Picea*, *Larix*, and *Pseudotsuga* sapwood, but are lacking in *Cedrus*, *Tsuga*, and *Abies* (Bannan, 1936). Higher quantities of resin are generally found in conifers than in hardwoods (Wise and Jahn, 1952). Resin is a heterogeneous mixture of resin acids, fatty acids, acid esters, sterols, alcohols, waxes, and hydrocarbons (Wise and Jahn, 1952). The outer stem is covered by bark, which includes inner living phloem arising from the vascular cambium, and the outer periderm arising from the phellogen or cork cambium.

Phloem is a conducting tissue through which dissolved sugars and other assimilates move (Salisbury and Ross, 1978). Conifer phloem is comprised of elongated sieve cells and closely associated albuminous cells, both of which function in transport. Parenchyma cells facilitate storage and lateral transport of solutes and water, and phloem fibers provide support to the tissue. Phloem (including the thin cambium layer) is the principal food source for the developing larvae, and also provides nutrition for adults. The nutritional content of this region has been analyzed for several tree species, and is typically sufficient in starch and sugars, but very low in nitrogen relative to foliage and reproductive substrates (Shrimpton, 1973; Mattson, 1980; Miller and Berryman, 1986; Haack and Slansky, 1987; Popp *et al.*, 1989). Compared to other woody substrates available to subcortical insects, however, the phloem is relatively nutritious (Schowalter *et al.*, 1992). For example, sapwood, heartwood, and outer bark are particularly low in nitrogen, minerals and (where known) carbohydrates. Thus, the most favorable substrate is restricted to a very narrow band (Haack and Slansky, 1987). Seasonal variation is much more pronounced in phloem than other woody tissues, however, with starch, nitrogen, lipid, and mineral content generally highest in autumn and winter, and declining rapidly in spring to low levels (Haack and Slansky, 1987). This variation in nutritional quality could reduce the reproductive potential of bark beetles, especially northern species that are limited to narrow periods of flight and development, even in the absence of tree defenses (see Chapter 5). In summary, the very scattered, patchy, and uncertain environment confronting bark beetles is not limited to between-tree availability, but extends to within tree spatial and temporal patterns as well (see also Chapter 5).

6.2.2 Enhancement of the food base by microorganisms

Scolytids rely on a wide array of fungi, yeasts, bacteria, and protozoans to facilitate colonization of their host substrate and obtain the nutrients necessary for brood development and survival (Whitney, 1982; Chapters 3 and 7). These symbiotic relationships are relatively species-specific, and in most cases mutualistic. Examples include nitrogen-fixing bacteria that facilitate *Dendroctonus terebrans* larval nutrition (Bridges, 1981), and yeasts that aid in the transmission of the fungi from *Ips avulsus* adults to larvae (Gouger, 1972). These non-phytopathogenic microorganisms usually are vectored by the beetle, but also may be transported by other associates, such as nematodes and mites that are phoretic or parasitic on the beetles.

The fungi carried by ambrosia beetles include *Ambrosiella, Fusarium, Monilia* and *Cephalosporium.* These fungi provide the beetles with their sole source of food (Francke-Grosmann, 1967) and derive their nutrients from host sapwood (Brues, 1946). They are neither pathogenic nor wood-decaying. Without continued tending by the beetles, the fungi could not survive in the galleries (Batra, 1966).

6.2.3 Chemical and histological barriers to host colonization

Despite its favorable nutritional and environmental attributes, the subcortical environment of living trees can be extremely inhospitable to invading organisms. The requirement of phloem death for beetle reproduction has placed strong selective pressures on conifers for effective defense systems. Host chemistry includes a broad array of toxic monoterpenes, resin acids, and phenolics that can kill the beetles at all stages of development and also inhibit fungal growth (Smith, 1963; Cobb *et al.*, 1968; Coyne and Lott, 1976; Raffa *et al.*, 1985). Trees generally can withstand low density attacks and prevent beetle reproduction, but can be overwhelmed by a mass attack (Raffa and Berryman, 1983a).

The complex set of interactions that operate among several tree defense mechanisms, multiple attacks, and beetle behavior are illustrated in a simplified form in Fig. 6.1. Tree chemistry, at least among healthy individuals, discourages most beetles from entering (Fig. 6.1a) (Hynum and Berryman, 1980; Moeck *et al.*, 1981). This can be viewed as a form of tree resistance, avoidance by the beetle of an unfavorable environment (see section 6.4.3), or both, depending on one's perspective.

If landing is followed by bark penetration, the tree responds to attack by secreting resin directly into the beetle's path (Chapters 5 and 8). The rapid flow of resin from severed ducts may physically impede beetle progress, interfere with the flow of beetle-produced pheromones into the atmosphere, and/or repel beetles from further attack (Figs 6.1a and 6.1c). Resin volume, flow and crystallization rate may partially explain the greater resistance of some trees to beetle colonization (Hodges *et al.*, 1979; Chapter 5). Cates and Alexander (1982) found higher resin flow rates in *P. ponderosa* that were not colonized by *D. ponderosae* than in colonized trees. Popp *et al.* (1991) reported that the beetle vectored fungi *Ophiostoma minus* and *O. ips* stimulated resin flow in *P. elliottii* and *P. taeda.* However, adult bark beetles and fungi vectored by the beetles generally can tolerate the consitutive resin chemistry of their normal hosts (Smith, 1963, 1966; Cobb *et al.*, 1968; Shrimpton and Whitney, 1968).

Once beetles contact phloem tissue, the tree responds to fungal metabolites and physical wounding (Chapter 8). This response includes cellular desiccation and necrosis, starch loss and accumulation of translocated and/or newly synthesized toxins. Monoterpene and phenolic concentrations increase while resin flow impedes beetle colonization (Fig. 6.1e). Host allelochemicals can reach adulticidal, ovicidal, and fungistatic concentrations within a few days of initial attack (Shrimpton and Whitney, 1968; Raffa and Berryman, 1983b; Raffa *et*

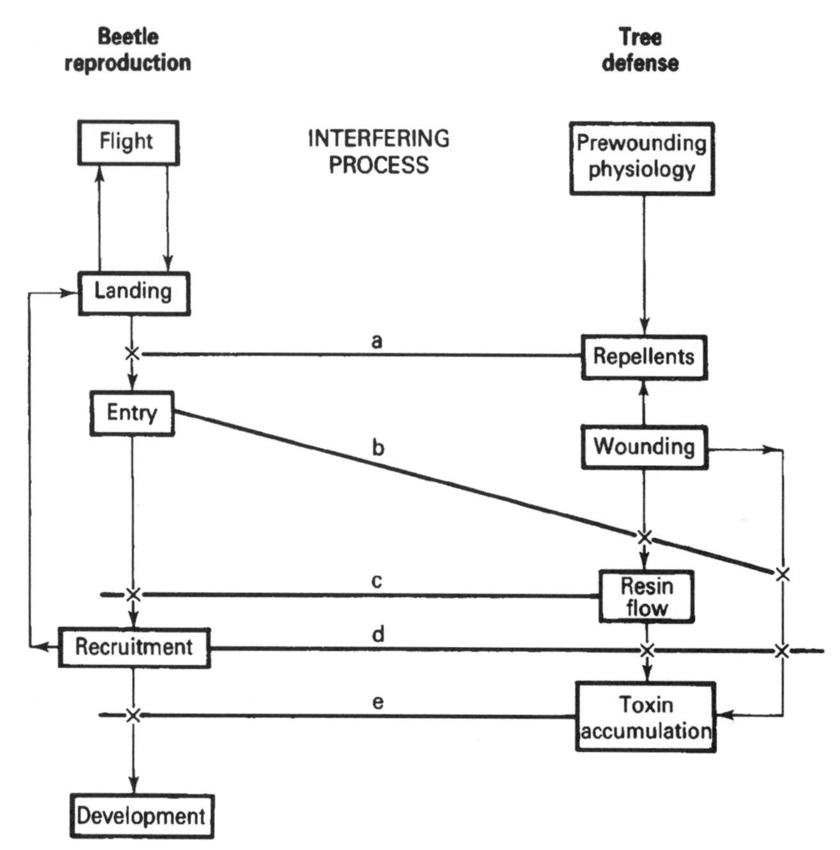

Fig. 6.1. Simplified schematic of conifer defenses against bark beetles and fungal associates. The sequence of events leading to either successful beetle colonization or tree defense is depicted in the left and right columns, respectively. Each sequence is connected by a series of narrow lines. Interactions where one agent interferes with a process of the other are illustrated with thick lines (see text for explanation of letters). These lines originate from an attribute or action (shown in a box) of the interfering agent, and interrupt the flow from one process to another where indicated by "x." For a more mechanistic version, raw data from supported experiments and references, see Raffa (1991a) and Raffa and Klepzig (1992) and references therein.

al., 1985). Unless these induced defenses are immobilized by the attacking beetle cohort, or occur too slowly to preclude mass attack and colonization, reproduction and brood development cannot proceed. A simultaneous, pheromone-mediated attack by many beetles and their vectored fungi can deplete both the constitutive and induced host resins (Figs 6.1b and 6.1d). For example, resin flow rates from *P. contorta* during the third day of aggregation by *D. ponderosae* were only 33% of those prior to attack (Raffa and Berryman, 1983a). Natural mass attacks and simulated multiple fungal inoculations can prevent the increase in monoterpene concentrations that otherwise would occur (Horntvedt *et al.*, 1983; Raffa and Berryman, 1983b).

Host entry by beetles, however, does not guarantee that mass attacks will result under field conditions. If it did, then all trees could be killed, regardless of resistance. The likelihood that an entered tree will become a focus of attraction depends on the number of beetles simultaneously engaged in the early stage of attack, the population of dispersing beetles, weather, and the rate of resin flow from wounds (Raffa and Berryman, 1983a). For example, among *P.*

contorta first entered by single "pioneer" *D. ponderosae* (as opposed to beetles diverted from mass attacks on nearby trees), the likelihood of mass attack and subsequent tree death was inversely related to primary resin flow rate (Raffa and Berryman, 1982a). However, resin flow rate was not different among trees that subsequently survived or died during this outbreak (Raffa and Berryman, 1982a). Resistant trees did show higher rates of monoterpene accumulation in response to simulated attack than did trees that subsequently were killed (Raffa and Berryman, 1982a). Thus, constitutive and rapidly induced defenses must interact as a cohesive system to defend most healthy trees under most conditions (Berryman, 1972).

Because tree resistance can be overcome only at high densities of beetles, aggregation of sufficient beetles must be balanced against limited food availability for larvae. Consequently, behavioral responses that simply orient bark beetles to trees they can kill are not necessarily adaptive. If the number of colonizers required to overcome resistance exceeds the number of offspring that can be produced within the finite substrate, then replacement rate falls below 1.0. Bark beetles have evolved strategies and mechanisms for regulating densities of ovipositing adults.

6.3 SCOLYTID STRATEGIES

6.3.1 Range of host relationships

Scolytid species can be characterized according to the physiological condition of trees that they colonize (Rudinsky, 1962; Stark 1982). Most species will enter only dead trees or dead tree parts. Others can attack stressed trees, with different beetle species falling along a gradient of host stress levels required to elicit entry. A very few species can colonize and kill healthy trees. Species that can colonize healthy trees are termed "aggressive," as compared to "non-aggressive" species that colonize dying or dead trees. The terms "primary" or "secondary" have been used to describe the time during the sequence of infestation that a particular species usually arrives. This generally, but not entirely, corresponds to the aggressive versus non-aggressive gradient. Ecologically, scolytids can be classified as "saprophytic" (feeding only on dead plant material), "facultatively parasitic" (normally saprophytic but capable of parasitizing stressed live plants) or "nearly obligate parasitic" (usually colonizing living plant material).

This pattern is complicated, however, by at least three factors. First, although each species can be characterized by an upper limit of host vigor that they can overcome, almost all will colonize trees of inferior condition. For example, even the aggressive species can colonize fallen trees. Second, certain species exhibit a range of behaviors through time, in which they are limited to severely stressed trees during periods of low population density, but colonize healthy trees during periods of high density. Third, different stresses affect trees differently, making strict comparisons difficult.

6.3.2 Aggressive and non-aggressive species

The specialized ability of a few species to colonize living trees was facilitated by two evolutionary trends: the development of sophisticated chemical communication systems and association with phytopathogenic fungi. Both non-aggressive and aggressive species use aggregation pheromones to optimize host finding, mate finding, and utilization of a scarce resource (Atkins, 1966). Aggressive species, however, also use pheromones to recruit conspecifics that overwhelm host defenses, so that the tree or tree section can be killed, a necessary requirement for beetle survival (Wood, 1972). Host monoterpenes function as kairomones that synergize attraction to beetle-produced pheromones in tree-killing species (Borden, 1982, 1985; D.L. Wood, 1982).

Different life history strategies between aggressive and non-aggressive scolytids are reflected in their host colonization mechanisms. For example, most scolytids incorporate both tree- and insect-produced volatiles in their host-finding repertoire, but their roles vary. Many non-aggressive species are attracted to volatiles, such as ethanol and/or monoterpenes, from dead or dying trees (Vité and Gara, 1962; Moeck, 1970, Rudinsky, 1966; Klimetzek *et al.*, 1986). These "primary" cues attract and or arrest beetles, which then may produce "secondary," i.e., pheromonal, attractants as well. Use of host attractants by pioneer beetles of the aggressive species *Dendroctonus brevicomis, D. ponderosae,* and *D. frontalis,* however, has not been demonstrated under controlled conditions despite numerous attempts. Prior to the onset of pheromone emission, landing rates by aggressive species on resistant and susceptible trees are equal (Payne *et al.*, 1978; Hynum and Berryman, 1980; Moeck *et al.*, 1981), and appear more strongly influenced by visual cues (Shepherd, 1966; Pitman and Vité, 1969). Host volatiles are essential synergists to pheromone plumes by aggressive species (D. Wood, 1982), however, and also may interact with visual cues (Lanier *et al.*, 1976; Payne and Coulson, 1985; Tilden *et al.*, 1979).

Establishment of attacking beetles in living hosts is augmented by the introduction of symbiotic fungi (Whitney, 1982; Horntvedt *et al.*, 1983). The pathogenicities of the fungi typically associated with different beetle species vary. Some species carry relatively virulent fungal symbionts classified as *Ophiostoma* spp. (=*Ceratocystis*) and its anamorph, *Leptographium*, of the class Ascomycetes, that invade the host's vascular system and secrete toxins, thereby hastening tree death (Chapter 8). Most fungal associates, however, are only moderately parasitic and generally are better suited for a saprophytic mode.

The extent to which different fungi rely on beetles for transport also varies (Chapters 2 and 3). Fungi that are carried in specialized pitlike depressions or pouches in the exoskeleton, termed mycangia, as well as many of the externally vectored *Ophiostoma* spp., rely entirely on beetles for between-tree movement. In some cases, phoretic mites and/or specialized beetle behaviors are required for recontamination of the emerging brood (Moser, 1985; Chapter 7). Other fungi appear to rely on beetles for long-distance movement, but can move between trees through root grafts. For example, the highly pathogenic *Leptographium wageneri*, *L. terebrantis*, and *L. procerum* are introduced into stands first by several root and lower-stem feeding beetles, but subsequently can progress through the stand without insect vectors (Goheen and Cobb, 1978; Witcosky *et al.*, 1986; Chapters 3 and 9). Mechanisms of spread include infection through root grafts, movement across roots externally in contact, and short-distance progression through soil. These fungi often predispose neighboring trees to bark beetle attack (Goheen and Cobb, 1980; Hunt and Morrison, 1980). The newly killed trees then become colonized by root-feeding beetle vectors of *Leptographium*, such as *Hylastes*, *Hylobius*, and *Dendroctonus*, initiating a chain reaction (Klepzig *et al.*, 1991).

6.4 MECHANISMS OF HOST COLONIZATION BY BARK BEETLES

Host colonization by bark beetles consists of four phases (Wood, 1972): (1) dispersal, the movement of new adults away from natal or hibernation sites; (2) host selection, including orientation to a potential host, landing, tasting or evaluating the host, and sustained feeding; (3) concentration, the recruitment of additional conspecifics (i.e. "mass attack") through pheromones and/or liberated host attractants; and (4) establishment, the elimination of host resistance followed by successful oviposition and fungal growth in the host substrate. It must be emphasized, however, that while these designations may facilitate conceptualization and experimentation, they are in fact part of a behavioral continuum.

6.4.1 Dispersal

Newly emerged adults tend to be positively phototactic and negatively geotropic when environmental conditions are suitable for leaving a breeding site or overwintering harborage (Graham, 1959; Atkins, 1966). Marked *D. brevicomis* may disperse up to 3 km, and sometimes as far as 14 km (Miller and Keen, 1960). Birch *et al.* (1981) reported that *Scolytus multistriatus* were caught on pheromone-baited traps in a California desert 8 km from any *Ulmus* host. Despite the capacity for long range dispersal, most individuals probably travel much shorter distances, especially if attractive sources are nearby. For example, Salom and McLean (1989) captured 25% of marked *Trypodendron lineatum* in 28 pheromone traps within a 100 m radius of the release point. Gara (1965) routinely recaptured about 25% of marked *Ips paraconfusus* (=*confusus*) at a small number (usually five) of attractant traps within 3–25 m from the release point. In growing infestations of *D. frontalis*, 97% of parent adult beetles re-emerged following oviposition and gallery construction to engage in new attacks, mostly within the immediate infestation (Coulson *et al.*, 1978). Many brood adults also are believed to fly to nearby trees as the infestation grows. In general, if attractive sources are close to emergence sites, substantial numbers of beetles will terminate dispersal after a relatively short flight.

The change from dispersal to host- and/or mate-directed orientation appears to be under physiological control. Newly emerged *T. lineatum* are strongly photopositive, and flight exercise is required for individuals to respond to attractant odors (Graham, 1959). Lipid content of *Dendroctonus pseudotsugae* is negatively correlated with the propensity to respond to olfactory stimuli, and increased flight time reduces lipid content and increases responsiveness (Atkins, 1966, 1969). Different qualities and quantities of food during larval development may contribute to physiological and behavioral variation among individuals within a population (Borden *et al.*, 1986). *Ips paraconfusus* that emerge early from logs or originate from low density populations are more prone to cease dispersal and respond to pheromones than are late-emerging beetles and those from high density populations (Gara, 1965).

Seasonal variation in the physiological composition of scolytid populations may add to this complexity. Adult *D. frontalis* fat contents are high in spring and fall generations, when long-range dispersal is more likely (Hedden and Billings, 1977). Conversely, they are low during the summer months, when beetles often expand local infestations by making short flights from brood trees to new attacks (Nijholt, 1965; Gara, 1967).

One proposed adaptive benefit to long distance dispersal is increased genetic mixing from outbreeding (Borden, 1982; D. Wood, 1982). Although inbreeding often is assumed to be deleterious, a moderate level of inbreeding is probably the rule rather than the exception for many species of animals and plants (Shields, 1982). If long-distance dispersal of bark beetles fosters gene flow among localities, then theory would predict homogeneity of gene frequencies across broad areas. Yet studies of electrophoretically detectable allozyme markers have found local samples of scolytids within regions to represent genetically distinct mating populations (Stock and Guenther, 1979). Sympatric populations of *D. ponderosae* from different species of host trees were genetically divergent at some electrophoretic loci (Sturgeon and Mitton, 1982), suggesting that host-related biotype formation may occur in bark beetles (Langor *et al.*, 1990). Variable dispersal distances may be more important as an optimal strategy for exploiting temporary, patchy resources in an uncertain environment (Atkins, 1966). If a female produces progeny that include both long and short distance dispersers, she may balance the risk of not having host trees available nearby with the risk of emigration to unsuitable habitats among her descendants.

6.4.2 Host selection

6.4.2.1 Initial landing

Recognizing that tree death by bark beetles was non-random throughout a forest, Person (1931) proposed that susceptible pines were selected by directed orientation to chemical cues associated with stress physiology. He rejected the possibility of random search for weakened trees based on the rarity of trees displaying unsuccessful attacks. Studies with some secondary and non-aggressive species have supported this view. *Dendroctonus valens* are strongly attracted to volatile terpenes in oleoresin that is released upon injury or stress (Vité and Gara, 1962), and many species are attracted to resin, individual monoterpenes, and alcohols (Chapman, 1962; Rudinsky, 1966). Ethanol, a common volatile from stressed and fermenting plants, attracts ambrosia beetles (Moeck, 1970), and terpenes and ethanol can synergistically attract many species that colonize stressed or deteriorating hosts (Phillips *et al.*, 1988).

Tree-killing species, however, appear to land in response to visual cues (Shepherd, 1966) and subsequently select or reject a tree following some close-range evaluation. Studies of *S. ventralis* found that 68–88% of the trees examined were attacked at least once during the early dispersal period, even though only 3% were successfully colonized (Berryman and Ashraf, 1970). Likewise, pre-aggregation landing rates by *D. ponderosae* on *Pinus contorta*, as measured by barrier traps on tree boles, were randomly distributed in relation to which trees ultimately succumbed to attack (Hynum and Berryman, 1980). By screening trees to prevent attacks and eliminate the chances of pheromonal attraction, Moeck *et al.* (1981) found no significant differences in landing by tree-killing scolytids among healthy, root diseased, artificially water stressed, mechanically injured, and chemically injured *P. ponderosa*. To date, no studies using this precaution have demonstrated primary attraction among aggressive scolytids.

While random landing may seem a risky, even counter-intuitive, strategy for an insect that seeks a limited and patchy resource, the combination of land-and-choose behavior with secondary attraction at suitable hosts may be highly effective. Raffa and Berryman (1980) found that the annual proportion of trees that were contacted by *D. ponderosae* ranged from 89% to 100% over a 4-year period. Anderbrant *et al.* (1988) found that over 90% of *Picea* trees in a stand that never became attacked by *Ips typographus* were visited by these beetles. It is important to note that these studies used non-attractive traps that covered less than 1% of the bark area. Moeck *et al.* (1981) calculated a visitation rate of one beetle per tree per day for *D. brevicomis* in stands of *P. ponderosa*. Therefore, if every tree is likely to be visited randomly and sampled by bark beetles on a daily basis, it is very likely that susceptible trees will be located readily, pioneer beetles will produce pheromones, and an exponentially larger number of diffusely dispersed beetles will be directed to the resource. Given that suppression of host resistance requires mass attack, the creation of attractive host foci and concentration of dispersing beetles would be quite rapid under epidemic, compared to endemic, situations.

A note of caution is necessary, however, because the physiological conditions of beetles collected in passive traps cannot be assessed. Insect responsiveness is known to vary as a result of internal regulatory mechanisms. If a large proportion of captured beetles are not in "attack" mode (Moeck *et al.*, 1981), these data would be confounded. However, until such a condition is physiologically defined and characterized, the results of controlled experiments and field observations do not allow rejection of the hypothesis that landing is not chemically directed. As with many aspects of bark beetle behavior, the possibility of a flexible strategy governed by beetle population density also cannot be dismissed.

6.4.2.2 Post landing

Once on the bark, beetles may either enter the tree or resume dispersal. Many beetles resume flight even if the tree is susceptible and undergoing mass attack (Hynum, 1978; Bunt *et al.*, 1980). Assessment may take place in response to short distance olfaction, chemical cues perceived on contact, or biting. Together, the high landing rates and low number of abandoned pitch tubes suggest that sustained feeding is normally not necessary for host assessment. Bark beetles apparently can taste the outer phloem and accept or reject the host prior to severing xylem resin ducts.

Beetles are positively thigmotactic (orient toward contact with a solid surface) and tend to initiate feeding in bark crevices or between bark plates (Norris and Baker, 1967; Elkinton and Wood, 1980). Initiation and maintenance of feeding is governed by chemical feeding stimulants. *Ips paraconfusus* males bored through the outer bark of both a host and non-host conifer, but rejected the non-host after they contacted phloem tissue (Elkinton and Wood, 1980). Elkinton *et al.* (1981) subsequently demonstrated that certain phloem extracts stimulated tunneling in this species. Feeding by *D. ponderosae* females is incited by both non-polar and polar extracts of outer bark from *P. contorta*, but only the latter stimulates sustained feeding (Raffa and Berryman, 1982b). Feeding on extracts was not related to whether trees subsequently survived or were killed by natural populations, even though beetle densities were high enough to ensure that all trees were tasted. Stimulants were not required for continued excavation once in the phloem, but repellents induced host abandonment (Raffa and Berryman, 1982b).

Although the presence of feeding incitants, stimulants, and repellents has been documented, most of these chemicals have not been identified in most systems. Ethanol, which is commonly associated with stress physiology (Kramer and Kozlowski, 1979), may be involved in various systems (Klimetzek *et al.*, 1986). Treatment of *Picea engelmannii* bark with 95% ethanol induced attacks by *Dendroctonus rufipennis* on resistant trees, suggesting that ethanol might serve as a feeding stimulant (Moeck, 1981). Norris and Baker (1969) found that ethanol was required as a feeding stimulant for *Xyleborus ferrugineus* on an artificial diet. However, we know of no studies in which ethanol levels have been quantified in standing conifers susceptible to bark beetle attack.

6.4.3 Concentration

6.4.3.1 Pheromone biosynthesis

Pheromones can be synthesized by direct bioconversion of an acquired host compound or from precursors with structurally dissimilar carbon skeletons. These pathways are represented by terpene alcohols and by an array of bicyclic ketals, hydrocarbons, and oxygenated short-chain, cyclic, or aromatic compounds, respectively. Examples of different types of bark beetle pheromones, that alternatively may be classified by chemical structures (Francke, 1988), are shown in Fig. 6.2. Beetles also may emit unmodified plant compounds that are sequestered and subsequently released in the plume. For example, both sexes of *D. pseudotsugae* release limonene, which synergizes the beetle-produced pheromones (Rudinsky *et al.*, 1977).

Conversion of host monoterpenes to terpene alcohols and ketones is very common in pine bark beetles, and includes such pheromones as *cis*- and *trans*-verbenol, myrtenol, and verbenone derived from *alpha*-pinene (Vanderwel and Oehlschlager, 1987). The terpene alcohol ipsdienol is derived directly from the host monoterpene myrcene through oxidation

Fig. 6.2. Molecular structure for selected examples of bark beetle pheromones (see text for references). Those derived from plant precursors include ipsdienol from myrcene, *trans*-verbenol and verbenone from alpha-pinene. Pheromones presumed not derived from direct conversion of plant compounds include the bicyclic ketals *exo*-brevicomin and multistriatin, the tricyclic ketal lineatin, sulcatol, 3,2-MCH, *cis*-pityol, the spiroacetal chalcogran, and the methyl ester of a carboxylic acid, (*E*,*Z*)-2,4-methyldecadienoate.

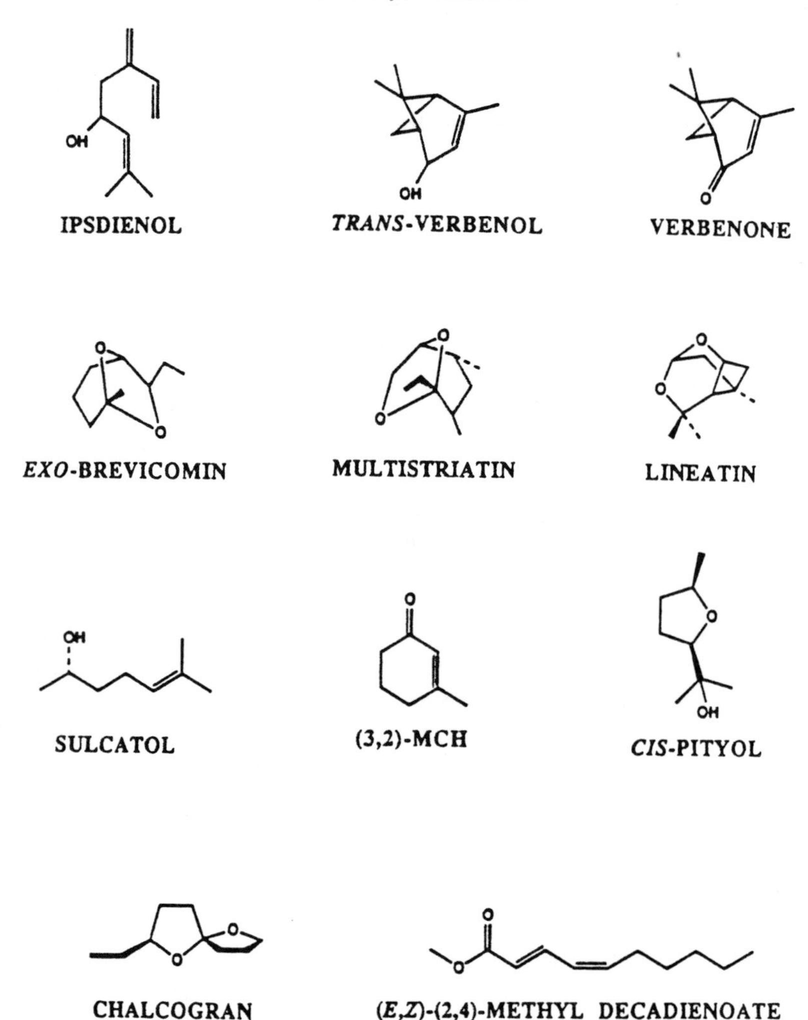

of the allylic carbon (Hughes, 1974; Hendry *et al.*, 1980). Use of ipsdienone as an intermediary may facilitate chiral specificity (Fish *et al.*, 1984). Ipsdienol is converted to ipsenol, another pheromone in the multicomponent blend of *I. paraconfusus*, via saturation of a double bond (Hughes, 1974; Fish *et al.*, 1984). Byers (1981) noted that the amount of myrcene present in the nuptial chambers of male *I. paraconfusus* in *P. ponderosa* was not high enough to account for the amount of ipsdienol present in male guts, suggesting that substantial myrcene may be sequestered during larval development. Alternatively, myrcene-derived ipsdienol may represent just one of several biosynthetic pathways (Hendry *et al.*, 1980; Byers and Birgersson, 1990). Several other terpene alcohols that have no known behavioral activity are produced by pine bark beetles upon exposure to host monoterpenes. Terpene alcohol production probably represents a detoxification mechanism, the products of which have been selected through evolutionary time as pheromones.

Other pheromones, unique to bark beetles and structurally dissimilar from host tree compounds, appear to arise from *de novo* synthesis, although labeled precursor studies are limited (Borden, 1985; Vanderwel and Oehlschlager, 1987). These include the cyclic ketals and acetals frontalin, *exo*-brevicomin, *endo*-brevicomin, multistriatin, lineatin, and chalcogran. Other less complex pheromone molecules are probably of isoprene origin and could arise from precursors in either the beetles or plants. These include simple cyclic alcohols and ketones, and branched short-chain alcohols. Franke *et al.* (1987) speculated that 6-methyl-5-hepten-2-one, a component of pine resin and probably an intermediate in insect isoprene metabolism, undergoes reduction to sulcatol, an ambrosia beetle pheromone, and also could be biotransformed to yield the bicyclic ketal, frontalin. At least one fatty acid, the methyl ester of *E*-,*Z*-2,4-decadienoate, serves as a scolytid pheromone (Byers *et al.*, 1988), and "lanierone" is the first record of a carotenoid ring compound among scolytid pheromones (Teale *et al.*, 1991).

The occurrence of scolytid pheromones appears to reflect phylogenies to some extent. For example, the myrcene-derived pheromones ipsdienol and ipsenol predominate among the Ipini (Vité *et al.*, 1972; Phillips *et al.*, 1989) and related Dryocetini (Klimetzek *et al.*, 1989). However, closely related compounds are not always restricted to certain clades. Ipsdienol is also produced by *D. brevicomis* (Byers, 1982) and *D. ponderosae* (D. Hunt *et al.*, 1986), species distantly related to *Ips*. Moreover, unrelated *Dendroctonus* and *Dryocoetes* species use the bicyclic ketal *exo*-brevicomin (Borden, 1982; Borden *et al.*, 1987a,b).

Control of pheromone production and release is generally sex-specific and regulated by physiological and endocrine mechanisms. For example, application of juvenile hormone (JH) stimulates pheromone production in male *I. paraconfusus* and female *D. brevicomis* (Borden *et al.*, 1969; Hughes and Renwick, 1977a), but not female *D. frontalis* (Bridges, 1982). Feeding in *I. paraconfusus* males presumably removes inhibition of the corpus allatum and triggers release of JH, that then activates brain neurosecretory cells to release a stimulatory brain hormone that drives pheromone synthesis (Hughes and Renwick, 1977b). Frontalin, a highly volatile bicyclic ketal and potent male attractant, is present in unfed and in-flight females of *D. frontalis* and *D. terebrans* (Coster and Vité, 1972; Phillips *et al.*, 1989) and declines rapidly following feeding during gallery initiation. Terpene alcohol production is presumed to proceed via inhalation of precursor vapors, oxidation in the fat body by mixed function oxidases, and concentration or deposition in the hindgut region (Byers, 1981; Borden, 1985).

Microbial associates also produce scolytid pheromones, but their significance under natural conditions is unclear. The bacterium *Bacillus cereus* from the guts of *I. paraconfusus* produces *cis*- and *trans*-verbenol upon *in vitro* exposure to *alpha*-pinene (Brand *et al.*,

1975). Production of ipsdienol and ipsenol in *I. paraconfusus* can be inhibited by amending beetle diet with the antibiotic streptomycin (Byers and Wood, 1981). However, axenic *I. paraconfusus* and *D. ponderosae* produce most terpene alcohol pheromones at levels equal to or slightly higher than those of wild beetles upon exposure to precursors (Conn *et al.*, 1984; Hunt and Borden, 1989). Hunt and Borden (1989) suggested that transovarially transmitted microbes may affect pheromone production. Microbes also may be important in reducing interspecific competition among *D. ponderosae*, by oxidizing *trans*-verbenol to the anti-aggregation pheromone verbenone, either within beetles or in gallery walls (Borden *et al.*, 1986). Although microbes may modulate pheromone synthesis, most scolytids probably can produce some pheromones in their absence (Hunt and Borden, 1989; Vanderwel and Oehlschlager, 1987).

6.4.3.2 Behavioral responses to pheromones

Pheromones are perceived by antennal chemoreceptors (Borden and Wood, 1966) consisting of compound- and enantiomer-specific receptor cells (Payne, 1970; Mustaparta *et al.*, 1979). The number of beetles caught in traps, the most commonly used index of scolytid pheromone activity, reflects a combination of behaviors such as chemotaxis (Akers and Wood, 1989), anemotaxis (Choudhury and Kennedy, 1980; Byers, 1988), and arrestment (Bennett and Borden, 1971). Integration by the central nervous system, and the details of these taxes and kineses, are not well understood.

Bark beetle pheromones are usually behaviorally active as blends with other pheromones and/or host odors, typically monoterpenes, thereby yielding complex and species-specific signals (Bedard *et al.* 1969, Renwick and Vité, 1972; Phillips *et al.*, 1989). Such pheromone combinations facilitate reproductive isolation and niche partitioning among sympatric species. In California, for example, *Ips pini* and *I. paraconfusus* both use ipsdienol, but only *I. paraconfusus* incorporates ipsenol and *cis*-verbenol into its pheromone blend (Wood *et al.*, 1968; Birch *et al.*, 1980a). *Ips latidens* is attracted to ipsenol and *cis*-verbenol, but not in the presence of ipsdienol (Wood *et al.*, 1967; Seybold *et al.*, 1988). Enantiomers and positional isomers further increase the effective number of pheromones from the same compound. For example, *Gnathotrichus retusus* use S-(+)-suicatol as a pheromone, whereas *G. sulcatus* use a 65:35 blend of S-(+)- and R-(-)-sulcatol (Borden *et al.*, 1976). Addition of small amounts of the (-) enantiomer reduces *G. retusus* response (Borden *et al.*, 1980a). Interspecific inhibition also can occur across genera, such as between *D. brevicomis* and *I. paraconfusus* (Byers and Wood, 1980) and between *D. ponderosae* and *I. pini* (Hunt and Borden, 1988).

Interspecific attraction also occurs among species that exploit different parts of the same tree. Among a complex of scolytids colonizing *P. taeda* in the southeastern US, for example, *I. avulsus* and *I. calligraphus* are mutually cross-attractive, and volatiles from *I. grandicollis* increase attraction of *I. avulsus* to its pheromone (Hedden *et al.*, 1976; Birch *et al.*, 1980b). Cross-attraction probably occurs between *D. frontalis* and *D. terebrans* (Payne *et al.*, 1987; Phillips *et al.*, 1989, 1990). Predatory or cohabiting species from other families also are attracted to scolytid pheromones (D. Wood, 1982). Different species that share the same host tissues, species, and physiological status have a variety of mechanisms for maintaining niche separation in time and space (Dixon and Payne, 1980; Phillips, 1988, 1990).

6.4.3.3 Aggregation

Regardless how pioneer beetles make initial contact, the overwhelming majority of beetles arrive at trees in response to pheromones. Once several individuals of the host-selecting sex bore into the host and produce attractants, numerous conspecifics of both sexes may arrive. As attacks accumulate and pathogenic fungi are inoculated into multiple infection courts along the stem, tree defenses are exhausted and successful brood establishment becomes possible. Some tree-killing beetles apparently can optimize attack densities so that maximal brood production is balanced with a minimum number of colonizers needed to overcome host resistance (Renwick and Vité, 1970; Byers, 1983; Raffa and Berryman, 1983a; Birgersson and Bergström, 1989).

Pheromone release and response must be relatively rapid in tree-killing species, so that enough beetles can be recruited to overcome host resistance before induced wound responses are manifested (see Section 6.2.3). Females of various *Dendroctonus* spp. contain pheromones in their hindguts upon emergence or immediately upon contacting a host tree (D. Wood, 1982). However, not all entries into trees generate attraction under natural conditions (Raffa and Berryman, 1983a).

The concentration of host odors in the plume can affect the sex ratio of responders (Renwick and Vité, 1970). For example, the first group of *D. ponderosae* to arrive in response to attacks by female beetles consists of about 85% females, and the proportion of males increases steadily thereafter (Raffa and Berryman, 1983a). This, and a related system in *D. brevicomis* (Vité and Pitman, 1969), appears to result from gender-specific responses to pheromone:host monoterpene ratios. For example, a low ratio of *trans*-verbenol to *alpha*-pinene attracts primarily females, whereas mostly males respond to high *trans*-verbenol:*alpha*-pinene ratios. This presumably reflects the rapidly changing host environment during the dynamics of colonization (Renwick and Vité, 1970).

Among tree-killing *Dendroctonus* species, an increase in certain male-produced pheromones and other semiochemicals at high attack densities facilitates termination of aggregation, spacing and short-range dispersal to new trees. For example, the responses of both sexes of *D. ponderosae* are reduced by high concentrations of *exo*-brevicomin and frontalin released by a critical density of males during mass attack (Borden *et al.*, 1987b). Similarly, males of *D. frontalis* produce (-)-*endo*-brevicomin that interrupts the responses of aggregating beetles to the female-produced pheromone (Payne *et al* ., 1978 ; Vité *et al* . 1985) . Verbenone, an oxidation product of *alpha*-pinene, is produced by both sexes of many *Dendroctonus* species after several days in a host, by beetle-associated microorganisms, and by simple autoxidation (Borden *et al.*, 1986; Hunt and Borden, 1990). Verbenone has an inhibitory or interruptive effect on most species tested so far at high concentrations, but attracts female *D. frontalis* at low concentrations (D. Wood, 1982; Borden *et al.*, 1986).

Host colonization by less aggressive species terminates in several ways. *Ips* and other Ipini males apparently cease production of attractant pheromone as their harems become full (Borden, 1982). Attraction to *S. multistriatus* females declines following pairing with males due to a reduction of one pheromone component relative to another (Lanier *et al.*, 1977). *Dendroctonus pseudotsugae* utilizes a complex system for density regulation in which stridulatory sounds produced by males induce females to release an anti-aggregation pheromone that "masks" the attractive effect of other pheromones (Rudinsky, 1969).

Among the aggressive species, the termination of colonization on a selected tree often is followed by "switching" of repelled beetles onto neighboring trees (Gara and Coster, 1968; Geizler *et al.*, 1980). This increases the chances of initiating successful aggregation on surrounding trees (Raffa and Berryman, 1983a).

Pheromone communication can be affected by stand structure. Mass attack and tree mortality to *D ponderosae* and *D. frontalis* are reduced at inter-tree distances greater than 6 m (Sartwell and Stevens, 1975; Schowalter *et al.*, 1981). Tree spacing directly reduces host susceptibility to bark beetles (Chapter 5), increases air flow and convection, disrupting pheromone plumes (Fares *et al.*, 1980; Amman *et al.*, 1988), and perhaps affects visual orientation.

6.4.4 Establishment

Where attack densities are sufficient to exhaust tree defenses, very little (or no) manifestation of the tree's potential to resist attack is expressed in terms of adult or brood mortality, or microbial inhibition. The major factors that prevent a cohort from achieving its full reproductive potential are competition, natural enemies, and climate. Effects of climate and natural enemies are discussed in Chapters 4 and 7.

Once the potential effects of host resistance have been removed, any mutual benefits among members of a colonizing cohort are eliminated. Larvae compete directly through cannibalism and competition for food and space (Cole, 1971). Colonizing females can reduce the effects of crowding on their progeny, however, by adjusting oviposition rates to attack density, and then re-emerging to attack new trees into which the remainder of their eggs are allocated (Coulson *et al.*, 1978). Thus, the degree of competition among the brood is less than would be expected from a particular cohort if full oviposition occurred. This reproductive adjustment confers higher larval survivorship at the individual level (Raffa and Berryman, 1987) and more efficient allocation of the resource at the population level (Coulson, 1979), provided that additional susceptible hosts are available.

6.5 INTEGRATION OF SCOLYTID STRATEGIES AND MECHANISMS

6.5.1 Aggressive versus non-aggressive strategies

Tree killing appears to be a specialized or derived ecological strategy (Fig. 6.3). For example, regardless of the host's condition when it is attacked by adult beetles, all scolytids require non-resistant, recently dead tissue for brood production. Also, only three species in North America are regular tree killers, falling in the class of near-obligate parasites: *D. ponderosae*, *D. brevicomis*, and *D. frontalis* (including the sibling species *D. vitei* and *D. mexicanus*). Other *Dendroctonus*, such as *D. rufipennis*, *D. pseudotsugae*, *D. simplex*, and some *Ips* and *Scolytus* species comprise the class of facultative parasites that normally colonize hosts that are fallen, stressed, injured, or under attack by other insects or pathogens. Healthy trees also can be colonized following a population buildup in nearby weakened trees, but these outbreaks are usually less expansive and persistent than those of the more aggressive species. The herbivore/saprophyte group comprises the overwhelming majority of Scolytidae, with many species competing for trees that are initially killed by other scolytids or other predisposing agents.

Each strategy has reproductive advantages and disadvantages. Concentrating on severely stressed trees, as do most scolytids, allows beetles to avoid host defense mechanisms. However, it also restricts beetles to a relatively rare and unpredictable resource. In addition, these trees often have thin or poor-quality phloem, and are available to many interspecific competitors. Specializing on standing trees may represent an evolutionary "flight" from superior competitors as much as an "invasion" of a new habitat. Attacking healthy trees, however, incurs the risk of being killed by host defenses and can require such large cohorts that intraspecific competition is severe.

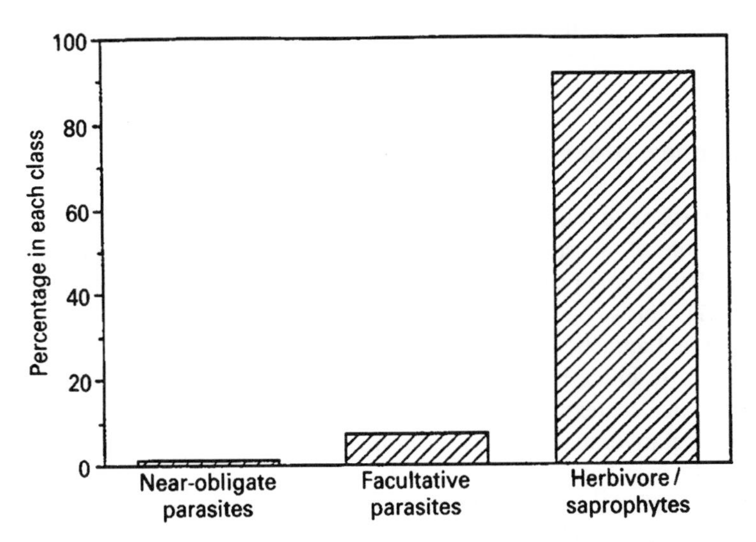

Fig. 6.3. The distribution of phloem-feeding conifer bark beetles in the US and Canada into three life history classes (see text for explanation). Only species that colonize stems and branches were considered in the analysis. N = 251. Data from Bright (1976, 1981), Bright and Stark (1973) and S. Wood (1982).

The extent to which aggressive behavior is advantageous over evolutionary time depends largely on the ecology of the host plant (Raffa and Berryman, 1987). For example, the ecological tendency of many temperate pines to grow in pure, even-aged, early successional stands following fire may render them more prone to area-wide tree killing than other tree groups. In some highly co-evolved systems, these outbreaks may favor long-term reproductive success of the host (Raffa and Berryman, 1987). In other systems large outbreaks may be maladaptive to both host and parasite if killed hosts are replaced by non-hosts (Raffa and Berryman, 1987; Chapter 9).

Only near-obligate tree-killers can substantially increase the number of available hosts as a function of their own population densities (Berryman, 1982). However, these individuals are relatively unfit at competing with members of the less aggressive species in killed trees. This can pose problems for tree-killing species during low density periods. To a certain extent, their dependence on high population levels makes "feast or famine" cycles inevitable, whereas facultative parasites and herbivore/saprophytes may be better equipped to locate dead and severely stressed trees each season. The ability of the aggressive species to concentrate within the most weakened and diseased trees can facilitate reproduction during non-outbreak periods (Cobb *et al.*, 1974; Chapter 9).

These tradeoffs also raise the possibility that individuals within a population may employ different and/or flexible strategies. For example, Raffa and Berryman (1983a, 1987) have speculated that behavioral responses of beetles during the host recognition phase may be subjected to density-dependent selective pressures, with relatively "discriminating" individuals favored during low population periods and relatively "non-discriminating" individuals favored during outbreaks. Likewise, individual variation in pheromone production has been documented, raising the possibility that a portion of the population may not enter trees as "pioneers," but instead "cheat" by responding only to aggregation pheromones, not producing large amounts of pheromones once entered, or various combinations thereof (Birgersson *et al.*, 1988). There is, to date, no experimental evidence for strict reliance on such strategies, and a number of factors may preclude their evolution. However, flexible systems in which individuals possess multiple strategies, or the frequencies of various behaviors vary with insect density and/or environmental conditions, seem plausible (Schlyter and Birgersson, 1989).

6.5.2 Pheromones, cooperation and exploitation

The complex emission of multiple pheromone components and host-released volatiles arising from the actions of one sex usually affects both sexes. However, most of the individual components of bark beetle pheromones are sex specific in both their production and initial effect (Table 6.1). This suggests that, for most scolytid species, pheromones serve primarily in mate finding and attraction, and aggregation may result merely from the perception by others that a rare resource has been located. If so, then scolytid pheromones probably evolved in the sexual contexts of courtship and male-male competition, and were elaborated further into cooperative attack messages in tree killing species where both senders and responders benefit.

The relative benefits to senders and receivers depend on the sexual combinations and host plant conditions involved. For example, female signalers benefit from male responders through fertilization. Moreover, by releasing sex pheromones after arriving at hosts, females recruit males who have survived the dispersal phase, and in so doing are perhaps practicing a form of mate choice and thereby increasing their chances of mating with superior males. The male recruits also assist females in gallery construction and provide some protection from and/or an alternate food source to natural enemies (Kirkendall, 1983). Response to female-produced chemical signals benefits males in mate finding. It is probably advantageous for males to mate as often as possible. Last-male sperm precedence (i.e. fertilization is most likely by sperm from the most recent male in multiply-mated females) may occur in bark beetles as in many other insect groups. Mating at host trees could be especially important in systems where gallery-constructing females may have mated previously with siblings prior to emergence (Witanahchi, 1980; Phillips, 1988), with unrelated males outside trees (Svhira and Clark, 1980), or in previously attacked trees (Coulson *et al.*, 1978).

Female signalers of near-obligate parasites may benefit from the arrival of other females through their combined effect in exhausting host resistance. Individual fitness increases up to an optimal attack density, and declines thereafter with crowding (Raffa and Berryman, 1983a; Cook and Hain, 1987). If the host is dead or dying, however, there probably is little benefit, and perhaps a net detriment, to female signalers who attract other females. For saprophytic species, and for aggressive species during colonization of dead trees, individual female fitness declines with any (or a very slight) increase in attack density (Raffa and Berryman, 1987; Kirkendall, 1989; Anderbrant, 1990). Such interactions could be considered exploitation rather than reciprocal communication (Alcock, 1982; Raffa and Berryman, 1987). In cases of both cooperation and exploitation, female responders to another female's pheromone benefit by locating breeding sites and food.

Females benefit from male pheromone signals in male-initiated, harem polygynous gallery systems, such as in *Ips*, by locating both a suitable resource and a successful male. Male signalers benefit in procuring mates that have a high likelihood of successfully establishing brood, because the early selection and colonization phases have been completed. The benefits conferred by male-male communication vary with the system and context. In male-initiated gallery systems, signaling males of aggressive species such as *I. typographus* may benefit from the arrival of recruits, but among less aggressive species host resistance does not provide a threat. In the latter species, which are clearly more common, responding males exploit

Table 6.1. Comparison of male and female production of and attraction to individual pheromone components among selected conifer-infesting scolytids

Species (Refs)	Pheromone	Producing sex	Relative responses of males and females[a]
Dendroctonus frontalis (1, 2, 3, 4)	Frontalin	Female	Predominantly males attracted to frontalin–turpentine mixture
D. brevicomis (3, 5, 6)	*Exo*-brevicomin	Female	Predominantly males attracted when released with host synergist
	Frontalin	Male	Predominantly females attracted when released with host synergist
D. ponderosae (7, 8, 23)	*Trans*-verbenol	Female, minor produced by male	Single component not active. (-)-Enantiomer attracts mostly males when released with myrcene. Low *trans*-verbenol:resin ratio attracts mostly females; high ratio mostly males
	Exo-brevicomin	Male	Attracts females when released at low levels with *trans*-verbenol and myrcene
D. rufipennis (9)	Frontalin	Female	Both sexes attracted equally
D. pseudotsugae (10, 11)	Frontalin	Both	Both
	Trans-verbenol	Female	
D. terebrans (12, 13)	Frontalin	Female	Predominantly males attracted when released with host turpentine
	Exo-brevicomin	Male	Predominantly females attracted when released with low host turpentine levels
Ips pini (14)	Ipsdienol	Male	Predominantly females attracted
I. calligraphus (15)	*Cis*- and *trans*-verbenol	Both	Predominantly females attracted
	Ipsdienol	Male	
I. paraconfusus (16, 17)	(+)-*Cis*-verbenol	Both	Predominantly females attracted
	Ipsenol	Male	
	Ipsdienol	Male	
I. grandicollis (4, 11, 18, 19)	Ipsenol	Male	Both sexes attracted
Trypodendron lineatum (20)	4,6,6-Lineatin	Female	Predominantly males attracted
Gnathotrichus sulcatus (21, 22)	65% (+):35(-)-Sulcatol	Male	Predominantly females attracted
G. retusus (22)	(+)-Sulcatol	Male	Predominantly females attracted

(1) Payne *et al.* (1978); (2) Moser and Browne (1978); (3) Renwick *et al.* (1973); (4) Van Der Wel and Oehlschlager (1987); (5) Wood *et al.* (1976); (6) Bedard *et al.* (1980); (7) Libbey *et al.* (1985); (8) Renwick and Vitè (1970); (9) Dyer (1973); (10) Borden (1982); (11) D. Wood (1982); (12) Phillips *et al.*, (1989); (13) Phillips *et al.* (1990); (14) Lanier *et al.* (1980); (15) Renwick and Vitè (1972); (16) Renwick *et al.* (1976); (17) Byers *et al.* (1979); (18) Vitè and Renwick (1971); (19) Hughes (1974); (20) Borden *et al.* (1982); (21) Byrne *et al.* (1974); (22) Borden *et al.* (1980); (23) Borden *et al.* (1987).

[a]In most cases, both sexes are attracted to some extent.

the mating signal of first-attacking males to locate host material, and possibly add competitive costs to their neighbors. "Anti-aggregation" or inhibitory pheromones may function as "pre-rivalry" signals in male-male interactions, in that both signaler and responder benefit by avoiding rivalry. High release levels due to the presence of many males could signal a lack of available females or food to incoming males. The resulting deflection may orient responding males to other parts of the tree in species with male-initiated galleries (Byers, 1983), or to females "switching" to other trees in female-initiated gallery systems (Gara, 1967).

A largely sexual and/or exploitative evolutionary context of scolytid pheromones can modify our interpretation of aggregation. Given the common occurrence of aggregations on host plants by other phytophagous, non-cooperating insects, and the saprophytic mode of most scolytids (Alcock, 1982), aggregations by most species of bark beetles may not be mutually adaptive to individuals, but rather are a consequence of discovering a rare and patchily distributed resource.

6.6 CONCLUSIONS

An ability to exploit the selective pressures on bark beetles ultimately will help us progress from a "crisis management" to a more preventive approach (Chapter 11). The brief period during which these insects are without the protection of the subcortical environment and must find new hosts is particularly tenuous (Wood, 1979).

Understanding the cues and mechanisms used by beetles during host recognition may improve our ability to assay tree susceptibility and thereby predict outbreaks, devise management schemes for enhancing stand resistance, and provide guidance to tree breeding programs. Although stand susceptibility has been associated with a variety of site and physiological factors (Chapters 4 and 5), simple techniques for correlating tree factors with susceptibility are not available.

The chemical communication systems of bark beetles also offer opportunities for management tactics (Borden, 1989). For example, estimates of bark beetle population trends are critical for effective pest management (Bakke, 1985), especially when efforts can be focused on stands accurately characterized as susceptible to attack. Use of semiochemicals for the direct removal or disruption of flying beetle populations also has potential for reducing losses (Bakke *et al.*, 1983; Gray and Borden, 1989; Payne and Billings, 1989; Chapter 11). These tactics can be integrated with other direct strategies, such as sanitation or insecticide application (Borden *et al.*, 1986) and biological control (Raffa, 1991b). Further research is needed to make these approaches more reliable.

REFERENCES

Akers, R.P. and Wood, D.L. (1989). Olfactory orientation responses by walking female *Ips paraconfusus* bark beetles: I. chemotaxis assay. *J. Chem. Ecol.* **15**, 3–24.

Alcock, J. (1982). Natural selection and communication among bark beetles. Florida. *Entomol.* **65**, 17–32.

Amman, G.D., McGregor, M.D., Schmitz, R.F. and Oakes, R.D. (1988). Susceptibility of lodgepole pine to infestation by mountain pine beetles following partial cutting of stands. *Can. J. For. Res.* **18**, 688–695.

Anderbrant, O. (1990). Gallery construction and oviposition of the bark beetle *Ips typographus* (Coleoptera: Scolytidae) at different breeding densities. *Ecol. Entomol.* **15**, 1–8

Anderbrandt, O., Schlyter, F. and J. Löfqvist. (1988). Dynamics of tree attack in the bark beetle *Ips typographus* under semi-epidemic conditions. *In* "Integrated Control of Scolytid Bark Beetles" (T.L. Payne and H. Saarenmaa, eds), pp.35–51. Virginia Polytechnic Institute, Blacksburg, VA.

Atkins, M.D. (1966). Behavioral variation among scolytids in relation to their habitat. *Can. Entomol.* **98**, 285–288.

Atkins, M.D. (1969). Lipid loss with flight in the Douglas-fir beetle (*Dendroctonus pseudotsugae*). *Can. Entomol.* **101**, 164–165.

Bakke, A. (1985). Deploying pheromone-baited traps for monitoring *Ips typographus* populations. Z. *Ang. Entomol.* **99**, 33–39.

Bakke, A., Saether, T. and Kvamme, T. (1983). Mass trapping of the spruce bark beetle *Ips typographus*: Pheromone and trap technology. *Medd. Nor. Inst. Skogforsk.* **38**, 1–35.

Bannan, M.W. (1936). Vertical resin ducts in the secondary wood of the Abietinae. *Nem. Phytol.* **35**, 11–46.

Batra, L.R. (1966). Ambrosia fungi: extent of specificity to ambrosia beetles. *Science* **153**, 193–195.

Bedard, W.D., Jr., Tilden, P.E., Wood, D.L., Silverstein, R.M., Brownslee, R.G. and Rodin, J.O. (1969). Western pine beetle: field response to its sex pheromone and a synergistic host terpene, myrcene. *Science* **164**, 1284–1285.

Bedard, W.D., Jr., Wood, D.L., Tilden, P.E., Lindahl, K.O., Jr., Silverstein, R.M. and Rodin, J.O. (1980). Field responses of the western pine beetle and one of its predators to host- and beetle-produced compounds. *J. Chem. Ecol.* **6**, 625–641.

Bennett, R.B. and Borden, J.N. (1971). Flight arrestment of tethered *Dendroctonus pseudotsugae* and *Trypodendron lineatum* (Coleoptera: Scolytidae) in response to olfactory stimuli. *Ann. Entomol. Soc. Am.* **64**, 1273–1286.

Berryman, A.A. (1972). Resistance of conifers to invasion by bark beetle-fungus associations. *BioScience* **22**, 598–602.

Berryman, A.A. (1982). Population dynamics of bark beetles. *In* "Bark Beetles in North American Conifers" (J.B. Mitton and K.B. Sturgeon, eds), pp. 264–314. University of Texas Press, Austin, TX.

Berryman, A.A. and Ashraf, M. (1970). Effects of *Abies grandis* resin on the attack behavior and brood survival of *Scolytus ventralis* (Coleoptera: Scolytidae). *Can. Entomol.* **102**, 1229–1236.

Birch, M.C., Light, D.M., Wood, D.L., Browne, L.E., Silverstein, R.M., Bergot, B.J., Ohloff, G., West, J.R. and Young, J.C. (1980a). Pheromonal attraction and allomonal interruption of *Ips pini* in California by the two enantiomers of ipsdienol. *J. Chem. Ecol.* **6**, 703–717.

Birch, M.C., Svihra, P., Paine, T.D. and Miller, J.C. (1980b). Influence of chemically mediated behavior on host tree colonization by four cohabiting species of bark beetles. *J. Chem. Ecol.* **6**, 395–414.

Birch, M.C., Paine, T.D. and Miller, J.C. (1981). Effectiveness of pheromone micro-trapping of the smaller European elm bark beetle. *Calif. Agric* **35**, 6–7.

Birgersson, G. and Bergström, G. (1989). Volatiles released from individual spruce bark beetle entrance holes: Quantitative variations during the first week of attack. *J. Chem. Ecol.* **15**, 2465–2483.

Birgersson, G., Schlyter, F., Bergström, G. and Löfqvist, J. (1988). Individual variation in the aggregation pheromone content of the spruce bark beetle *Ips typographus*. *J. Chem. Ecol.* **14**, 1737–1761.

Borden, J.H. (1982). Aggregation pheromones. *In* "Bark Beetles in North American Conifers" (J.B. Mitton and K.B. Sturgeon, eds), pp.74–139. University of Texas Press, Austin, TX.

Borden, J.H. (1985). Aggregation Pheromones. *In* "Behavior" (G.A. Kerkut, ed.), pp. 257–285. Pergamon Press, Oxford.

Borden, J.H. (1989). Semiochemicals and bark beetle populations: exploitation of natural phenomena by pest management strategies. *Hol. Ecol.* **12**, 501–510.

Borden, J.H. and Wood, D.L. (1966). The antennal receptors and olfactory response of *Ips confusus* (Coleoptera: Scolytidae) to male sex attractant in the laboratory. *Ann. Entomol. Soc. Am.* **59**, 253–261.

Borden, J.H., Nair, K.K. and Slater, C.C. (1969). Synthetic juvenile hormone: induction of sex pheromone production in *Ips confusus*. *Science.* **1166**, 1626–1627.

Borden, J.H., Chong, L.J., McLean, J.A., Slessor, K.N. and Mori, K. (1976). *Gnathotrichus sulcatus* synergistic response to enantiomers of the aggregation pheromone sulcatol. *Science.* **192**, 894–896.

Borden, J.H., Handley, J.R., Johnston, B.D., MacConnell, J.G., Silverstein, R.M., Slessor, K.N., Swigar, A.A. and Wong, D.T.W. (1979). Synthesis and field testing of 4, 6, 6-liniatin, the aggregation pheromone of *Trypodendrum lineatum* (Coleoptera: Scolytidae). *J. Chem. Ecol.* **5**, 681–689.

Borden, J.H., Handley, J.R., McLean, J.A., Silverstein, R.M., Chong, L., Slessor, K.N., Johnston, B.D. and Schuler, H.R. (1980a). Enantiomer-based specificity in pheromone communication by two sympatric *Gnathotrichus* species (Coleoptera: Scolytidae). *J. Chem. Ecol.* **6**, 445–456.

Borden, J.H., Lindgren, B.S. and Chong, L.J. (1980b). Ethanol and alpha-pinene as synergists for the aggregation pheromones of two *Gnathotrichus Can. J. For. Res.* **10**, 290–232.

Borden, J.H., King, C.J., Lindgren, B.S., Chong, L., Gray, D.R., Oehlschlager, A.C., Slessor, K.N. and

Pierce, H.D., Jr. (1982). Variations in response of *Trypodendron lineatum* from two continents to semiochemicals and trap form. *Environ. Entomol.* **11**, 403–408.
Borden, J.H., Hunt, D.W.A., Miller, D.R. and Slessor, K.N. (1986). Orientation in forest Coleoptera: an uncertain outcome to responses by individual beetles to variable stimuli. *In* "Mechanisms in Insect Olfaction" (T.L. Payne, M.C. Birch and C.E.J. Kennedy, eds) pp. 97–109. Oxford University Press, Oxford.
Borden, J.H., Pierce, A.M., Pierce, H.D., Jr., Chong, L.J., Stock, A.J. and Oehlschlager, A.C. (1987a). Semiochemicals produced by western balsam bark beetle, *Dryocoetes confusus* Swaine (Coleoptera: Scolytidae). *J. Chem. Ecol.* **13**, 823–836.
Borden, J.H., Ryker, L.C., Chong, J.L., Pierce, H.D., Johnston, B.D. and Oehlschlager, A.C. (1987b). Response of the mountain pine beetle, *Dendroctonus ponderosae* Hopkins (Coleoptera: Scolytidae), to five semiochemicals in British Columbia lodgepole pine forests. *Can. J. For. Res.* **17**, 118–128.
Brand, J.M., Bracke, J.W., Wood, D.L., Browne, L.E. (1975). Production of verbenol pheromone by a bacterium isolated from bark beetles. *Nature* **254**, 136–137.
Bridges, J.R. (1981). N-fixing bacteria associated with bark beetles. *Microb. Ecol.* **7**, 131–137.
Bridges, J.R. (1982). Effects of juvenile hormone on pheromone synthesis in *Dendroctonus frontalis. Environ. Entomol.* **11**, 417–420.
Bright, D.E., Jr. (1976). The insects and arachnids of Canada. Part 2. The bark beetles of Canada and Alaska (Coleoptera: Scolytidae). Publ. 1576. Canada Dept. Agric., Ottawa, 241 pp.
Bright, D.E., Jr. (1981). Taxonomic monograph of the genus *Pityophthorus* Eichhoff in North and Central America (Coleoptera: Scolytidae). Entomol. Soc. Canada Memoirs 118, 378 pp.
Bright, D.E., Jr. and Stark, R.W. (1973). The bark and ambrosia beetles of California (Coleoptera: Scolytidae and Platypodidae). Calif. Insect Surv. Bull. 16. University of California Press, Berkeley, 169 pp.
Brues. C.T. (1946). "Insect Dietary. An Account of the Food Habits in Insects." Harvard University Press, Cambridge.
Bunt, W.D., Coster, J.E. and Johnson, P.C. (1980). Behavior of the southern pine beetle on the bark of host trees during mass attack. *Ann. Entomol.Soc. Am.* **73**, 647–652.
Byers, J.A. (1981). Pheromone biosynthesis in the bark beetle *Ips paraconfusus* during feeding or exposure to vapors of host plant precursors. *Insect Biochem.* **11**, 563–569.
Byers, J.A. (1982). Male-specific conversion of the host plant compound myrcene, to the pheromone, (+)-ipsdienol, in the bark beetle, *Dendroctonus brevicomis. J. Chem. Ecol.* **82**, 363–372.
Byers, J.A. (1983). Sex-specific responses to aggregation pheromone: regulation of colonization density in the bark beetle *Ips paraconfusus. J. Chem. Ecol.* **9**, 129–142.
Byers, J.A. (1988). Upwind flight orientation to pheromone in western pine beetle tested with rotating windvane traps. *J. Chem. Ecol.* **14**, 189–198.
Byers, J.A. and Birgersson, G. (1990). Pheromone production in a bark beetle independent of myrcene precursor in host pine species. *Naturwissenschaften* **77**, 385–387.
Byers, J.A. and Wood, D.L. (1980). Interspecific inhibition of the response of the bark beetles, *Dendroctonus brevicomis* and *Ips paraconfusus*, to their pheromones in the field. *J. Chem. Ecol.* **6**, 149–164.
Byers, J.A. and Wood, D.L. (1981). Antibiotic-induced inhibition of pheromone synthesis in a bark beetle. *Science.* **213**, 763–764.
Byers, J.A., Wood, D.L., Browne, L.E., Fish, R.H., Piatek, B. and Hendry, L.B. (1979). Relationship between a host plant compound, myrcene and pheromone production in the bark beetle, *Ips paraconfusus. J. Insect Physiol.* **25**, 477–482.
Byers, J.A., Birgersson, G., Löfqvist, J. and Bergström, G. (1988). Synergistic pheromones and monoterpenes enable aggregation and host recognition by a bark beetle, *Pityogenes chalcographus. Naturwissenschaften* **75**, 153–155.
Byrne, K.J., Swiggar, A.A., Silverstein, R.M., Borden, J.H. and E. Stokkink. (1974). Sulcatol: population aggregation pheromone in the scolytid *Gnathotrichus sulcatus. J. Insect Physiol.* **20**, 1895–1900.
Cates, R.G. and Alexander, H. (1982). Host resistance and susceptibility. *In* "Bark Beetles in North American Conifers: A System for the Study of Evolutionary Biology" (J.B. Mitton and K.B. Sturgeon, eds), pp. 212–263. University of Texas Press, Austin, TX.
Chapman, J.A. (1962). Field studies on attack flight and log selection by the ambrosia beetle *Trypodendron lineatum* (Oliv.) (Coleoptera: Scolytidae). *Can. Entomol.* **94**, 74–92.

Choudhury, J.H. and Kennedy, J.S. (1980). Light versus pheromone-bearing wind in the control of flight direction by bark beetles, *Scolytus multistriatus*. *Physiol. Entomol.* **5**, 207–214.

Cobb, F.W., Jr., Krstic, M., Zavarin, E. and Barber, H.W., Jr. (1968). Inhibitory effects of volatile oleoresin components on *Fomes annosus* and four *Ceratocystis* species. *Phytopathology* **58**, 327–335.

Cobb, F.W., Jr., Parmeter, J.R., Jr., Wood, D.L. and Stark, R.W. (1974). Root pathogens as agents predisposing ponderosa pine and white fir to bark beetles. *In* "Proc. Fourth International Conference on *Fomes annosus*," pp. 8–15. USDA Forest Serv., USDA Forest Serv., Washington, DC.

Cole, W.E. (1971). Crowding effects among single age larvae of the mountain pine beetle. *Environ. Entomol.* **2**, 285–293.

Conn, J.E., Borden, J.H., Hunt, D.W.A., Holme, J., Whitney, H.S., Spanier, O.J., Pierce, H.D. and Oehlschlager, A.C. (1984). Pheromone production by axenically reared *Dendroctonus ponderosae* and *Ips paraconfusus* (Coleoptera: Scolytidae). *J. Chem. Ecol.* **10**, 281–290.

Cook, S.P. and Hain, F.P. (1987). Susceptibility of trees to southern pine beetle, *Dendroctonus frontalis* (Coleoptera: Scolytidae). *Environ.Entomol.* **16**, 9–14.

Coster, J.E. and Vité, J.P. (1972). Effects of feeding and mating in pheromone release in the southern pine beetle. *Ann. Entomol. Soc. Am.* **65**, 263–266.

Coulson, R.N. (1979). Population dynamics of bark beetles. *Annu. Rev. Entomol.* **24**, 417–447.

Coulson, R.N., Fargo, W.S., Pulley, P.E., Foltz, J.L., Pope, D.N., Richerson, J.V. and Payne, T.L. (1978). Evaluation of the re-emergence of parent adult *Dendroctonus frontalis* (Coleoptera:Scolytidae). *Can. Entomol.* **110**, 475–486.

Coyne, J.F. and Lott, L.H. (1976). Toxicity of substances in pine oleoresin to southern pine beetles. *J. Georgia Entomol. Soc.* **11**, 301–305.

Dixon, W.N. and Payne, T.L. (1980). Attraction of entomophagous and associate insects of the southern pine beetle to beetle and host tree-produced volatiles. *J. Georgia Entomol. Soc.* **15**, 378–389.

Dyer, E.D.A. (1973). Spruce beetle aggregated by the synthetic pheromone frontalin. *Can. J. For. Res.* **3**, 486–494.

Elkinton, J.W. and Wood, D.L. (1980). Feeding and boring behavior of the bark beetle *Ips paraconfusus* (Coleoptera Scolytidae) on the bark of a host and non-host tree species. *Can. Entomol.* **112**, 797–809.

Elkinton, J.W., Wood, D.L. and Browne, L.E. (1981). Feeding and boring behavior of the bark beetle, *Ips paraconfusus*, in extracts of ponderosa pine phloem. *J. Chem. Ecol.* **7**, 209–220.

Fares, Y., Sharpe, P.J.H. and Magnusen, C.E. (1980). Pheromone dispersion in forests. *J. Theor. Biol.* **84**, 335–359.

Fish, R.H., Browne, L.E. and Bergot, B.J. (1984). Pheromone biosynthetic pathways: conversion of ipsdienone to (-)-ipsdienol, a mechanism for enantioselective reduction in the male bark beetle, *Ips paraconfusus*. *J. Chem. Ecol.* **10**, 1057–1064.

Francke, W. (1988). Identification and synthesis of new pheromones. *Agric. Ecosystems Environ.* **21**, 21–30.

Francke, W., Pan, M.L., Konig, W.A., Mori, K., Puapoomchareon, P., Heuer, H. and Vité, J.P. (1987). Identification of "pityol" and "grandisol" as pheromone components of the bark beetle, *Pityophthorus pityographus*. *Naturwissenschaften.* **74**, 343–345.

Francke-Grosmann, H. (1967). Ectosymbiosis in wood-inhabiting insects. *In* "Symbiosis," Vol. II (S.M. Henry, ed) pp. 141–205. Academic Press, New York.

Gara, R.I. (1965). Studies on the flight behavior of *Ips confusus* (Lec.) (Coleoptera: Scolytidae) in response to attractants. PhD Dissertation, Oregon State University, Corvallis, 141 pp.

Gara, R. I. (1967). Studies on the attack behavior of the southern pine beetle. I. The spreading and collapse of outbreaks. *Contrib. Boyce Thompson Inst.* **23**, 349–354 .

Gara, R.I. and Coster, J.E. (1968). Studies on the attack behavior of the southern pine beetle. III. Sequence of tree infestation within stands. *Contrib. Boyce Thompson Inst.* **24**, 69–79 .

Geizler, D.R., Gallucci, V.F. and Gara, R.I. (1980). Modeling the dynamics of mountain pine beetle aggregation in a lodgepole pine stand. *Oceologia (Berl.)* **46**, 244–253.

Goheen, D.J. and Cobb, F.W., Jr. (1978). Occurrence of *Verticicladiella wagenerii* and its perfect state, *Certocystis wagenerii* sp. nov., in insect galleries. *Phytopathology* **68**, 1192–1195 .

Goheen, D.J. and Cobb, F.W., Jr. (1980). Infestation of *Ceratocystis wageneri*-infected ponderosa pine by bark beetles (Coleoptera: Scolytidae) in the central Sierra Nevada. *Can. Entomol.* **112**, 725–730.

Gouger, R. J . (1972) . Interrelationships of *Ips avulsus* (Eichh.) and associated fungi. *Dissert. Abst.* **32**, 6453–B.

Graham, K. (1959) Release by flight exercise of the chemotropic response from photopositive domination in a scolytid beetle. *Nature* **184**, 282–284.

Gray, D.R. and Borden, J.H. (1989). Containment and concentration of mountain pine beetle (Coleoptera: Scolytidae) infestations with semiochemicals: validation by sampling of baited and surrounding zones. *J. Econ. Entomol.* **82**, 1399–1405.

Haack, R.A. and Slansky, F., Jr. (1987). Nutritional ecology of wood-feeding Coleoptera, Lepidoptera, and Hymenoptera. *In* "The Nutritional Ecology of Insects, Mites, Spiders, and Related Invertebrates" (F. Slansky, Jr. and J.G. Rodriguez, eds), pp. 449–486. Wiley, New York.

Hedden, R.L. and Billings, R.F. (1977). Seasonal variations in fat content and size of the southern pine beetle in East Texas. *Ann. Entomol. Soc. Am.* **70**, 876–880.

Hedden, R.L., Vité, J.P. and Mori, K. (1976). Synergistic effect of a pheromone and a kairomone on host selection and colonization by *Ips avulsus. Nature* **261**, 696–697.

Hendry, L.B., Piatek, B., Browne, L.E., Wood, D.L., Byers, J.A., Fish, R.H. and Hicks, R.A. (1980). *In vivo* conversion of a labelled host plant chemical to pheromones of the bark beetle *Ips paraconfusus. Nature* **284**, 485.

Hodges, J.D., Elam, W.W., Watson, W.F. and Nebeker, T.E. (1979). Oleoresin characteristics and susceptibility of four southern pines to southern pine beetle (Coleoptera: Scolytidae) attacks. *Can. Entomol.* **111**, 889–896.

Horntvedt, R., Christiansen, E., Solheim, H. and Wang, S. (1983). Artificial inoculation with *Ips typographus*-associated blue-stain fungi can kill healthy Norway spruce trees. *Med. Norsk Inst. Skogforskn.* **38**, 1–20.

Hughes, P.R. (1974). Myrcene: a precursor of pheromones in *Ips* beetles. *J. Insect Physiol.* **20**, 1271–1275.

Hughes, P.R. and Renwick, J.A.A. (1977a). Hormonal and host factors stimulating pheromone synthesis in female western pine beetles *Dendroctonus brevicomis. Physiol. Entomol.* **2**, 289–292.

Hughes, P.R. and Renwick, J.A.A. (1977b). Neural and hormonal control of pheromone biosynthesis in the bark beetle, *Ips paraconfusus. Physiol. Entomol.* **2**, 117–123.

Hunt, D.W.A. and Borden, J.H. (1988). Response of mountain pine beetle, *Dendroctonus ponderosae* Hopkins, and pine engraver, *Ips Pini* (Say), to ipsdienol in southwestern British Columbia. *J. Chem. Ecol.* **14**, 277–293.

Hunt, D.W.A. and Borden, J.H. (1989). Terpene alcohol pheromone production by *Dendroctonus ponderosae* and *Ips paraconfusus* (Coleoptera: Scolytidae) in the absence of readily culturable microorganisms. *J. Chem. Ecol.* **15**, 1433–1463.

Hunt, D.W.A. and Borden, J.H. (1990). Conversion of verbenols to verbenone by yeasts isolated from *Dendroctonus ponderosae* (Coleoptera: Scolytidae). *J. Chem. Ecol.* **16**, 1385–1397.

Hunt, D.W.A., Borden, J.H., Pierce, H.D., Jr., Slessor, K.N., King, G.G.S. and Csyzewska, E.K. (1986). Sex-specific production of ipsdienol and myrcenol by *Dendroctonus ponderosae* (Coleoptera: Scolytidae) exposed to myrcene vapors. *J. Chem. Ecol.* **12**, 1579–1586.

Hunt, R.S. and Morrison, D.J. (1980). Black stain root disease in British Columbia. Canadian Forestry Serv. For. Pest. Leafl. No. 67. Forestry Canada, Pacific Forestry Centre, Victoria, BC.

Hynum, B.G. (1978). Migration phase interactions between the mountain pine beetle and lodgepole pine. PhD Dissertation, Washington State University.

Hynum, B.G. and Berryman, A.A. (1980). *Dendroctonus ponderosae* (Coleoptera: Scolytidae); pre-aggregation landing and gallery initiation on lodgepole pine. *Can. Entomol.* **112**, 185–191.

Kirkendall, L.R. (1983). The evolution of mating systems in bark and ambrosia beetles (Coleoptera: Scolytidae and Platypodidae). *Zool. J. Linn. Soc.* **77**, 293–352.

Kirkendall, L.R. (1989). Within-harem competition among *Ips* females, an overlooked component of density-dependent larval mortality. *Hol. Ecol.* **12**, 477–487.

Klepzig, K.D., Raffa, K.F. and Smalley, E.B. (1991). Association of insect-fungal complexes with red pine decline in the Lake States. *Forest Sci.* **37**, 1119–1139.

Klimetzek, D., Köhler, J., Vité, J.P. and Kohnle, U. (1986). Dosage response to ethanol mediates host selection by "secondary" bark beetles. *Naturwissenschaften* **73**, 270–272.

Klimetzek, D., Köhler, J. Krohn, S. and Francke, W. (1989). Das Pheromon-System des Waldreben-Brokenkafers, *Xylocleptis bispinus* Duft. (Col., Scolytidae). *J. Appl. Entomol.* **107**, 304–309.

Kramer, P.J. and Kozlowski, T.T. (1979). "Physiology of Woody Plants" Academic Press, New York.

Langor, D.W., Spence, J.R. and Pohl, G.R. (1990). Host effects on fertility and reproductive success of *Dendroctonus ponderosae* Hopkins (Coleoptera: Scolytidae). *Evolution* **44**, 609–618.

Lanier, G.N., Silverstein, R.M. and Peacock, J.W. (1976). Attractant pheromone of the European elm bark beetle (*Scolytus multistriatus*): isolation, identification, synthesis, and utilization studies. *In* "Perspectives in Forest Entomology" (J.T. Anderson and H.K. Kaya, eds), pp. 149–175. Academic Press, New York.

Lanier, G.N., Gore, W.E., Pearce, G.T., Peacock, J.W. and Silverstein, R.M. (1977). Response of the European bark beetle, *Scolytus multistriatus* (Coleoptera Scolytidae) to isomers and components of its pheromone. *J. Chem. Ecol.* **3**, 1–8.

Lanier, G.N., Classon, A., Stewart, T., Piston, J.J. and Silverstein, R.M. (1980). *Ips pini*: the basis for interpopulational differences in pheromone biology. *J. Chem. Ecol.* **6**, 677–687.

Libbey, L.M., Ryker, L.C. and Yandell, K.L. (1985). Laboratory and field studies of volatiles released by *Dendroctonus ponderosae* Hopkins (Coleoptera: Scolytidae). *Z. Ang. Entomol.* **100**, 381–392.

Mattson, W.J., Jr. (1980). Herbivory in relation to plant nitrogen content. *Annu. Rev. Ecol. Syst.* **11**, 119–161.

Miller, J.M. and Keen, F.P. (1960). "Biology and Control of the Western Pine Beetle." USDA Forest Serv. Misc. Publ. 800. USDA Forest Serv., Washington DC, 381 pp.

Miller, R.H. and Berryman, A.A. (1986). Carbohydrate allocation and mountain pine beetle attack in girdled lodgepole pines. *Can. J. For. Res.* **16**, 1036–1040.

Moeck, H.A. (1970). Ethanol as the primary attractant for the ambrosia beetle *Trypodendron lineatum* (Coleoptera: Scolytidae). *Can. Entomol.* **102**, 985–995.

Moeck, H.A. (1981). Ethanol induces attack on trees by spruce beetles, *Dendroctonus rufipennis*, (Coleoptera: Scolytidae). *Can. Entomol.* **113**, 939–942.

Moeck, H.A., Wood, D.L., and Lindahl, K.Q., Jr. (1981). Host selection behavior of bark beetles (Coleoptera: Scolytidae) attacking *Pinus ponderosae*, with special emphasis on the western pine beetle, *Dendroctonus brevicomis*. *J. Chem. Ecol.* **7**, 49–83.

Moser, J.C. (1985). Use of sporothecae by phoretic *Tarsonemus* mites to transport ascospores of coniferous bluestain fungi. *Trans. Br. Mycol. Soc.* **84**, 750–753.

Moser, J.C. and Browne, L.E. (1978). A nondestructive trap for *Dendroctonus frontalis* Zimmermann (Coleoptera: Scolytidae). *J. Chem. Ecol.* **4**, 1–7.

Mustaparta, H., Angst, M.E. and Lanier, G.N. (1979). Specialization of olfactory cells to insect- and host-produced in the bark beetle *Ips pini* (Say). *J. Chem. Ecol.* **5**, 109–123.

Nijholt, W.W. (1965). Moisture and fat content in the ambrosia beetle *Trypodendron lineatum* (Oliv.). *Entomol. Soc. Br. Col. Proc.* **62**, 16–18.

Norris, D.M., and Baker, J.M. (1969). Nutrition of *Xyleborus ferrugineus*, I: ethanol in diets as a tunneling (feeding) stimulant. *Ann. Entomol. Soc. Am.* **62**, 592–594.

Payne, T.L. (1970). Electrophysiological investigations in response to pheromones in bark beetles. *Contrib. Boyce Thomp. Instit.* **24**, 275–282.

Payne, T.L. and Billings, R.F. (1989). Evaluation of (S)-verbenone applications for suppressing southern pine beetle (Coleoptera: Scolytidae) infestations. *J. Econ. Entomol.* **82**, 1702–1708.

Payne, T.L. and Coulson, R.N. (1985). Role of visual and olfactory stimuli in host selection and aggregation behavior by *Dendroctonus frontalis*. *In* "The Role of the Host in the Population Dynamics of Forest Insects" (L. Safranyik, ed.), pp. 73–82. Forestry Canada, Pacific Forestry Centre, Victoria, BC.

Payne, T.L., Coster, J.E., Richerson, J.V., Edson, L.J. and Hart, E.R. (1978). Field response of the southern pine beetle to behavioral chemicals. *Environ. Entomol.* **7**, 578–582.

Payne, T.L., Billings, R.F., Delorme, J.D., Andryszak, N.A., Bartels, J., Franke, W. and Vité, J.P. (1987). Kairomonal-pheromonal system in the black turpentine beetle, *Dendroctonus terebrans* (Ol.). *J. Appl. Entomol.* **103**, 15–22.

Person, H.L. (1931). Theory in explanation of the selection of certain trees by the western pine beetle. *J. Forestry* **29**, 696–699.

Phillips, T.W. (1988). Chemical signals and male mating strategies in Scolytidae. *In* "Proc. XVII International Congress of Entomology", p. 430., University of British Columbia, Vancouver, BC.

Phillips, T.W. (1990). Attraction of *Hylobius pales* (Herbst) (Coleoptera: Curculionidae) to pheromones of bark beetles (Coleoptera: Scolytidae). *Can. Entomol.* **122**, 423–427.

Phillips, T.W., Wilkening, A.J., Atkinson, T.H., Nation, J.L., Wilkinson, R.C. and Foltz, J.L. (1988). Synergism of turpentine and ethanol as attractants for certain pine-infesting beetles (Coleoptera). *Environ. Entomol.* **17**, 456–462.

Phillips, T.W., Nation, J.L., Wilkinson, R.C. and Foltz, J.L. (1989). Secondary attraction and field activity of beetle-produced volatiles in *Dendroctonus terebrans*. *J. Chem. Ecol.* **15**, 1513–1533.
Phillips, T.W., Nation, J.L., Wilkinson, R.C., Foltz, J.L., Pierce, H.D. and Oehlschlager, A.C. (1990). Response specificity of *Dendroctonus terebrans* (Coleoptera: Scolytidae) to enantiomers of its sex pheromones. *Ann. Entomol. Soc. Am.* **83**, 251–257.
Pitman, G.B. and J.P. Vité. (1969). Aggregation behavior of *Dendroctonus ponderosae* (Coleoptera: Scolytidae) in response to chemical messengers. *Can. Entomol.* **101**, 143–149.
Popp, M.P., Wilkinson, R.C., Jokela, E.J., Harding, R.B. and Phillips, T.W. (1989). Effects of slash pine nutrition on the reproductive performance of *Ips calligraphus* (Coleoptera: Scolytidae). *Environ. Entomol.* **18**, 795–789.
Popp, M.P., Johnson, J.D. and Massey, T.L. (1991). Stimulation of resin flow in slash and loblolly pine by bark beetle vectored fungi. *Can. J. For. Res.* **21**, 1124–1126.
Raffa, K.F. (1991a). Induced defensive reactions in conifer-bark beetle systems. *In* "Phytochemical Induction by Herbivores" (D.W. Tallamy and M.J. Raupp, eds). Academic Press, New York.
Raffa, K.F. (1991b). Temporal and spatial disparities among bark beetles, predators, and associates responding to synthetic bark beetle pheromones: implications for coevolution and pest management. *Environ. Entomol.* **20**, 1665–1679.
Raffa, K.F., and Berryman, A.A. (1980). Flight responses and host selection by bark beetles. *In* "Dispersal of Forest Insects: Evolution, Theory, and Management Implications." Proceedings of the Second International Union of Forestry Research Organizations (A.A. Berryman and L. Safranyik, eds), pp. 213–233. Washington State University Press, Pullman.
Raffa, K.F., and Berryman, A.A. (1982a). Physiological differences between lodgepole pines resistant and susceptible to the mountain pine beetle and associated microorganisms. *Environ. Entomol.* **11**, 486–492.
Raffa, K.F. and Berryman, A.A. (1982b). Gustatory cues in the orientation of *Dendroctonus ponderosae* (Coleoptera: Scolytidae) to host trees. *Can. Entomol.* **114**, 97–103.
Raffa, K.F. and Berryman, A.A. (1983a). The role of host plant resistance in the colonization behavior and ecology of bark beetle (Coleoptera: Scolytidae). *Ecol. Monogr.* **53**, 27–49.
Raffa, K.F. and Berryman, A.A. (1983b). Physiological aspects of lodgepole pine wound responses to a fungal symbiont of the mountain pine beetle, *Dendroctonus ponderosae* (Coleoptera: Scolytidae). *Can. Entomol.* **115**, 723–734.
Raffa, K.F. and Berryman, A.A. (1987). Interacting selective pressures in conifer-bark beetle systems: a basis for reciprocal adaptations? *Am. Nat.* **129**, 234–262.
Raffa, K.F. and Klepzig, K.D. (1992). Tree defense mechanisms against insect-associated fungi. *In* "Defense Mechanisms of Woody Plants Against Fungi" (R.A. Blanchette and A.R. Biggs, eds), pp. 354–390. Springer-Verlag, New York.
Raffa, K.F., Berryman, A.A., Simasko, J., Teal, W. and Wong, B.L. (1985). Effects of grand fir monoterpenes on the fir engraver, *Scolytus ventralis* (Coleoptera: Scolytidae), and its symbiotic fungus. *Environ. Entomol.* **14**, 552–556.
Renwick, J.A.A. and Vité, J.P. (1970). Systems of chemical communication in *Dendroctonus*. *Contrib. Boyce Thompson Instit.* **24**, 283–292.
Renwick, J.A.A., and Vité, J.P. (1972). Pheromones and host volatiles that govern aggregation of the six-spined engraver beetle, *Ips calligraphus*. *J. Insect Physiol.* **18**, 1215–1219.
Renwick, J.A.A., Hughes, P.R. and Ty, T.D. (1973). Oxidation productions of pinene in the bark beetle *Dendroctonus frontalis*. *J. Insect. Physiol.* **19**, 1735–1740.
Renwick, J.A.A., Hughes, P.R. and Krull, I.S. (1976). Selective production of cis- and trans-verbenol from (-) and (+)-alpha-pinene by a bark beetle. *Science* **191**, 199–200.
Rudinsky, J.A. (1962). Ecology of Scolytidae. *Annu. Rev. Entomol.* **7**, 327–348.
Rudinsky, J.A. (1966). Scolytid beetles associated with Douglas-fir: response to terpenes. *Science* **152**, 218–219.
Rudinsky, J.A. (1969). Masking of the aggregation pheromone in *Dendroctonus pseudotsugae* Hopk. *Science* **166**, 884–885.
Rudinsky, J.A., Morgan M.E., Libbey, L.M. and Putnam, T.B. (1977). Limonene released by the scolytid beetle *Dendroctonus pseudotsugae*. *Zeit. Ang. Entomol.* **82**, 376–380.
Salisbury, F.B. and Ross, C.W. (1978). "Plant Physiology." Wadsworth Publ. Co., Belmont, California.
Salom, S.M. and McLean, J.A. (1989). Influence of wind on the spring flight of *Trypodendron lineatum* (Coleoptera: Scolytidae) in a second-growth coniferous forest. *Can. Entomol.* **121**, 109–119.

Sartwell, C. and Stevens, R.E. (1975). Mountain pine beetle in ponderosa pine: prospects for silvicultural control in second-growth stands. *J. Forestry* **73**, 136–140.

Schlyter, F. and Birgersson, G. (1989). Individual variation of pheromone in bark beetles and moths — a comparison and an evolutionary background. *Hol. Ecol.* **12**, 457–465.

Schowalter, T.D., Pope, D.N., Coulson, R.N. and Fargo, W.S. (1981). Patterns of southern pine beetle (*Dendroctonus frontalis* Zimm.) infestation enlargement. *Forest Sci.* **27**, 837–849.

Schowalter, T.D., Caldwell, B.A., Carpenter, S.E., Griffiths, R.P., Harmon, M.E., Ingham, E.R., Kelsey R.G., Lattin, J.D. and Moldenke, A.R. (1992). Decomposition of fallen trees: effects of initial conditions and heterotroph colonization rates. *In* "Tropical Ecosystems: Ecology and Management" (K.P. Singh and J.S. Singh, eds), pp. 373–383. Wiley Eastern Ltd., New Delhi.

Seybold, S.J., Wood, D.L., Lindhal, K.O., Silverstein, R.M. and West, J.R. (1988). Aspects of the chemical ecology of *Ips latidens* (LeConte). *In* "Proc. XVII International Congress of Entomology," p.429. University of British Columbia, Vancouver, BC.

Shepherd, R.F. (1966). Factors influencing the orientation and rates of activity of *Dendroctonus ponderosae* (Coleoptera: Scolytidae). *Can. Entomol.* **98**, 507–518.

Shields, W.M. (1982). "Philopatry, Inbreeding, and the Evolution of Sex." State University of New York Press, Albany, 245 pp.

Shrimpton, D.M. (1973). Extractives associated with the wound response of lodgepole pine attacked by the mountain pine beetle and associated microorganisms. *Can. J. Bot.* **51**, 527–534.

Shrimpton, D.M. and Whitney, H.S. (1968). Inhibition of growth of blue stain fungi by wood extractives. *Can. J. Bot.* **46**, 757–761.

Smith, R.H. (1963). Toxicity of pine resin vapors to three species of *Dendroctonus* bark beetles. *J. Econ. Entomol.* **56**, 827–831.

Smith, R.H. (1966). Resin quality as a factor in the resistance of pines to bark beetles. *In* "Breeding Pest-resistant Trees" (H.D. Gerhold, R.E. McDermott, B.J.Schreiner and J.A. Winjeski, eds) pp. 189–196 Pergamon Press, New York.

Stark, R.W. (1982). Generalized ecology and life cycle of bark beetles. *In* "Bark Beetles in North American Conifers." (J.B. Mitton and K.B. Sturgeon, eds), pp. 21–45. University of Texas Press, Austin, TX.

Stock, M.W. and Guenther, J.D. (1979). Isozyme variation among mountain pine beetle (*Dendroctonus ponderosae*) populations in the Pacific Northwest. *Environ. Entomol.* **8**, 889–893.

Sturgeon, K.B. and Mitton, J.B. (1982;). Evolution of bark beetle communities. *In* "Bark Beetles in North American Conifers" (J.B. Mitton and K.B. Sturgeon, eds), pp. 350–384. University of Texas Press, Austin, TX, 527 pp.

Svhira, P. and Clark, J.K. (1980). The courtship of the elm bark beetle. *Calif. Agric.* **34**, 7–9.

Teale, S.A., Webster, F.X., Zhang, A. and Lanier, G.N. (1991). Lanierone: a new pheromone compound from *Ips pini* (Coleoptera: Scolytidae) in New York. *J. Chem. Ecol.* **17**, 1159–1176.

Tilden, P.E., Bedard, W.D. Jr., Wood, D.L., Lindahl, K.Q. and Rausch, P.A. (1979). Trapping the western pine beetle at and near a source of synthetic attractive pheromone: effects of trap size and position. *J. Chem. Ecol.* **5**, 579–531.

Vanderwel, D. and Oehlschlager, A.C. (1987). Biosynthesis of pheromones and endocrine regulation of pheromone production in Coleoptera. *In* "Pheromone Biochemistry" (G.D. Prestwich and G.J. Bloomquist, eds), pp. 175–215 Academic Press, New York.

Vité, J.P. and Gara, R.I. (1962). Volatile attractants from ponderosa pine attacked by bark beetles (Coleoptera: Scolytidae). *Contrib. Boyce Thompson Inst.* **21**, 251–254.

Vité, J.P. and Pitman, G.B. (1969). Aggregation behavior of *Dendroctonus brevicomis* in response to synthetic pheromones. *J. Insect Physiol.* **15**, 1617–1622.

Vité, J.P. and Renwick, J.A.A. (1971). Population aggregation pheromone in the bark beetle, *Ips grandicollis*. *J. Insect Physiol.* **17**, 1699–1704.

Vité, J.P., Bakke, A. and Renwick, J.A.A. (1972). Pheromones in *Ips* (Coleoptera: Scolytidae): occurrence and production. *Can. Entomol.* **104**, 1967–1975.

Vité, J.P., Billings, R.F., Ware, C.W. and Mori, K. (1985). Southern pine beetle: enhancement or inhibition of aggregation response mediated by enantiomers of endo-brevicomin. *Naturwissenshaften.* **72**, 99–100.

Whitney, H.S. (1982). Relationships between bark beetles and symbiotic organisms. *In* "Bark Beetles in North American Conifers" (J.B. Mitton and K.B. Sturgeon, eds), pp. 183–211. University of Texas Press, Austin, TX.

Wise, L.E. and Jahn, E.C. (1952). "Wood Chemistry," Vol. I. Van Nostrand Reinhold, Princeton, NJ.
Witanahchi, J.P. (1980). Evidence for pre-emergence mating among mature progeny of *Ips grandicollis* (Eichhoff). *J. Aust. Entomol. Soc.* **19**, 93–100.
Witcosky, J.J., Schowalter, T.D. and Hansen, E.M. (1986). *Hylastes nigrinus* (Coleoptera: Scolytidae), *Pissodes fasciatus*, and *Steremnius carinatus* (Coleoptera: Curculionidae) as vectors of black-stain root disease of Douglas-fir. *Environ. Entomol.* **15**, 1090–1095.
Wood, D.L. (1972). Selection and colonization of ponderosa pine by bark beetles. *Symp. Royal Entomol. Soc. London* **6**, 110–117.
Wood, D.L. (1979). Development of behavior modifying chemicals for use in forest pest management in the USA. *In* "Chemical Ecology: Odour Communication in Animals" (F.T. Ritter, ed.), pp. 261–279. Elsevier, Amsterdam.
Wood, D.L. (1982). The role of pheromones, kairomones, and allomones in the host selection behavior of bark beetles. *Annu. Rev. Entomol.* **27**, 411–446.
Wood, D.L., Stark, R.W., Silverstein, R.M. and Rodin, J.O. (1967). Unique synergistic effects produced by the sex attractant compounds of *Ips confusus* (LeConte) (Coleoptera: Scolytidae). *Nature* **215, 206**.
Wood, D.L., Browne, L.E., Bedard, W.D., Tilden, P.E., Silverstein, R.M. and Rodin, J.O. (1968). Response of *Ips confusus* to synthetic sex pheromone in nature. *Science* **159**, 1373–1374.
Wood, D.L., Browne, L.E., Ewing, B., Lindahl, K., Bedard, W.D., Tilden, P.E., Mori, K., Pitman, G.B. and Hughes, P.R. (1976). Western pine beetle: specificity among enantiomers of male and female components of an attractant pheromone. *Science* **192**, 896–898.
Wood, S.L. (1982). "The Bark and Ambrosia Beetles of North and Central America (Coleoptera: Scolytidae), A Taxonomic Monograph." Great Basin Nat. Mem. 6.

–7–
Invertebrate and Microbial Associates

F.M. STEPHEN,[1] C.W. BERISFORD,[2]
D.L. DAHLSTEN[3], P. FENN[4] and J.C. MOSER[5]
[1]*Department of Entomology, University of Arkansas, Fayetteville, AR, USA*
[2]*Department of Entomology, University of Georgia, Athens, GA, USA*
[3]*Division of Biological Control, University of California, Berkeley, CA, USA*
[4]*Department of Plant Pathology, University of Arkansas, Fayetteville, AR, USA*
[5]*USDA Forest Service, Southern Forest Exp. Stn., Pineville, LA, USA*

7.1 INTRODUCTION

Coincident with and immediately subsequent to bark penetration, colonization and establishment of bark beetle and pathogen populations in the host, a myriad of associated organisms that are intimately associated with the bark beetles arrives at and finds access to the subcortical environment of infested trees. Although many of these associated species have been identified and cataloged, relatively little is known about the biology or impact of most species. Evaluation of the effects of associated species is difficult for a variety of reasons, including the wide expanse of taxonomic categories (ranging from pathogenic bacteria to arthropod parasitoids to avian vertebrate predators) and ecological groupings that encompass all aspects of multiple species interactions. In addition, the cryptic habit of these organisms within their hosts confounds efforts to sample and even to observe. As a result of these research difficulties, the importance of associates in promotion or natural control of bark beetles and pathogens may be underestimated. The perception that associated species have little effect cannot be supported or rejected unless research in this area is encouraged. This

BEETLE–PATHOGEN INTERACTIONS IN CONIFER FORESTS
ISBN 0-12-628970-0

chapter focuses on the effects of associated species on the survival and reproduction of bark beetles and pathogens.

7.2 ORGANISMS INVOLVED

Researchers have been aware of the many associated organisms in bark beetle-infested trees for some time (Shelford, 1913; Blackman and Stage, 1924; Savely, 1939) and have noted that their importance relates to the variety of roles they play in the successional process of tree death and decomposition. Taxonomic inventories (some more extensive than others) of associates have been compiled for many of the primary bark beetles, including most species of *Dendroctonus* and some species of *Ips* and *Scolytus* (Table 7.1). Table 7.1 is not intended to be a complete inventory of all published literature, but does represent the majority of current North American information.

Some associated species are host-tree specific. However, the roles of most related species are similar across tree taxa, permitting a functional classification based on these roles. One approach to classification of the ecological groups of organisms associated with bark beetles in North American conifers is shown in Fig. 7.1. Characteristics of each group and examples of the organisms in each category are discussed below.

7.2.1 Bark beetle predators

Most information on bark beetle predators comes from studies of predation in beetle galleries or on the bark surface of beetle-infested trees (Dahlsten, 1982). Predation of bark beetles flying from tree to tree is poorly known.

Examination of the predaceous arthropod families and genera listed in Table 7.1 reveals that many are common to several of the bark beetles, and some to nearly all. Predaceous

Fig. 7.1. Bark beetle/associate ecological relationships, with examples of organisms comprising specific groups

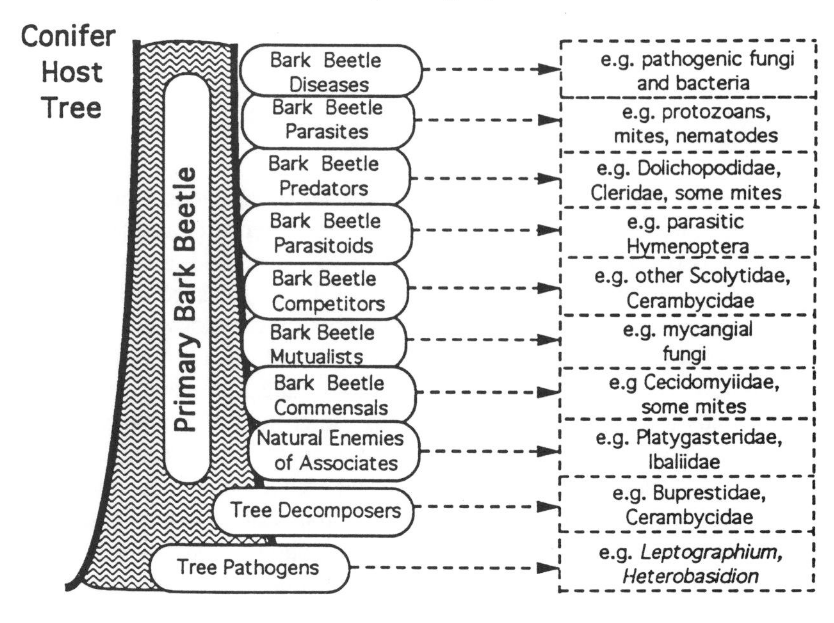

Table 7.1. Arthropod families and genera recorded as associates of conifer-infesting *Dendroctonus Scolytus*, and *Ips* bark beetles in North America

	Dendroctonus							*Scolytus*								*Ips*											
	Df[a]	Db	Dp	Da	Dps	Dr	Ds	Sa	Sl	Sp	Spr	Ss	St	Su	Sv	Ia	Ic	Ig	Ib	Ip	Ico	Iac	Ipi	Ipe	Il	Im	Ipl
Hemiptera																											
Anthocoridae																											
Lyctocoris	x	x		x	x																						
Scoloposcelis	x	x	x													x	x	x		x			x				
Aradidae																											
Aradus	x	x		x																							
Coleoptera																											
Histeridae				x	x																						
Plegaderus	x	x	x																	x						x	
Platysoma	x	x	x				x																				
Staphylinidae				x	x	x																					
Aleocharinae																				x							x
Nudobius	x	x	x																					x		x	x
Quedius																									x		
Buprestidae	x																										
Buprestis		x	x	x																							
Melanophila		x	x																								
Chrysobothris		x	x																								
Dermestidae	x		x																								
Ostomidae																											
Temnochila	x	x	x	x	x										x												
Tenebriodes	x	x		x	x																						
Corticotomus	x																										
Cleridae						x																					
Thanasimus	x		x		x	x																					
Enoclerus		x	x	x	x	x								x	x	x	x	x		x				x			x
Cymatodera		x	x																								
Nitidulidae																											
Epuraea	x		x	x																							
Glischrochilus			x	x																							
Soronia		x		x																							

Table 7.1 continued

	Dendroctonus							*Scolytus*								*Ips*											
	Df[a]	Db	Dp	Da	Dps	Dr	Ds	Sa	Sl	Sp	Spr	Ss	St	Su	Sv	Ia	Ic	Ig	Ib	Ip	Ico	Iac	Ipi	Ipe	Il	Im	Ipl
Bostrichidae			x																								
Rhizopgagidae																											
Rhizophagus	x	x	x	x		x	x							x													
Cucujidae																											
Cucujus		x	x																								
Silvanus	x	x	x																								
Lathridiidae	x																										
Corticaria		x	x																								
Colydiidae																											
Aulonium	x	x	x	x																							
Lasconotus	x	x	x	x	x									x	x					x							x
Colydium	x	x																									
Othniidae					x																						
Othnius		x	x	x																							
Tenebrionidae																											
Corticeus	x	x	x	x	x	x								x													
Melandryidae																											
Rushia		x	x	x	x																						
Serropalpus		x	x												x												
Throscidae																											
Triagus	x			x																							
Cerambycidae																											
Acanthocinus	x	x	x	x																							
Monochamus	x		x													x	x	x					x				
Asemum	x		x	x																							
Anoplodera			x																								
Stenocorus			x	x																							
Megasemum			x	x																							
Xylotrechus			x	x											x												
Spondylis		x	x																								
Callidium		x	x																								

	Dendroctonus							*Scolytus*								*Ips*											
	Df[a]	Db	Dp	Da	Dps	Dr	Ds	Sa	Sl	Sp	Spr	Ss	St	Su	Sv	Ia	Ic	Ig	Ib	Ip	Ico	Iac	Ipi	Ipe	Il	Im	Ipl
Curculionidae																											
Cossonus	x	x	x																								
Platypodidae																											
Platypus	x	x		x													x	x					x				
Scolytidae																											
Dendroctonus	x	x	x	x												x	x	x		x							
Scolytus															x												
Ips	x	x	x	x												x	x	x					x				
Pseudohylesinus															x												
Hylastes	x	x		x												x	x	x									
Pityokteines		x	x												x												
Orthotomicus			x	x																							
Pityogenes			x	x														x									
Hylurgops		x	x	x																							
Xyleborus		x		x																							
Pityophthorus	x	x	x												x												
Crypturgus	x		x												x			x									
Gnathotrichus	x	x		x												x	x	x					x				
Diptera																											
Sciaridae																											
Bradysia		x	x	x																							
Cecidomyiidae																											
Lestodiplosis	x		x																								
Winnertzia	x	x																									
Stratiomyidae					x																						
Zabrachia	x	x	x			x	x																				
Scenopinidae			x		x																						
Empididae					x																						
Drapetis	x	x	x																								
Tachypeza		x	x																								

Table 7.1 continued

	Dendroctonus							*Scolytus*									*Ips*										
	Df[a]	Db	Dp	Da	Dps	Dr	Ds	Sa	Sl	Sp	Spr	Ss	St	Su	Sv	Ia	Ic	Ig	Ib	Ip	Ico	Iac	Ipi	Ipe	Il	Im	Ipl
Dolichopodidae																											
Medetera	x	x	x		x	x	x							x	x			x		x			x	x			x
Phoridae																											
Megaselia		x	x												x												
Lonchaeidae																											
Lonchaea	x	x	x		x	x								x	x												
Xylophagidae																											
Xylophagus																								x			
Hymenoptera																											
Siricidae																											
Sirex		x	x	x														x									
Ichneumonidae																											
Gelis		x	x																								
Xorides															x												
Braconidae																											
Bracon																							x				
Coeloides	x	x	x	x	x	x								x	x	x	x	x		x	x		x	x			x
Atanycolus	x	x		x										x							x						
Dendrosoter	x	x	x	x						x						x		x		x	x		x				
Doryctes	x	x																									
Meteorus	x	x	x																								
Chelonus	x	x																									
Orgilus																					x						
Spathius	x						x	x	x	x			x	x	x	x		x		x			x		x		
Cosmophorus			x				x																				
Heterospilus											x					x		x					x				
Ecphylus									x			x		x	x												
Stephanidae																											
Schlettererius		x	x																								
Eulophidae																											
Tetrastichus	x	x																									

	Dendroctonus							Scolytus								Ips											
	Df[a]	Db	Dp	Da	Dps	Dr	Ds	Sa	Sl	Sp	Spr	Ss	St	Su	Sv	Ia	Ic	Ig	Ib	Ip	Ico	Iac	Ipi	Ipe	Il	Im	Ipl
Thysanidae																											
Signophora	x	x																									
Encrytidae																											
Avetianella		x	x																								
Eupelmidae																											
Eupelmus	x	x														x			x								
Lutnes	x		x																								
Torymidae																											
Crodontomerus																	x										
Roptrocerus	x	x	x		x	x	x								x	x	x	x		x	x		x	x			x
Pteromalidae																											
Tomicobia		x			x											x	x	x			x		x				
Cheiropachus		x			x					x				x	x												
Coelopisthia																								x			
Macromesus		x	x								x			x	x					x							
Amblymerus		x	x																		x		x				
Rhopalicus	x	x	x				x				x			x		x		x		x			x				x
Heydenia	x	x	x				x							x		x		x					x				
Dinotiscus	x	x	x	x	x	x								x	x	x	x	x		x	x		x	x			x
Eurytomidae																											
Eurytoma	x	x	x					x						x		x		x		x	x		x				
Ibaliidae																											
Ibalia		x	x																								
Scelionidae																											
Telenomus	x	x																									
Ceraphronidae																											
Ceraphron	x	x																									
Platygasteridae																											
Leptacis	x	x																									
Platygaster	x	x	x																								

Table 7.1 continued

	Dendroctonus							*Scolytus*								*Ips*											
	Df[a]	Db	Dp	Da	Dps	Dr	Ds	Sa	Sl	Sp	Spr	Ss	St	Su	Sv	Ia	Ic	Ig	Ib	Ip	Ico	Iac	Ipi	Ipe	Il	Im	Ipl
Formicidae																											
Camponotus	x	x	x	x																							
Leptothorax	x		x																								
Monomorium	x			x																							
Formica			x	x																							
Sphecidae																											
Passaloecus		x	x																								
TOTAL TAXA	61	75	74	41	21	12	9	2	2	2	3	1	1	16	21	18	13	22	1	16	9	0	19	7	1	2	10

[a]Df = *D. frontalis*, Db = *D. brevicomis*, Dp = *D. ponderosae*, Da = *D. adjunctus*, Dps = *D. pseudotsugae*, Dr = *D. rufipennis*, Ds = *D. simplex*, Sa = *S. abietis*, Sl = *S. laricis*, Sp = *S. piceae*, Spr = *S. praeceps*, Ss = *S. subscaber*, St = *S. tsugae*, Su = *S. unispinosus*, Sv = *S. ventralis*, Ia = *I. avulsus*, Ic = *I calligraphus*, Ig = *I grandicollis*, Ib = *I borealis*, Ip = *I. paraconfusus*, Ico = *I confusus*, Iac = *I. acuminatus*, Ipi = *I. pini*, Ipe = *I. perturbatus*, Il = *I. latidens*, Im = *I. mexicanus*, Ipl = *I. plastographus*.

insects known to feed on bark beetles are primarily members of the families Anthocoridae (Heteroptera), Formicidae (Hymenoptera), Histeridae, Staphylinidae, Ostomidae (=Trogositidae), Cleridae, Rhizophagidae, Cucujidae, Colydiidae, Othniidae, Tenebrionidae, Melandryidae (Coleoptera), Stratiomyidae, Empididae, Dolichopodidae and Lonchaeidae (Diptera). In addition to predaceous insects, birds, and perhaps spiders, are important predators of bark beetles prior to bark penetration (Dahlsten, 1982).

The degree of host specificity for most species in these families is uncertain. Many prey on other members of the vast complex of associated organisms in addition to the primary bark beetle host.

7.2.2 Bark beetle parasitoids

Parasitoids differ from predators in that they do not kill their prey directly but deposit eggs or larvae on the selected prey. The offspring subsequently consume and kill the prey during their development. Parasites, by contrast, usually do not kill their hosts (Dahlsten, 1982). All parasitoids of bark beetles are wasps (Hymenoptera). Parasitoid species most commonly associated with bark beetles are members of the families Braconidae, Eulophidae, Eupelmidae, Torymidae, Pteromalidae, and Eurytomidae. Parasitoids are an important and diverse group of bark beetle associates. The total number of associated parasitoid species is not known exactly for any of the bark beetles, because the host relationships of many of the uncommon potential parasitoids have not been investigated. Many of these species are quite host specific, whereas others attack both the primary bark beetle and scolytids that are potential competitors (Berisford, 1974).

Sorting out the biology of this parasitoid complex has been recognized as an extremely demanding and difficult task. For many parasitoid species the host relationships are uncertain, and there may be more species and/or varieties than previously thought (Espelie *et al.*, 1990). However, some genera are common to most bark beetle species, probably interacting similarly among hosts. Generally, the bark beetle species of greatest economic importance are those for which classification of the associates has been most thorough. As in-depth studies are conducted with the less well known beetles, undoubtedly a greater diversity of parasitoids will be encountered, and the consistency with which some parasitoid species are found on several beetle taxa will increase.

7.2.3 Bark beetle parasites and diseases

Organisms parasitic on bark beetles include nematodes, mites, and protozoans. Massey (1974) summarized the biology and taxonomy of nematodes associated with North American bark beetles. Reviews by Dahlsten (1982) and Mills (1983) provide a good source of literature relating to various parasites and their impact on bark beetles. Diseases, including those caused by bacteria, fungi, nematodes, protozoa and possibly viruses, may be important factors regulating populations of *D. frontalis* (Moore, 1971; Sikorowski *et al.*, 1979). Nematodes have been reported to reduce fertility and fecundity in *D. frontalis* and *D. pseudotsugae* (Thong and Webster, 1975; Kinn, 1980). Hoffard and Coster (1976) found four species of nematodes infecting three southern *Ips* spp. in Texas. Infection delayed emergence of adults but had little effect on reproduction. Mills (1983) noted that the fungal pathogen *Beauvaria bassiana* has been reported from *Ips amitinus* and *I. typographus* in Europe.

7.2.4 Commensals

Commensals (organisms that benefit from, but do not affect, their associate) are represented by a large and diverse assemblage of taxa whose biology and relationship to bark beetles, and the phloem-inhabiting guild as a whole, are poorly known. Many of the mites discussed below are commensal. Lindquist (1969) reviewed the tarsonemid mite associates of *Ips* and related bark beetles. Other commensals include fungivorus dipterans and coleopterans, many of whom are listed in Table 7.1.

A vast complex of mites is found with pine bark beetles in their host trees (Fig. 7.2). The role of different species varies extensively and is unknown for many. Unlike most insect associates of bark beetles, mites are wingless. This requires that individuals ride on beetle hosts or other flying associates (phoresy) in order to disperse to new hosts. A phoretic relationship ensures dispersal to new subcortical habitats but does not necessarily imply that the mite interacts with the beetle in any other way. The majority of mites have little or no effect on the bark beetles with which they are associated. Fig. 7.3, devised by Wilson (1980),

Fig. 7.2 A variety of predacious and saprophytic mites, such as the fungus-feeding acarids (probably *Histiogaster* sp.), are associated with bark beetle tunnels, commonly reaching new hosts by riding on dispersing beetles.

depicts a unique method for evaluating the effect of mites on *D. frontalis*. This relationship is based on data from Kinn and Witcosky (1978), Moser (1975, 1976), and Moser and Roton (1971). The figure shows that the closer the phoretic relationship between a mite and *D. frontalis* (as measured by the index of association), the less threat that mite is to the beetle. Thus, mite species with a low index of association possess high probabilities for predation on *D. frontalis,* whereas mites with a high index are benign. In fact, none of the mites phoretic on *D. frontalis* substantially harm its brood, except certain mite species when starved (Moser, 1975; Wilson, 1980).

The phoretic latitude (number of animal species that a mite will ride) varies with each particular mite species. Although mite parasitoids tend to be the most specific, practically all ride more than one host species. An exception to this may be *Pyemotes parviscolyti.* So far, this mite is known to ride only *Pityophthorus annectans* (called *P. bisulcatus* in Moser *et al.*, 1971), a common but inconspicuous bark beetle infesting small branches of southern pines. One other mite that appears to be specific to a single bark beetle is *Ereynetes sinescutulis* which is known to ride only *Ips pini* (Hunter *et al.*, 1989). However, like many other mites recorded from single species, the biology of this species is incompletely known, and more collecting may broaden its phoretic latitude. Many species of the egg-parasitic genus *Iponemus* generally ride only one species of *Ips* (Lindquist, 1969). However, the three species in the southern pine subcortical habitat each tend to make "mistakes," riding one of the other two *Ips* species (but never *D. frontalis*) about 5–10% of the time (J. C. Moser, unpublished).

Fig. 7.3. Relationship between the relative closeness of association of mites and *D. frontalis* and the observed degree of predation of those mites on *D. frontalis.* (J.C. Moser data, adapted from Wilson, 1980.)

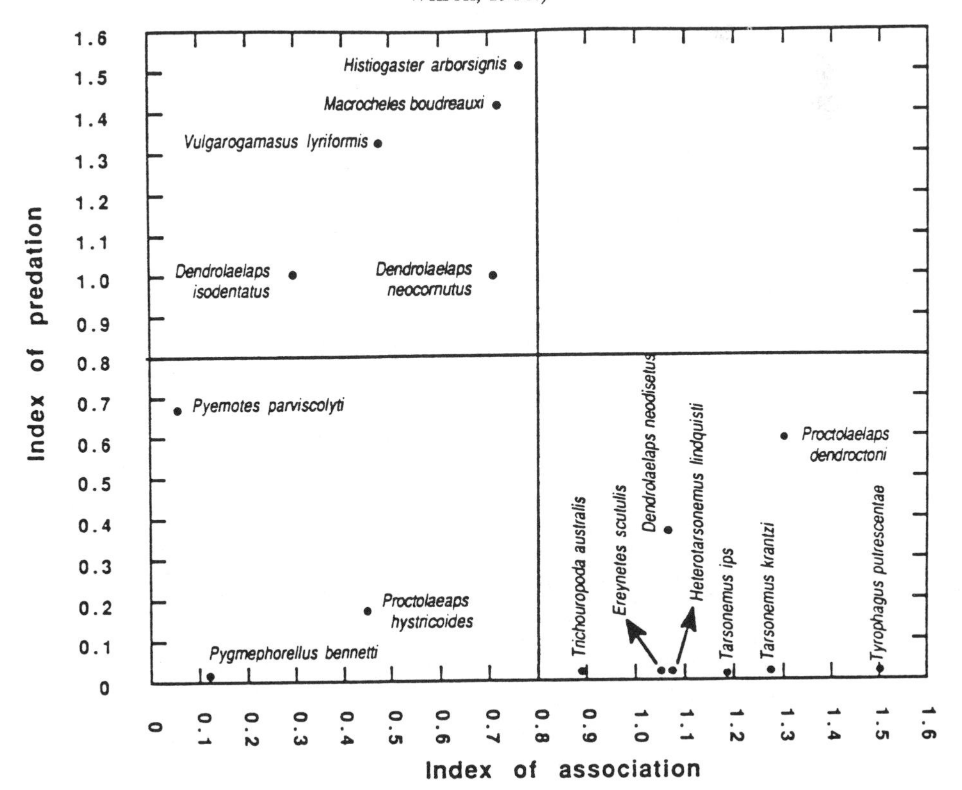

Other mites ride closely related species of bark beetles. *Tarsonemus krantzi* seems to be restricted to certain members of the *D. frontalis* species group (Lanier *et al.*, 1988), having been found so far on *D. frontalis, D. mexicanus*, and *D. adjunctus*, but not *D. brevicomis. Proctolaelaps hystrix, Histiostoma media*, and *Tarsonemus* "*terebrans*" are phoretic on the "turpentine" species group of *Dendroctonus*, i.e. *D. rhizophagus, D. terebrans*, and *D. valens. Elattoma* n. sp. #9 has a somewhat broader latitude, riding only *D. frontalis* and *D. terebrans* in the southern pine habitat, but never *Ips grandicollis*, one of the three species of *Ips* in this habitat.

The Holarctic species *Iponemus gaebleri* apparently rides all species of *Ips* that infest *Picea*, with one subspecies exploiting an *Ips* that attacks *Pinus* (Lindquist, 1969). *Pyemotes scolyti* is a classic example of a parasite specific to a particular genus of bark beetles, *Scolytus*. Thus *P. scolyti* can be found in subcortical habitats as diverse as *Ulmus, Pseudotsuga*, and *Prunus*. Members of the genus *Mucroseius* and perhaps species of a few other genera are phoretic only on cerambycid beetles (Kinn and Linit, 1989). *Cercoleipus coelonotus* is restricted to the genus *Ips*, but only the larger species. This is understandable because *C. coelonotus* is the largest mite associated with bark beetles, approaching the size of many tick deutonymphs. In the southern pine subcortical habitat, *C. coelonotus* rides only *Ips calligraphus*, the largest of the three species of *Ips*. Under experimental conditions, *C. coelonotus* rode *D. ponderosae*, but under field conditions this was not seen (Kinn, 1971). In central Louisiana, at least, *Tarsonemus subcorticalis* rides *Ips* and *Monochamus*, but rejects *Dendroctonus* (Kinn and Linit 1989; Moser, unpublished data). The egg parasite, *Paracarophaenax ipidarius* is recorded to ride only under the thorax of *Ips typographus* in Europe. However, in North America (where *I. typographus* does not occur) it occupies this position on *Ips pini, I. paraconfusus* and *I. plastographus* and rides under the elytra of *D. brevicomis* (Kinn, 1971; Moser, unpublished data). As with many species whose biology is poorly known, this disparity of host and phoretic data suggests two or more sibling species.

The vast majority of mites seem to accept any subcortical habitat, riding many of the scolytids and their associates. Of these, the beetle associates carry the most. Two common groups of associates, *Corticeus* spp. (Tenebrionidae) and the Cleridae, may exceed the primary scolytids in the number of individuals and species of mites carried. These subcortical mites include *Histiogaster arborsignis, Histiostoma varia, Mexecheles virginiensis, Paraleius leontonychus, Pleuronectocelaeno drymocoetes, Proctolaelaps fiseri, P. hystricoides, Tarsonemus ips, T. subcorticalis, Trichouropoda australis, Vulgarogamasus lyriformis*, and many others (Moser and Roton, 1971; Moser unpublished data). Some of these mites such as *Histiogaster arborsignis* and *Histiostoma varia* stick to practically any animal under bark (including other mites).

Some of the above mite taxa apparently are restricted to particular regions, perhaps because they have not been introduced to other habitats. Examples include *Macrocheles boudreauxi* and *Proctolaelaps dendroctoni*, which are known only from the southern pine subcortical habitat. *Pyemotes giganticus* has been found with at least 16 species of bark beetles (and one *Corticeus* associate) in western conifers. This mite also rode *D. frontalis*, which is not native to western North America, under experimental conditions (Cross *et al.*, 1981; Moser, 1981).

Some mites commonly seen in subcortical habitats of *Pinus* also may be found in the subcortical habitats of nearby tree species. *Trichouropoda hirsuta* is a mite commonly seen in the southern pine subcortical habitat and is normally phoretic on cerambycids (Kinn and Linit, 1989). However, phoretic deutonymphs also have been found on adults of the tenebrionid *Alobates pennsylvanica* collected from well decomposed stumps of *Liquidambar styraciflua* (J. C. Moser, unpublished).

7.2.5 Competitors

Primary beetle and pathogen species benefit from being the initial organisms to colonize the nutrient-rich phloem tissue of freshly killed conifers. Other phloem-inhabiting species that rapidly locate and colonize this resource include other scolytids, species of the beetle families Cerambycidae and Buprestidae, and various competing fungi and other microorganisms. An example illustrating the structure and dynamics of the phloem-inhabiting insect guild, is seen in the southern *Pinus* community of *D. frontalis*, *Ips* spp. and *Monochamus titillator* (Birch *et al.*, 1980; Coulson *et al.*, 1980; Paine *et al.*, 1981; Miller, 1985; Wagner *et al.*, 1985; Flamm *et al.*, 1987, 1989). In this situation, larvae of the cerambycid *Monochamus* will kill bark beetle larvae by indiscriminate foraging through phloem inhabited by the scolytids, but mortality to the scolytids normally is minimized through niche partitioning, i.e. different colonization and development rates and phloem utilization strategies.

Competing fungi such as *Trichoderma* spp. and *Penicillium* spp. may inhibit colonization and spread of pathogenic fungi. For example, *Trichoderma* prevents growth of *Heterobasidion annosum* and perhaps other pathogenic fungi in conifer stumps (Goldfarb *et al.*, 1989). Competition also occurs among beetle-vectored pathogens. Parmeter *et al.* (1989) reported that coinoculation of *Pinus ponderosa* with isolates of *Leptographium terebrantis* and *Ophiostoma ips* significantly reduced the rate of sapwood penetration observed for *L. terebrantis* alone. Inhibition can occur through production of antibiotic substances that prevent pathogen establishment or growth or through rapid growth and depletion of available resources (Rayner and Todd, 1979).

7.2.6 Mutualists

Mutualists include some mites and many microorganisms. The phoretic mite *Dendrolaelaps neodisetus* benefits *Dendroctonus frontalis* by preying on parasitic nematodes (Kinn, 1980). Many species of bacteria, yeasts and mycelial fungi have been reported to benefit associated conifer-attacking bark beetles. Whitney (1982) lists over 100 examples of microorganisms associated with bark beetles or their habitats in conifers and suggests that many more remain to be discovered. Although the ecological relationships of many of these associations are unknown and remain fruitful topics of investigation, Whitney (1982) gives 12 examples of yeasts and mycelial fungi that are proven or suspected mutualists of conifer bark beetles. Other recent reviews of bark beetle–fungal relationships include Batra (1979) and Beaver (1989).

Dissemination is the primary benefit that the mites and microorganisms receive from their associated beetles. Indeed, many require the penetration of the bark barrier by the host beetles in order to colonize susceptible tree tissues (Schowalter *et al.*, 1991). The pleomorphic (many distinct life stages) growth habit and gelatinous spores of the fungi (Fig. 7.4) represent adaptations for insect transmission; insect vectors are the only known mechanisms for dispersal in some species (Webber and Gibbs, 1989; Chapter 3).

Most of the mutualistic fungi are vectored specifically by certain species of bark beetles, and most are adapted to be transported in the mycangia of their respective vectors. Beaver (1989) and Bright (Chapter 2) discuss the benefits of mutualism and the relationships of beetle mycangia and fungal transmission. Mycangia are not always necessary for a success-

Fig. 7.4. A diverse fungal flora is inoculated rapidly into the subcortical habitat of trees colonized by bark beetles

ful relationship between beetle and fungus. A number of beetle–fungal relationships appear successful without mycangia on the host beetle (Witcosky *et al.*, 1986; Chapter 2). However, at least some species of mycangial fungi are known to undergo differentiation and to reproduce within the mycangia, indicating that the beetle must supply nutrients and growth factors (Norris, 1979).

The plant-pathogenic *Ophiostoma* apparently are vectored non-specifically on the exterior of the adult beetle or, in the case of *D. frontalis*, by two species of phoretic mites in the genus *Tarsonemus*. These mites carry ascospores of *O. minus* in a special spore-carrying structure called a sporotheca (Moser, 1985). *Tarsonemus* were common in *D. frontalis* infestations where *O. minus* was abundant; conversely, significantly fewer *Tarsonemus* were seen in infestations where the pathogen was rare (Bridges and Moser, 1986).

Bark beetles benefit from their fungal mutualists in many ways. Some examples are reviewed here with respect to the similarities and differences among different fungus/bark beetle associations.

7.2.6.1 Death of the host tree

Inoculation experiments have demonstrated that *Ophiostoma* spp. can colonize extensively and kill their hosts (Basham, 1970; Owen *et al.*, 1987). Therefore, these fungi are presumed to be responsible for host death. However, tree mortality often occurs so rapidly that there is some question whether these fungi are solely responsible (Whitney and Cobb, 1972; Parmeter *et al.*, 1989). In several cases, the extent of colonization and penetration by *Ophiostoma* has been limited before and even after the host trees have died. This has been observed for *Pinus taeda* killed in *D. frontalis* infestations (Bridges *et al.*, 1985) and for *P. ponderosa* killed in *D. brevicomis* infestations (Whitney, 1982; Whitney and Cobb, 1972).

These observations suggest that factors other than extensive colonization may be important in causing mortality, for example, production of systemic toxins by the pathogens (Hemingway *et al.*, 1977; Chapter 8). Detailed investigations of the host–fungal pathogen interactions are needed to clarify the role(s) of these fungi and/or their metabolites in tree mortality.

Most other mutualistic fungi apparently are not involved in the death of the host tissues, a requirement for successful larval development and ultimately for beetle reproduction. Deep penetration by these mutualistic fungi into the sapwood has not been demonstrated. Rapid and deep invasion would be required to disrupt host water conduction, leading to death of the tree.

7.2.6.2 Requirement for gallery production

Most research has shown that mutualistic fungi carried in the mycangia of female beetles during initial attack on a suitable host are not important to egg-gallery mining and egg laying. Barras (1973) showed that the length of galleries and number of eggs per gallery length did not differ between *D. frontalis* that had or lacked mycangial fungi. Recently, however, work with both parent and progeny generations of *D. frontalis* by Goldhammer *et al.* (1990) indicated that mining and egg laying were decreased in the absence of mycangial fungi.

7.2.6.3 Conditioning of host tissues for brood development

Research with *D. frontalis*, in particular, has shown clearly that mycangial fungi improve larval mining, duration of brood development, survival of larvae to adults, and adult beetle size (Barras, 1970; Bridges, 1985; Bridges and Perry, 1985; Goldhammer *et al.*, 1990). The mechanism(s) by which these fungi affect reproductive parameters is not clear. The mycangial fungi are found along the egg galleries and larval mines, and in the pupal chambers where they presumably infest the newly emerging brood adults. The timing of fungal growth and development in relation to changes in host tissues during larval development have not been clearly defined.

By contrast, the importance of mycangial fungi and mutualistic yeasts in preparing *Pinus contorta* phloem for *D. ponderosae* larval mining is questionable. Whitney (1971) has shown that newly laid eggs are deposited and second-to-fourth instar larvae mine in phloem that lacks mutualistic fungi and yeasts. Mining and pupation occurred several millimeters ahead of the growing fungi in essentially axenic non-discolored phloem. The fungi eventually colonize the pupal chambers and reinfest the new brood adults.

7.2.6.4 Mutualistic fungi as a food source

The ambrosial growth habit of some mutualistic fungi within the galleries, mines and pupal chambers of conifer bark beetles has been interpreted to indicate their use as a food source for brood development. However, there is little evidence that fungivory is required by conifer bark beetles. Several non-aggregating *Dendroctonus* species (*D. micans, D. punctatus, D. terebrans* and *D. valens*) have no known fungal mutualists and feed on unaltered host tissues (Berryman, 1989). Some species can be raised *in vitro* on sterile phloem or wood bolts and show no requirement for specific fungi; however, yeast extract can enhance beetle development and survival *in vitro* (Bedard, 1966; Strongman, 1987).

In the mutualistic association between *D. ponderosae* and *O. montium, O. clavigerum* and yeasts, the second-to-fourth instar larvae mine phloem in advance of these fungi

(Whitney, 1971). This would preclude these fungi as a food source for the larvae. Observations on the development of pupae and teneral adults, however, suggested that consumption of these fungi occurred during the final stages of brood development.

7.2.6.5 Interactions between fungal mutualists and Ophiostoma

Observations and experiments have suggested that factors within tissues colonized by *Ophiostoma* are detrimental to beetle brood development. Larvae mine away from tissues colonized by *Ophiostoma* and expend more energy in producing longer, winding galleries (Barras, 1970; Franklin, 1970; Yearian *et al*., 1972). Female *D. frontalis* forced to mine and oviposit in *Ophiostoma*-infected tissues produced smaller broods with extended development times (Barras, 1970; Paine and Stephen, 1988). Other observations and a few experiments have shown that the mutualistic fungi may benefit the beetles by restricting the development of *Ophiostoma*. In southern pines and *P. ponderosa, Ophiostoma* spp. often are restricted in their invasion to sectors in the sapwood and associated phloem, while surrounding tissues are colonized heavily by the mutualistic fungi (Barras, 1970; Whitney and Cobb, 1972). In experimental work, *Ophiostoma* was restricted in pine bolts infested by *D. frontalis* carrying their mycangial fungi but was more extensive in tissues infested by female beetles deprived of mycangial fungi (Bridges and Perry, 1985). How the mycangial fungi restrict growth of *Ophiostoma* is unknown. Perhaps they out-compete the *Ophiostoma* for specific nutrients, but antibiosis does not appear to be involved. Differences in inoculum loads between *Ophiostoma* and mycangial mutualists may influence the initiation and speed of *D. frontalis* infestation growth.

7.2.6.6 Production of pheromones

Mutualistic fungi, yeasts and bacteria associated with conifer bark beetles have been shown to convert host tree terpenes into beetle aggregation pheromones (Chapter 6). Mutualistic symbionts also can produce various volatile products including alcohols that augment the effectiveness of the beetle pheromones. With *D ponderosae*, the mutualistic fungi can oxidize the aggregation pheromone *trans*-verbenol to the anti-aggregation pheromone verbenone and thus may signal the termination of a successful mass attack (Borden *et al*., 1986).

7.2.7 Natural enemies of associates

Many families and genera that are commonly reared from bark beetle-infested material (Table 7.1) are thought to be predaceous or parasitic on other associated species found with bark beetles. The Platygasteridae, for example, probably are parasitoids of the many fungus-feeding larvae of the dipteran families Sciaridae or Cecidomyiidae found at the phloem–sapwood interface in bark beetle-infested trees. Not only are insect natural enemies encountered, but the variety of mites, pseudoscorpions, and other arachnids that are predaceous on some of the associated community is vast. Unfortunately the biology and role of most are relatively unknown.

7.2.8 Tree decomposers

Large saprophages such as the Cerambycidae and Buprestidae function as wood degraders. These species feed on decomposing tissue and also vector wood-decomposing fungi that grow in this habitat. Exclusion of these insects from logs can slow decomposition significantly (Edmonds and Eglitis, 1989).

Various saprophytic invertebrates and spores of non-vectored fungi find easier access to decaying tissues under the bark following penetration by bark beetles and other wood boring insects (Käärik, 1974; Schowalter *et al.*, 1992). Pathogenic fungi often persist as saprophytes in the decaying tree. *Ophiostoma* spp., other ascomycetes and deuteromycetes, and some bacteria degrade cell contents and cause "soft rot"; *Heterobasidion annosum*, *Armillaria* spp., and other basidiomycete "decay" fungi are capable of enzymatic degradation of cellulose and/or lignin (Käärik, 1974; Rayner and Todd, 1979). Initial colonization by soft rot organisms can inhibit decay fungi through antibiosis and resource depletion or, with nitrogen-fixing bacteria, can stimulate decay fungi through provision of necessary nitrogen, vitamins and other resources and degradation of toxic substances (Käärik, 1974; Blanchette and Shaw, 1978; Barz and Weltring, 1985).

These organisms ultimately are responsible for the decomposition and mineralization of wood and cycling of nutrients from dead trees. Roots and mycorrhizal fungi infuse decaying logs and transport nutrients into living plants. Schowalter *et al.* (1992) provide more detailed discussion and current literature on this topic.

7.3 ARRIVAL SEQUENCE

Stephen and Dahlsten (1976a) noted two basic arrival/colonization patterns for *D. brevicomis* in *P. ponderosa.* The first pattern, exemplified in Fig. 7.5, reflects rapid mass attack during the first summer flight period when synchronized emergence of overwintering beetles results in large populations of adults available for colonization. A slower, more extended attack period was found during the second generation of *D. brevicomis*, when populations of adults are less dense due to lack of synchronization in emergence (Fig. 7.6). The type of arrival pattern exhibited in Fig. 7.5 also would be expected for bark beetle species, such as *D. frontalis* and *D. ponderosae* that require large population sizes to respond rapidly and overcome tree resistance (McCambridge, 1967; Gara and Coster, 1968; Dixon and Payne, 1979). Ashraf and Berryman (1969) found a somewhat slower, more extended colonization period with *Scolytus ventralis* in *Abies grandis*. Berisford and Franklin (1971) noted a rapid arrival pattern with *I. avulsus* (Fig. 7.5). *Ips grandicollis*, normally seen attacking weak hosts or slash, shows a different pattern of gradual arrival (Fig. 7.6). The amplitude and periodicity of the arrival curves for other bark beetles likely reflect temperatures, beetle population density, and the importance of mass arrival for successful colonization of a temporary resource.

In addition to the primary species that may be responsible for tree death, there are many secondary bark and ambrosia beetles that respond rapidly to the newly created habitat in the dead or dying tree. Many of these species respond to primary (plant-produced) or secondary (insect-produced) attractants during host selection and concentration (Borden, 1982). The role of microorganisms in attraction was proposed by Person (1931), subsequently discounted (Graham, 1967), and recently reconsidered (Borden, 1982; Dahlsten and Berisford, unpublished). Interspecific communication among bark beetle species that inhabit the same host has been established (Birch and Wood, 1975) and is important in the interactions among these species (Borden, 1982; Lewis and Cane, 1990).

Response of bark beetle predators and parasitoids to bark beetle aggregation pheromones was first demonstrated by Wood *et al.* (1968) and Bedard (1966), respectively. Since that

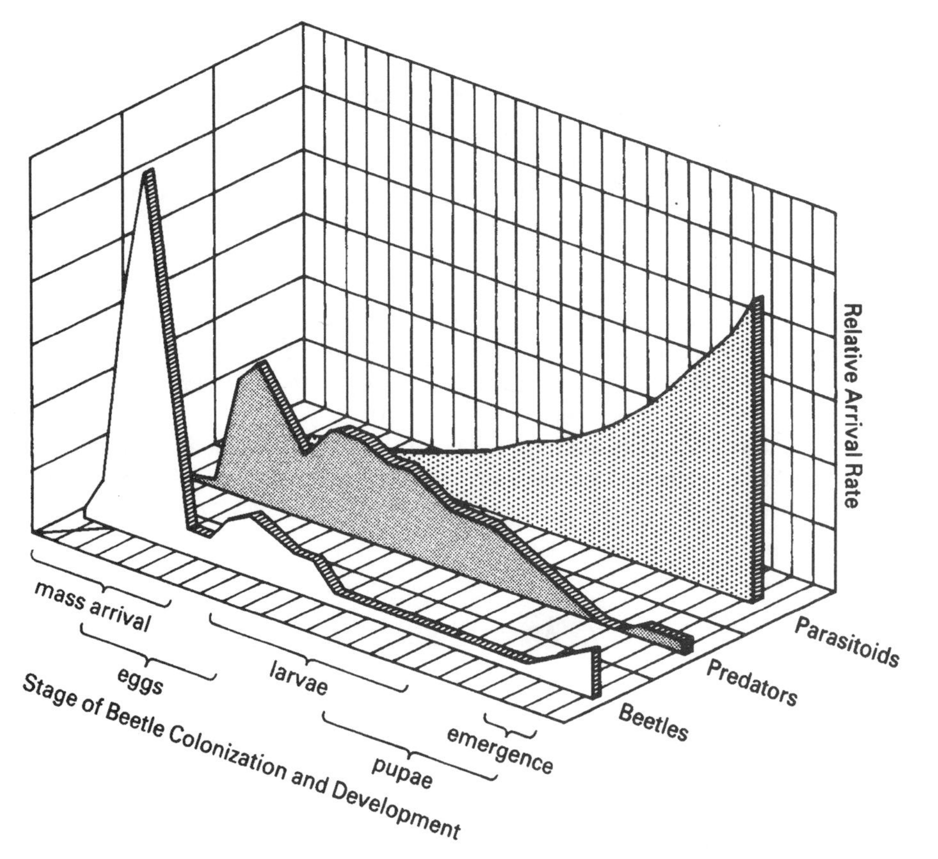

Fig. 7.5. Generalized pattern of primary bark beetle, predator and parasitoid arrival in relation to stage of the beetles' colonization and within-tree development.

time, many natural enemies now have been shown to respond to a wide variety of beetle and host-produced compounds. Payne (1989) provides examples of 29 species, from four insect orders, of natural enemies that rely on bark beetle-associated olfactory stimuli for host location.

The complex of natural enemies and other arthropod associates of conifer-infesting bark beetles arrives at its host tree in a predictable sequence that has been described in detail for at least two species of *Dendroctonus, D. brevicomis* (Stephen and Dahlsten, 1976b) and *D. frontalis* (Camors and Payne, 1973; Dixon and Payne, 1979), for *Ips* species (Berisford and Franklin, 1971; Berisford, 1974), and for *S. ventralis* (Ashraf and Berryman, 1969). Chemicals produced by these beetles, the host tree, and fungi interact to influence the predictable arrival of associated arthropods selecting suitable host locations. The sequence of arrival that has been observed for these bark beetles and their associates can be generalized as follows.

The arrival patterns of natural enemies and other associates is consistent for particular guilds (Stephen and Dahlsten, 1976b). Generalized patterns for guilds of natural enemies and other associates are proposed in Figs 7.5 and 7.6. Predators, primarily those that feed on attacking adult bark beetles, respond during and shortly after mass attack. In some species a second peak in predator response, perhaps associated with re-emerging parent adult bark beetles, has been observed. A more gradual arrival pattern is seen with the generalist preda-

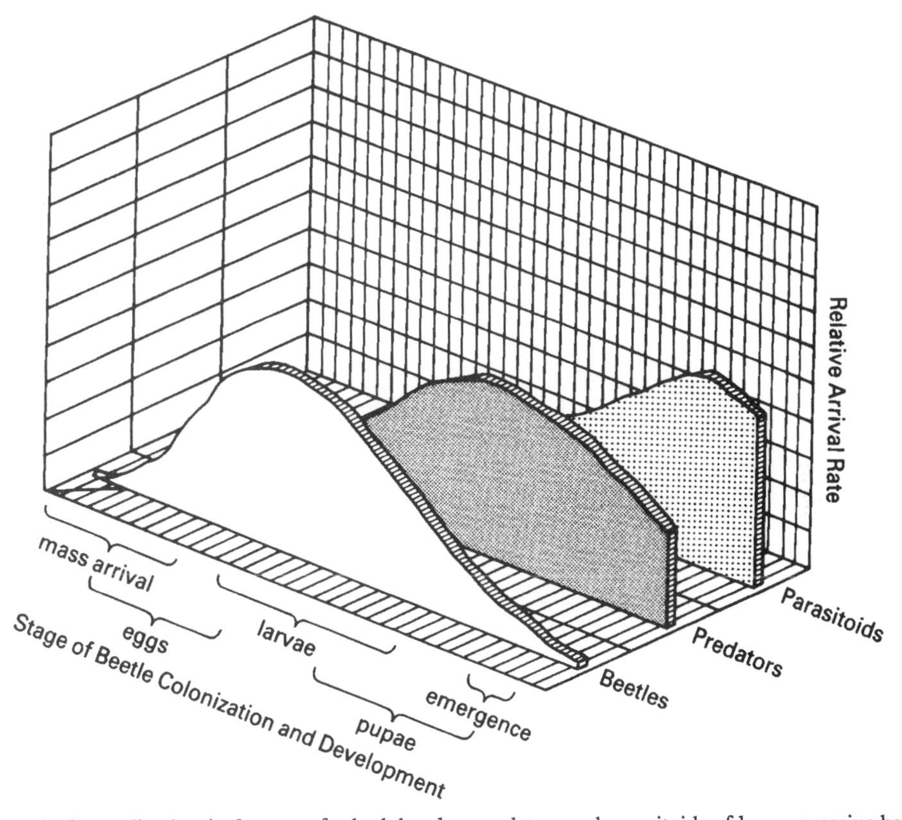

Fig. 7.6. Generalized arrival pattern for bark beetles, predators and parasitoids of less aggressive bark beetle species and those whose population densities are sufficiently low as to require a protracted period of mass arrival for successful colonization

tors, wood decomposers and fungivores. Bark beetle parasitoids, that must arrive when suitable stages of immature beetles are present, show a rapidly increasing response later in the beetle life cycle. The constancy of these overall patterns is seen for different parasitoid species with similar host requirements (Fig. 7.7) and for numerous groups of other arthropod associates sharing similar host requirements.

Recent research indicates the potential importance of a third trophic level interaction. Microorganisms play a role in producing chemicals to which parasites and perhaps other associates can respond. Most members of the parasitoid guilds of both *D. frontalis* and *D. brevicomis* are strongly attracted to billets infected with fungi and/or yeasts from these beetles (Dahlsten and Berisford, unpublished).

7.4 IMPACT OF NATURAL ENEMIES

Many species have been identified as natural enemies of the primary colonizers and the influence of these associates in natural control of bark beetle populations has been debated. With some exceptions (e.g. Hain and McClelland, 1979; Amman, 1984) most populations of bark beetles have been studied during outbreaks. In those situations it appeared that natural enemies had not been successful in preventing the outbreak. Natural enemies frequently are ascribed an important role at endemic, or low, population levels, but are credited

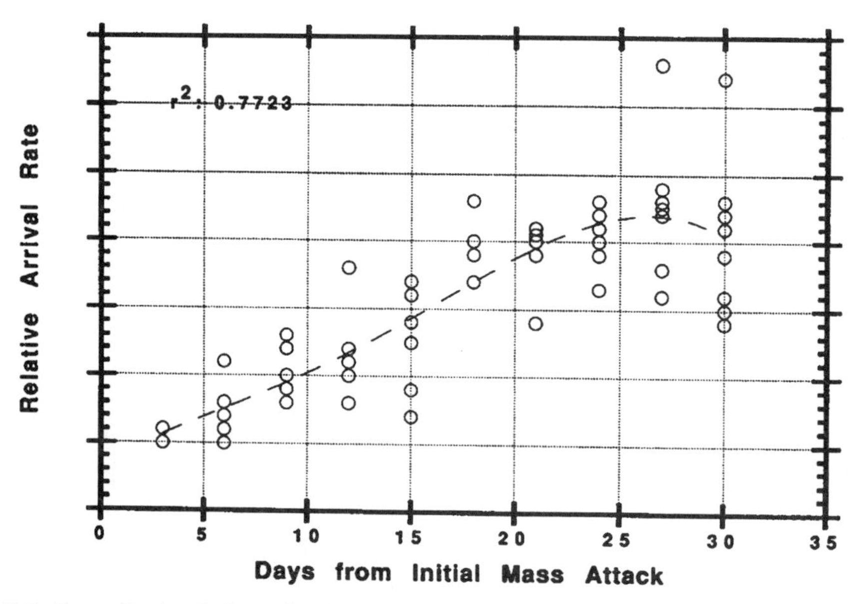

Fig. 7.7. Generalized arrival rate for eight species of *D. frontalis* parasitoids. (Data from Dixon and Payne, 1979.)

with minimal impact once bark beetle populations reach epidemic population levels. Unfortunately, little research has been designed to ascertain their actual importance. The literature on impact of arthropod parasitoids and predators on bark beetle population dynamics is scarce, and recent reviews by Berisford (1980), Dahlsten (1982), Mills (1983), Moeck and Safranyik (1984), and an edited book on the potential for biological control (Kulhavy and Miller, 1989) reveal little substantiative information to clarify their role.

Numerous authors have measured bark beetle mortality caused by individual species or complexes of natural enemies (see Dahlsten, 1982). Within-tree mortality caused by insect predators, parasitic mites, nematodes and insects averaged 17% in combined life tables for *S. ventralis* in *Abies grandis* (Berryman and Ferrell, 1988). Linit and Stephen (1983) and Moore (1972) reported about 25% within-tree mortality attributable to natural enemies of *D. frontalis*. Amman (1984) estimated that insect parasites and predators killed 8%, 33%, and 4% of *D. ponderosae* within trees in endemic, epidemic and postepidemic infestations, respectively.

Data from Stephen *et al.* (1989) revealed increased within-tree mortality from natural enemies of *D. frontalis* during the progression from endemic to epidemic to postepidemic population phases. Amman (1984) and Cole (1981), however, did not see similar responses with *D. ponderosae* populations. Recently Turchin (1990) and Turchin *et al.* (1991) have analyzed long-term population indices of *D. frontalis* abundance statistically and concluded that delayed density-dependent processes are responsible for regulating *D. frontalis* populations. They propose that natural enemies may provide this regulation. The variation in mortality and its relationship to host density must be understood if the role of natural enemies is to be evaluated properly.

Research to evaluate the importance of natural enemies in the dynamics of bark beetle populations should be designed to determine the extent to which natural enemy populations exhibit density dependent responses to their bark beetle host. This could be accomplished by measuring the absolute population densities of the bark beetle and its natural enemies over time. This seems a simple task in concept, but the effort needed to provide adequate estimations of populations of all the organisms involved is expensive and difficult. Adequate sampling protocols have been studied in depth for some bark beetles species, but not for others. However, adequate sampling techniques for estimation of natural enemy populations rarely have been considered, particularly in relationship to the host beetle populations involved. Stephen and Taha (1976) found that within-tree populations of parasitoids and predators were more highly aggregated than *D. frontalis*. Thus, accurate estimation of natural enemy density required larger sample units and sample size than did bark beetle life stages.

7.5 CONCLUSIONS

An extensive complex of organisms, primarily arthropods and fungi, rapidly colonize trees that are attacked and killed by bark beetles and pathogenic fungi. The composition and arrival sequence of this community are predictable in space and time, and appear similar in terms of community dynamics among bark beetle–conifer associations. Interactions among key elements in this complex, arthropod natural enemies and mutualistic fungi, can affect reproduction and survival of bark beetles and pathogenic fungi significantly. However, these interactions have not been studied in sufficient detail to assess their role in regulating bark beetle or pathogen epidemiologies.

REFERENCES

Amman, G.D. (1984). Mountain pine beetle (Coleoptera: Scolytidae) mortality in three types of infestations. *Environ. Entomol.* **13**, 184–191.

Ashraf, M. and Berryman, A.A. (1969). Biology of *Scolytus ventralis* (Coleoptera: Scolytidae) attacking *Abies grandis* in northern Idaho. *Melanderia* **2**, 1–23.

Barras, S.J. (1970). Antagonism between *Dendroctonus frontalis* and the fungus *Ceratocystis minor*. *Ann. Entomol. Soc. Am.* **63**, 1187–1190.

Barras, S.J. (1973). Reduction of progeny and development in the southern pine beetle following removal of symbiotic fungi. *Can. Entomol.* **105**, 1295–1299.

Barz, W. and Weltring, K.-M. (1985). Biodegradation of aromatic extractives of wood. *In* "Biosynthesis and Biodegradation of Wood Components" (T. Higuchi, ed.), pp. 607–666. Academic Press, Orlando, FL.

Basham, H.G. (1970). Wilt of loblolly pine inoculated with blue-stain fungi of the genus *Ceratocystis*. *Phytopathology* **60**, 750–754.

Batra, L.R., ed. (1979). "Insect–Fungus Symbiosis: Nutrition, Mutualism, and Commensalism". Allanheld, Osmund & Co., Montclair, NJ, 276 pp.

Beaver, R.A. (1989). Insect–fungus relationships in the bark and ambrosia beetles. *In* "Insect–Fungus Interactions" (N. Wilding, N.M. Collins, P.M. Hammond and J.F. Webber, eds), pp. 121–143. Academic Press, London.

Bedard, W.D. (1966). A ground phloem medium for rearing immature bark beetles (Scolytidae). *Ann. Entomol. Soc. Am.* **59**, 931–938.

Berisford, C.W. (1974). Parasite abundance in *Ips* spp. infestations as influenced by the southern pine beetle. *Environ. Entomol.* **3**, 695–696.

Berisford, C.W. (1980). Natural enemies and associated organisms. *In* "The Southern Pine Beetle" (R.C. Thatcher, J.L. Searcy, J.E. Coster and G.D. Hertel, eds), pp. 31–52. USDA Forest Serv. Tech. Bull. 1631, Washington, DC.

Berisford, C.W. and Franklin, R.T. (1971). Attack patterns of *Ips avulsus* and *I. grandicollis* (Coleoptera: Scolytidae) on four species of southern pines. *Ann. Entomol. Soc. Am.* **64**, 894–897.

Berryman, A.A. (1989). Adaptive pathways in scolytid–fungus associations. *In* "Insect–Fungus Interactions" (N. Wilding, N.M. Collins, P.M. Hammond and J.F. Webber, eds), pp. 145–159. Academic Press, London.

Berryman, A.A. and Ferrell, G.T. (1988). The fir engraver beetle in western states. *In* "Dynamics of Forest Insect Populations: Patterns, Causes, Implications" (A.A. Berryman, ed.), pp. 555–577. Plenum Press, New York.

Birch, M.C. and Wood, D.L. (1975). Mutual inhibition of the attractant pheromone reponse by two species of *Ips* (Coleoptera: Scolytidae). *J. Chem. Ecol.* **1**, 101–113.

Birch, M.C., Svihra, P. and Miller, J.C. (1980). Influence of chemically medicated behaviour on host tree colonization by four cohabiting species of bark beetles. *J. Chem. Ecol.* **6**, 395–414.

Blackman, M.W. and Stage, H.H. (1924). On the succession of insects living in the bark and wood of dying, dead and decaying hickory. *In* NY State College of Forestry Tech. Publ. 17, pp. 3–268. Syrecuse, NY.

Blanchette, R.A. and Shaw, C.G. (1978). Associations among bacteria, yeasts, and basidiomycetes during wood decay. *Phytopathology* **68**, 631–637.

Borden, J.H. (1982). Aggregation pheromones. *In* "Bark Beetles in North American Conifers" (J.B. Mitton and K.B. Sturgeon, eds), pp. 74–139. University of Texas Press, Austin, TX.

Borden, J.H., Hunt, D.W.A., Miller, D.R. and Slessor, K.N. (1986). Orientation in forest Coleoptera: an uncertain outcome to responses by individual beetles to variable stimuli. *In* "Mechanisms in Insect Olfaction" (T.L. Payne, M.C. Birch and C.E.J. Kennedy, eds), pp. 97–109. Oxford University Press, Oxford.

Bridges, J.R. (1985). Relationship of symbiotic fungi to southern pine beetle population trends. *In* "Integrated Pest Management Research Symposium: the Proceedings" (S.J. Branham and R.C. Thatcher, eds), pp. 127–135. USDA Forest Serv. Gen. Tech. Rpt. SO-56. USDA Forest Serv., Southern Forest Exp. Stn., New Orleans, LA.

Bridges, J.R. and Moser, J.C. (1986). Relationship of phoretic mites (Acari: Tarsonemidae) to the bluestaining fungus, *Ceratocystis minor*, in trees infested by southern pine beetle (Coleoptera: Scolytidae). *Environ. Entomol.* **15**, 951–953.

Bridges, J.R. and Perry, T.J. (1985). Effects of mycangial fungi on gallery construction and distribution of blue-stain in southern pine beetle infested pine bolts. *J. Entomol. Sci.* **20**, 271–275.

Bridges, J.R., Nettleton, W.A. and Conner, M.D. (1985). Southern pine beetle (Coleoptera: Scolytidae) infestations without the bluestain fungus, *Ceratocystis minor*. *J. Econ. Entomol.* **78**, 325–327.

Camors, F.B., Jr. and Payne, T.L. (1973). Sequence of arrival of entomophagous insects to trees infested with the southern pine beetle. *Environ. Entomol.* **2**, 267–270.

Cole, W.E. (1981). Some risks and causes of mortality in mountain pine beetle populations: a long-term analysis. *Res. Popul. Ecol.* **23**, 116–144.

Coulson, R.N., Pope, D.N., Gagne, J.A., Fargo, W.S., Pulley, P.E., Edson, L.J. and Wagner, T.L. (1980). Impact of foraging by *Monochamus titillator* (Col.: Cerambycidae) on within-tree populations of *Dendroctonus frontalis* (Col.: Scolytidae). *Entomophaga* **25**, 155–170.

Cross, E.A., Moser, J.C. and Rack, G. (1981). Some new forms of *Pyemotes* (Acarina: Pyemotidae) from forest insects with remarks on polymorphism. *Int. J. Acarol.* **7**, 179–196.

Dahlsten, D.L. (1982). Relationships between bark beetles and their natural enemies. *In* "Bark Beetles in North American Conifers" (J.B. Mitton and K.B. Sturgeon, eds), pp. 140–182. University of Texas Press, Austin, TX.

Dixon, W.N. and Payne, T.L. (1979). Sequence of arrival and spatial distribution of entomophagous and associate insects on southern pine beetle-infested trees. Texas A&M University Agric. Exp. Stn. Bull. MP-1432, College Station, TX.

Edmonds, R.L. and Eglitis, A. (1989). The role of the Douglas-fir beetle and wood borers in the decomposition and nutrient release from Douglas-fir logs. *Can. J. For. Res.* **19**, 853–859.

Espelie, K.E., Berisford, C.W. and Dahlsten, D.L. (1990). Cuticular hydrocarbons of geographically isolated populations of *Rhopalicus pulchripennis* (Hymenoptera: Pteromalidae): evidence for two species. *Comp. Biochem. Physiol.* **968**, 305–308.

Flamm, R.O., Wagner, T.L., Cook, S.P., Pulley, P.E., Coulson, R.N. and McArdle, T.M. (1987). Host colonization by cohabiting *Dendroctonus frontalis*, *Ips avulsus*, and *I. calligraphus* (Coleoptera: Scolytidae). *Environ. Entomol.* **16**, 390–399.

Flamm, R.O., Coulson, R.N., Beckley, P., Pulley, P.E. and Wagner, T.L. (1989). Maintenance of a phloem-inhabiting guild. *Environ. Entomol.* **18**, 381–387.

Franklin, R.T. (1970). Observations on the blue stain–southern pine beetle relationship. *J. Georgia Entomol. Soc.* **5**, 53–57.
Gara, R.I. and Coster, J.E. (1968). Studies on the attack behavior of the southern pine beetle. III. Sequence of tree infestation within stands. *Contrib. Boyce Thompson Inst.* **24**, 77–85.
Goldfarb, B., Nelson, E.E. and Hansen, E.M. (1989) *Trichoderma* species from Douglas-fir stumps and roots infested with *Phellinus weirii* in the western Cascade Mountains of Oregon. *Mycologia* **81**, 134–138.
Goldhammer, D.S., Stephen, F.M. and Paine, T.D. (1990). The effect of the fungi *Ceratocystis minor* (Hedgecock) Hunt, *Ceratocystis minor* (Hedgecock) Hunt var. *barrasii* Taylor, and SJB 122 on reproduction of the southern pine beetle, *Dendroctonus frontalis* Zimmermann (Coleoptera: Scolytidae). *Can. Entomol.* **122**, 407–418.
Graham, K. (1967). Fungal–insect mutualism in trees and timber. *Annu. Rev. Entomol.* **12**, 105–126.
Hain, F.P. and McClelland, W.T. (1979). Studies of declining and low level populations of the southern pine beetle in North Carolina. *In* "Proceedings of the Population Dynamics of Forest Insects at Low Levels" (F.P. Hain, ed.), pp. 9–26. North Carolina State Univ., Raleigh, NC.
Hemingway, R.W., McGraw, G.W. and Barras, S.J. (1977). Polyphenols in *Ceratocystis minor* -infected *Pinus taeda*: fungal metabolites, phloem and xylem phenols. *J. Agric. Food Chem.* **25**, 717–720.
Hoffard, W.H. and Coster, J.E. (1976). Endoparasitic nematodes of *Ips* bark beetles in Eastern Texas. *Environ. Entomol.* **5**, 128–132.
Hunter, P.E., Rosario, R.M.T. and Moser, J.C. (1989). Two new species of *Ereynetes* (Acari: Prostigmata: Ereynetidae) associated with bark beetles. *J. Entomol. Sci.* **24**, 16–20.
Käärik, A.A. (1974). Decomposition of wood. *In* "Biology of Plant Litter Decomposition" (C.H. Dickinson and G.J. F. Pugh, eds), pp. 129–174. Academic Press, London.
Kinn, D.N. (1971). The life cycle and behavior of *Cercoleipus coelonotus* (Acarina: Mesostigmata). Univ. Calif. Publ. Entomol. 65, 1–66. University of California Press, Berkeley, CA.
Kinn, D.N. (1980). Mutualism between *Dendrolaelaps neodisetus* and *Dendroctonus frontalis*. *Environ. Entomol.* **9**, 756–758.
Kinn, D.N. and Linit, M.J. (1989). A key to phoretic mites commonly found on long-horned beetles emerging from southern pines. USDA Forest Serv. Res. Note S0-357, USDA Forest Serv., Southern Forest Exp. Stn., New Orleans, LA, 8 pp.
Kinn, D.N. and Witcosky, J.J. (1978). The life cycle and behavior of *Macrocheles boudreauxi* Krantz. *Z. Ang. Entomol.* **84**, 136–144.
Kulhavy, D.L. and Miller, M.C. (1989). "Potential for Biological Control of *Dendroctonus* and *Ips* Bark Beetles." Stephen F. Austin State University Press, Nacogdoches, TX, 255 pp.
Lanier, G.N., Hendrichs, J.P. and Flores, J.E. (1988). Biosystematics of the *Dendroctonus frontalis* (Coleoptera: Scolytidae) complex. *Ann. Entomol. Soc. Am.* **81**, 403–418.
Lewis, E.E. and Cane, J.H. (1990). Pheromonal specificity of southeastern *Ips* pine bark beetles reflects phylogenetic divergence (Coleoptera: Scolytidae). *Can. Entomol.* **12**, 1235–1238.
Lindquist, E.E. (1969). Review of Holarctic tarsonemid mites (Acarina: Prostigmata) parasitizing eggs of Ipine bark beetles. *Mem. Entomol. Soc. Can.* **60**, 1–111.
Linit, M.J. and Stephen, F.M. (1983). Parasite and predator component of within-tree southern pine beetle (Coleoptera: Scolytidae) mortality. *Can. Entomol.* **115**, 679–688.
Massey, C.L. (1974). Biology and taxonomy of nematode parasites and associates of bark beetles in the United States. USDA Forest Serv. Agric. Handbook 446. USDA Forest Serv., Washington, DC, 233 pp.
McCambridge, W.F. (1967). Nature of induced attacks by the Black Hills beetle, *Dendroctonus ponderosae* (Coleoptera: Scolytidae). *Ann. Entomol. Soc. Am.* **60**, 920–928.
Miller, M.C. (1985). The effect of *Monochamus titillator* (F.)(Col., Cerambycidae) foraging on the emergence of *Ips calligraphus* (Germ.) (Coleoptera: Scolytidae) insect associates. *Z. Ang. Entomol.* **100**, 189–197.
Mills, N.J. (1983). The natural enemies of scolytids infesting conifer bark in Europe in relation to the biological control of *Dendroctonus* spp. in Canada. CIBC Biocontrol News Info. **4**, 305–328.
Moeck, H.A. and Safranyik, L. (1984). Assessment of predator and parasitoid control of bark beetles. Canadian Forestry Serv. Info. Rept. BC-X-248. Pacific Forest Research Centre, Victoria, BC.
Moore, G.E. (1971). Mortality factors caused by pathogenic bacteria and fungi of the southern pine beetle in North Carolina. *J. Invert. Pathol.* **17**, 28–37.
Moore, G.E. (1972). Southern pine beetle mortality in North Carolina caused by parasites and predators. *Environ. Entomol.* **1**, 58–65.

Moser, J.C. (1975). Mite predators of the southern pine beetle. *Ann. Entomol. Soc. Am.* **68,** 1113–1116.
Moser, J.C. (1976). Surveying mites (Acarina) phoretic on the southern pine beetle (Coleoptera: Scolytidae) with sticky traps. *Can. Entomol.* **108,** 809–813.
Moser, J.C. (1981). Transfer of a *Pyemotes* egg parasite phoretic on western pine bark beetles to the southern pine beetle. *Int. J. Acarol.* **7,** 197–202.
Moser, J.C. (1985). Use of sporothecae by phoretic *Tarsonemus* mites to transport ascospores of coniferous bluestain fungi. *Trans. Br. Mycol. Soc.* **84,** 750–753.
Moser, J.C. and Roton, L.M. (1971). Mites associated with southern pine bark beetles in Allen Parish, Louisiana. *Can. Entomol.* **103,** 1775–1798.
Moser, J.C., Cross, E.A. and Roton, L.M. (1971). Biology of *Pyemotes parviscolyti* (Acarina: Pyemotidae). *Entomophaga* **16,** 367–379.
Norris, D.M. (1979). The mutualistic fungi of Xyleborini beetles. *In* "Insect–Fungus Symbiosis: Nutrition, Mutualism and Commensalism" (L.R. Batra, ed.), pp. 53–63. Allanheld, Osmun & Co., Montclair, NJ.
Owen, D.R., Lindahl, K.Q., Jr., Wood, D.L. and Parmeter, J.R., Jr. (1987). Pathogenicity of fungi isolated from *Dendroctonus valens, D. brevicomis*, and *D. ponderosae* to ponderosa pine seedlings. *Phytopathology* **77,** 631–636.
Paine, T.D. and Stephen, F.M. (1988). Induced defenses of loblolly pine, *Pinus taeda*: potential impact on *Dendroctonus frontalis* within-tree mortality. *Entomol. Exp. Appl.* **46,** 39–46.
Paine, T.D., Birch, M.C. and Svihra, P. (1981). Niche breadth and resource partitioning by four sympatric species of bark beetles (Coleoptera: Scolytidae). *Oecologia* **48,** 1–6.
Parmeter, J.R., Slaughter, G.W., Chen, M., Wood, D.L. and Stubbs, H.A. (1989). Single and mixed inoculations of ponderosa pine with fungal associates of *Dendroctonus* spp. *Phytopathology* **79,** 768–772.
Payne, T.L. (1989). Olfactory basis for insect enemies of allied species. *In* "Potential for Biological Control of *Dendroctonus* and *Ips* Bark Beetles" (D.L. Kulhavy and M.C. Miller, eds), pp. 55–69. Stephen F. Austin State University Press, Nacogdoches, TX.
Person, H.L. (1931). Theory in explanation of the selection of certain trees by the western pine beetle. *J. Forestry* **31,** 696–699.
Rayner, A.D.M. and Todd, N.K. (1979). Population and community structure and dynamics of fungi in decaying wood. *Adv. Bot. Res.* **7,** 333–420.
Savely, H.E. (1939). Ecological relations of certain animals in dead pine and oak logs. *Ecol. Monogr.* **9,** 321–385.
Schowalter, T.D., Caldwell, B.A., Carpenter, S.E., Griffiths, R.P., Harmon, M.E., Ingham, E.R., Kelsey, R.G., Lattin, J.D. and Moldenke, A.R. (1992). Decomposition of fallen trees: effects of initial conditions and heterotroph colonization rates. *In* "Tropical Ecosystems: Ecology and Management" (K.P. Singh and J.S. Singh, eds), pp. 373–383. Wiley Eastern Ltd., New Delhi.
Shelford, V.E. (1913). Animal communities in temperate America. *Bull. Geo. Soc. Chicago* **5,** 362.
Sikorowski, P.P., Pabst, G.S. and Tomson, O. (1979). "The impact of diseases on southern pine beetle in Mississippi." Mississippi State Agric. Exp. Stn. Tech. Bull. 99. Mississippi State, MS. 9 pp.
Stephen, F.M. and Dahlsten, D.L. (1976a). The temporal and spatial arrival pattern of *Dendroctonus brevicomis* Le Conte in ponderosa pine. *Can. Entomol.* **108,** 271–282.
Stephen, F.M. and Dahlsten, D.L. (1976b). The arrival sequence of the arthropod complex following attack by *Dendroctonus brevicomis* (Coleoptera: Scolytidae) in ponderosa pine. *Can. Entomol.* **108,** 283–304.
Stephen, F.M. and Taha, H.A. (1976). Optimization of sampling effort for within-tree-populations of southern pine beetle and its natural enemies. *Environ. Entomol.* **5,** 1001–1007.
Stephen, F.M., Lih, M.P. and Wallis, G.W. (1989). Impact of arthropod natural enemies on *Dendroctonus frontalis* (Coleoptera: Scolytidae) mortality and their potential role in infestation growth. *In* "Biological Control of *Dendroctonus* and *Ips* Bark Beetles" (D.L. Kulhavy and M.C. Miller, eds), pp. 169–185. Stephen F. Austin State University Press, Nacogdoches, TX.
Strongman, D.B. (1987). A method for rearing *Dendroctonus ponderosae* Hopk. (Coleoptera: Scolytidae) from eggs to pupae on host tissue with or without a fungal complement. *Can. Entomol.* **119,** 207–208.
Thong, C.H.S. and Webster J.M. (1975). Effects of the bark beetle nematode, *Contortylenchus reversus*, on gallery construction, fecundity, and egg viability of the Douglas-fir beetle *Dendroctonus pseudotsugae* (Coleoptera: Scolytidae). *J. Invert. Pathol.* **26,** 235–238.

Turchin, P. (1990). Rarity of density dependence or population regulation with lags? *Nature* **344**, 660–663.

Turchin, P., Lorio, P.L., Jr., Taylor, A.D. and Billings, R.F. (1991). Why do populations of southern pine beetles (Coleoptera: Scolytidae) fluctuate? *Environ. Entomol.* **20**, 401–409.

Wagner, T.L., Flamm, R.O. and Coulson, R.N. (1985). Strategies for cohabitation among the southern pine bark beetle species: comparisons of life-process biologies. *In* "Integrated Pest Management Research Symposium: the Proceedings" (S.J. Branham and R.C. Thatcher, eds) pp. 87–101. USDA Forest Serv. Gen. Tech. Rpt. SO-56, USDA Forest Serv., Southern Forest Exp. Stn., New Orleans, LA.

Webber, J.F. and Gibbs, J.N. (1989). Insect dissemination of fungal pathogens of trees. *In* "Insect–Fungus Interactions" (N. Wilding, N.M. Collins, P.M. Hammond and J.F. Webber, eds), pp. 161–193. Academic Press, London.

Whitney, H.S. (1971). Association of *Dendroctonus ponderosae* (Coleoptera: Scolytidae) with blue stain fungi and yeasts during brood development in lodgepole pine. *Can. Entomol.* **103**, 1495–1503.

Whitney, H.S. (1982). Relationships between bark beetles and symbiotic organisms. *In* "Bark Beetles in North American Conifers" (J.B. Mitton and K.B. Sturgeon, eds), pp. 183–211. University of Texas Press, Austin, TX.

Whitney, H.S. and Cobb, F.W., Jr. (1972). Non-staining fungi associated with the bark beetle *Dendroctonus brevicomis* (Coleoptera: Scolytidae) on *Pinus ponderosa*. *Can. J. Bot.* **50**, 1943–1945.

Wilson, D.S. (1980). "The Natural Selection of Populations and Communities". Benjamin/Cummings Publishing Co., Inc., Menlo Park, CA. 186 pp.

Witcosky, J.J., Schowalter, T.D. and Hansen, E.M. (1986). *Hylastes nigrinus* (Coleoptera: Scolytidae), *Pissodes fasciatus* and *Steremnius carinatus* (Coleoptera: Curculionidae) as vectors of black-stain root disease of Douglas-fir. *Environ. Entomol.* **15**, 1090–1095.

Wood, D.L., Browne, L.E., Bedard, W.D., Tilden, P.E., Silverstein, R.M., Rodin, J.O. (1968). Response of *Ips confusus* to synthetic sex pheromones in nature. *Science* **159**, 1373–1374.

Yearian, W.C., Gouger, R.J. and Wilkinson, R.C. (1972). Effects of the bluestain fungus, *Ceratocystis ips*, on the development of *Ips* bark beetles in pine bolts. *Ann. Entomol. Soc. Am.* **65**, 481–487.

PART IV
Effects of Interactions

–8–

Host Response to Bark Beetle and Pathogen Colonization

T. E. NEBEKER,[1] J. D. HODGES[2] and C. A. BLANCHE[3]
Department of [1]Entomology and [2]Forestry, Mississippi State University, MS, USA
[3]Department of Plant and Soil Sciences, Southern University and A & M College, Baton Rouge, LA, USA

8.1 INTRODUCTION

Bark beetles must identify suitable hosts to colonize. Colonization then requires overcoming the resistance mechanisms of the host tree in order to successfully utilize the resource, but this can be accomplished only by recruitment of a critical minimum number of beetles (Wood, 1972; Hodges *et al.*, 1979, 1985; Chapter 6). The mechanism by which the host attempts to resist the invasion of bark beetles and their associated fungi has two recognized components: (1) the preformed or constitutive (primary) resin system and (2) the induced hypersensitive response. The oleoresin system in pines traditionally has been considered the primary defense against attack by bark beetles (Vité, 1961; Rudinsky, 1966; Reid *et al.*, 1967; Smith, 1975; Hodges *et al.*, 1979). This chapter evaluates these two components in relation to the colonization process and its effect on the survival or death of the host tree.

8.2 RESPONSE TO INITIAL COLONIZATION

Host selection has been attributed to the efforts of the first beetles to arrive. Colonization of the potential host begins with the biting process that is stimulated by chemical and physical cues (Chapter 6). This process is not particularly well understood. The cues used in host finding may also be stimuli for host selection and colonization. Thomas *et al.* (1981) assayed for biting responses of *Dendroctonus frontalis* to bark extracts of different polarities. The greatest number of biting responses was elicited when outer bark extracts were tested. By contrast, *Ips paraconfusus* distinguished a host, *Pinus ponderosa*, from a non-

BEETLE–PATHOGEN INTERACTIONS IN CONIFER FORESTS
ISBN 0-12-628970-0

host, *Abies concolor*, only after beetles had bored through the outer bark into the phloem (Elkinton and Wood, 1980). Tunneling by *D. frontalis* was influenced by diethyl ether and methanol extracts of inner and outer bark of *Pinus taeda* (White, 1981). White (1981) viewed these responses as indicative of responses to gustatory stimulants and deterrents. Considerable variation was observed and attributed to the variability in the hosts.

Raffa and Berryman (1982a), investigating the gustatory cues in the orientation of *D. ponderosae* to host trees, found that both benzene and methanol–water extracts increased feeding activity. Non-polar host compounds exhibited greater incitant (initiation of feeding) properties, while polar compounds were more powerful stimulants for the continuation of feeding.

Cues other than chemical signals also may be important in initiating the colonization process. Mattson and Haack (1987) hypothesized that some bark- and wood-boring species might utilize (orient toward) drought-induced acoustic signals during host selection and colonization. Such ultrasonic emissions produced as a result of water columns breaking in the xylem of stressed trees may serve as short-range stimuli to trigger the biting response and initiate the boring process. They may also act synergistically with other cues, such as bark texture (Elkinton and Wood, 1980) to enhance the behavioral response.

Once the adults have cut through the outer bark and into the inner bark and cambial tissue, resin ducts are severed and resin flow usually begins. At this point the adults spend time removing the resin from the entrance hole as it flows (Fig. 8.1). The adults must cope with both the preformed and induced defense systems in order to colonize.

8.2.1 Preformed defensive system

Many coniferous species have a well defined resin duct system, while others do not (Chapter 5). The preformed resin system is most highly developed in coniferous species that are capable of mobilizing large quantities of oleoresin to pitch out large numbers of attacking beetles (Christiansen *et al.*, 1987). In *P. taeda*, which has a well-formed resin duct system, formation of the radial resin ducts and their associated secretory cells is influenced by the age and growth rate of the annual ring in which the ducts are formed (DeAngelis *et al.*, 1986; Chapter 5). These radial resin ducts and associated epithelial cells serve two functions in the defense of the tree: (1) synthesis of new resin, and (2) delivery of resin to attack sites. Resin flow is the most important characteristic of this system for colonizing bark beetles and is partially under genetic control (Hodges *et al.*, 1977, 1979; Nebeker and Hodges, 1983; Nebeker *et al.*, 1988). Chapters 5 and 6 provide information on resin production and contribution to tree defense. This section will focus on the effects of disturbance and successful bark beetle and pathogen colonization on resin flow and chemistry.

Varying amounts of resin may be stored in this system and available upon demand to resist attack by bark beetles. The amount and rate of delivery of preformed resin, in *P. taeda*, are determined by (1) volume of the resin-producing system, (2) size of the resin pool at the time of attack, (3) physiological state of the system at time of attack, and (4) resin viscosity (DeAngelis *et al.*, 1986).

Resin flow can be stimulated or blocked by disturbance. Blanche *et al.* (1985a) reported that no oleoresin exudation could be detected in a lightning-struck *P. taeda* for at least 3 days following the strike. Eleven days after strike the flow was about 10% of that before the strike. Twenty-one days after the strike the flow had increased to about 2.3 times pre-strike levels. Resin flow from the unstruck control tree remained at about the same level during the period of observation.

Fig. 8.1. Adult *Dendroctonus frontalis* working a pitch tube on *Pinus taeda.* (Photo courtesy of the US Forest Serv., Southern Forest Exp. Stn., Pineville, LA.)

Resin flow also is altered in root pruned and basally wounded trees. Blanche *et al.* (1985b) reported that root pruning on two sides of *P. taeda* significantly reduced total resin flow 2 weeks after treatment. Normal flow was restored 7 months later. On the other hand, basal wounding stimulated resin flow for 10 months. Apparently, basal wounding stimulates the accumulation of resin precursors, resulting in an increase in available resin.

If the tree is successfully attacked, resin flow drops off rapidly (Christiansen *et al.*, 1987; Hodges *et al.*, 1989). Hodges *et al.* (1989) noted that the decrease in flow begins below and well above the attack zone and takes place within the first 2 days after initial attack. However, resin flow is evident at a reduced level up to 6 days after initial attack in the upper section of the infested *P. taeda* bole.

Resin chemistry also plays a role in the ability of a tree to resist attack by bark beetles and associated fungi. Toxicity tests of substances in pine oleoresin demonstrated that several monoterpenes are toxic to *D. frontalis* (Coyne and Lott, 1976). Those found to be toxic were limonene, *alpha*-pinene, *beta*-pinene, myrcene and *delta*-3-carene. Smith (1966) found the same monoterpenes to be toxic to other *Dendroctonus* spp. but in a slightly different order. Popp *et al.* (1991) reported that *Ophiostoma minus* and *0. ips* stimulated resin flow in *P. elliottii* and *P. taeda.* It appears that monoterpenes have antibiotic and repellent properties (Callaham, 1966; Bordasch and Berryman, 1977).

Trees struck by lightning exhibit dramatic changes in levels of specific monoterpenes (Blanche *et al.*, 1985a). Three weeks after being struck by lightning, the *alpha*-pinene level in *P. taeda* had increased 71%. Myrcene and camphene followed the same pattern of increase, with myrcene increasing more than three-fold. However, *beta*-pinene levels declined significantly, 37 times lower than before the strike. Limonene and *beta*-phellandrene also declined, but in an attenuated manner.

In light of current knowledge, the observation that lightning-struck trees almost invariably serve as focal points of bark beetle infestations (Coulson *et al.*, 1983, 1985) may be the result of one or all of the following conditions (Blanche *et al.*, 1983, 1985a):

(1) Fermentation of the phloem, either by anaerobic cellular respiration or by microorganisms, could result from wounding of *P. ponderosa* by lightning. Volatile odors from fermented phloem attract newly emerged beetle adults (Johnson, 1966).
(2) Certain host volatiles are released from the exposed wood and phloem of lightning-struck pines. Some of these volatiles are attractive to *Ips* spp. (Anderson and Anderson, 1968).
(3) The sudden release of ozone following a lightning strike attracts beetles (Howe *et al.*, 1971).
(4) Microorganisms invading a lightning wound produce chemicals attractive to bark beetles (Howe *et al.*, 1971).
(5) *D. terebrans* may respond to an attractant(s), produced as a result of the strike, and in turn produce a secondary attractant responsible for attack by *D. frontalis* (Hodges and Pickard, 1971).
(6) The ejection and deposition of the debris shower from the struck tree on neighboring trees cause a short-term oleoresin release that enhances the probability of discovery and attack by pioneer beetles (Taylor, 1974).
(7) Trees become more susceptible after lightning strike owing to the absence of, or reduction in, xylem resin flow immediately after a tree is struck by lightning (Blanche *et al.*, 1985a).
(8) Increased attractiveness results from quantitative changes in the volatile monoterpene fractions after the tree is struck by lightning (Blanche *et al.*, 1985a).
(9) Trees become more suitable after lightning strike, due to carbohydrate alteration and reduction in the relative water content of the bark (Hodges and Pickard, 1971).

From this we can see that there are a number of ways in which a tree can become more attractive or suitable when struck by lightning. Blanche *et al.* (1983) discussed the merits of each hypothesis and reviewed the supporting literature. Given the general evolutionary success of the bark beetles, they appear capable of responding to these chemical changes and may even detoxify defensive compounds, if they occur in abundance, in order to utilize resources such as lightning struck trees.

8.2.2 Induced defensive system

The second line of defense against attacks by bark beetles and associated fungi is known as the induced defense system (induced response). By its very nature, this induced response must have an instantaneous source of stored reserves to be effective in containing invasion. This system has been referred to as secondary resinosis (Reid *et al.*, 1967), dynamic reaction zone (Shain, 1967), hypersensitive response (Berryman, 1972), and wound response (Shrimpton, 1978). Basically, this response consists of: (a) localized autolysis of parenchyma cells accompanied by rapid cellular dessication, (b) tissue necrosis (lesion formation with accumulation of toxic or allellochemic compounds) (Fig. 8.2), (c) secondary resinosis by adjacent secretory and parenchyma cells (Fig. 8.3), and (d) formation of wound periderm to physically isolate the lesion from the mainstream of metabolism and transport (see Chapter 6).

The process works by chemically and physically containing and nutritionally depriving the invading organisms. Because this process of defense involves rapid and intense mobilization and utilization of storage reserves, particularly carbohydrates and other precursors of defensive chemicals, this constitutes a substantial drain of energy that otherwise could be channelled to growth and reproduction. For instance, a 30% decrease in soluble sugars and a 15% decrease in reducing sugars in the reaction zone of *Ophiostoma clavigerum*-inoculated *P. contorta* have been reported (Miller and Berryman, 1985). Declines of 36%, 80%, and 300% in amino nitrogen, carbohydrates, and triglyceride fatty acids, respectively, have been observed during a 5-week incubation period in *P. taeda* inoculated with *O. minus* (Richmond *et al.*, 1970). Barras and Hodges (1969) reported a drastic reduction in reducing sugars, but starch remained unchanged, in the inner bark of *P. taeda* treated with *D. frontalis*-microorganism complex and two beetle-associated fungi.

Conifers that lack well developed resin duct systems, i.e. species in the genera *Abies*, *Tsuga*, *Cedrus*, and *Pseudolarix* (Cates and Alexander, 1982), rely primarily on this induced defense strategy to contain invasion, whereas those with a well developed resin duct system, such as *Pinus*, defend themselves against invading organisms primarily through the preformed or constitutive defense system. However, pines also possess the ability to defend themselves through the induced defense system. The occurrence of hypersensitive response in *P. taeda* has been documented (Shain, 1967; Hain *et al.*, 1983; Paine *et al.*, 198S). Whether this induced defense system is effective in preventing successful host colonization by bark beetles has not been demonstrated unequivocally. There are observations of dead beetles within a fully formed hypersensitive lesion (Berryman, 1969; Christiansen and Horntvedt, 1983; Stephen and Paine, 1985), but it is not known if this was a result of resinosis or other factors.

The induced response is considered to be a more universal defensive system than the preformed defense system (Cates and Alexander, 1982). We suggest, however, that it probably has a minor role in the defense of *Pinus* spp. against bark beetle–fungus invasion in the southern US, largely due to the rapid mass attack behavior of *D. frontalis*. This certainly merits further investigation.

At present there are no hard data comparing the efficacies of these two defense systems with respect to bark beetle–microorganism invasion. However, Russell and Berryman

Fig. 8.2. Gallery construction by *Dendroctonus frontalis* without interference (center) from the host. Resinosis within the gallery and darkened tissue around the galleries to right and left of center are evidence that the induced defensive system is being activated. (Photo courtesy of the US Forest Serv., Southern Forest Exp. Stn., Pineville, LA.)

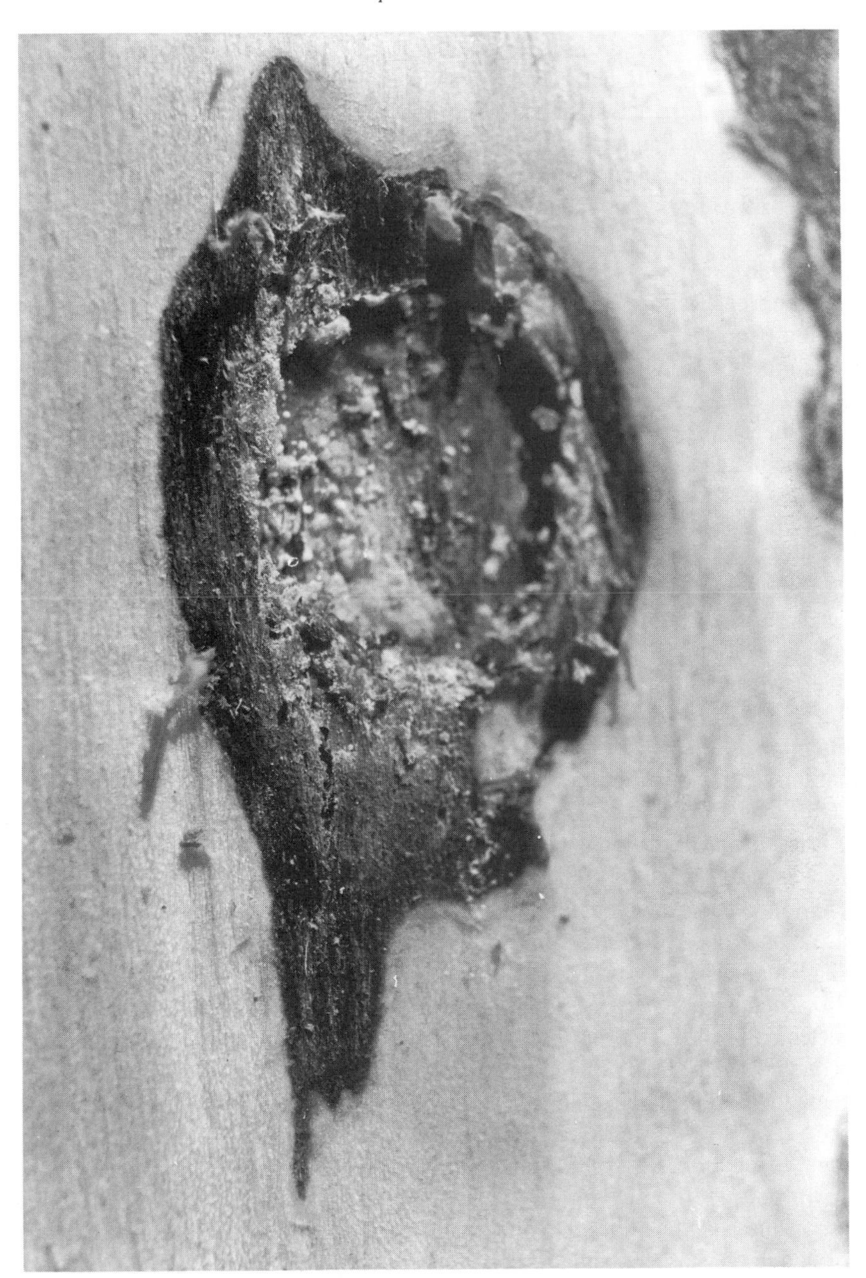

Fig. 8.3. Wound on *Pinus taeda* showing the surrounding lesion characteristic of the induced (hypersensitive) response. (Photo courtesy of R.A. Tisdale.)

(1976) reported that the induced system of *Abies grandis*, a conifer that lacks a well developed resin duct system, was more effective than the preformed resin system in preventing infection by *Trichosporium symbioticum*, a fungal associate of *Scolytus ventralis*. Also, Christiansen *et al.* (1987) concluded from the literature that the impregnation of the necrotic area with resinous and phenolic compounds prevents beetle gallery construction and fungal proliferation (Reid *et al.*, 1967; Berryman, 1969; Richmond *et al.*, 1970; Shrimpton, 1973).

In *Picea* the first response is the quick death of phloem tissue in an elliptical area adjacent to the egg gallery and secretion of resin (Safranyik, 1988). The dead tissue eventually is sealed off by wound periderm and sloughed. Within 3–6 weeks, traumatic resin ducts form on the xylem side of the cambium adjacent to the wound and completely surround the dead tissue, which becomes resin soaked (Safranyik, 1988).

Hypersensitive responses induced by bark beetle–associated–fungi have been reported to vary with season (Paine, 1984; Stephen and Paine, 1985; Cook *et al.*, 1986; Raffa and Smalley, 1988), tree species (Cook and Hain, 1987; Raffa and Smalley, 1988), general vigor of the tree (Raffa and Berryman, 1982b; Paine and Stephen, 1987a,b), and time of inoculation (Cook and Hain, 1987; Owen *et al.*, 1987; Raffa and Smalley, 1988). Larger lesions are formed from February to October than during the remainder of the year in *Pinus taeda* and *P. echinata*, with the largest lesions forming in August or September in *P. taeda* and September in *P. echinata* (Stephen and Paine, 1985; Cook *et al.*, 1986). The period during which larger lesions are formed coincides with the growing period for both species. This implies that the induced hypersensitive response depends on the physiological status or activity of the host, which in turn is influenced largely by temperature. Lesion size is also positively correlated with temperature (Cook *et al.*, 1986). Interestingly enough, the time of peak lesion formation (Stephen and Paine, 1985; Cook *et al.*, 1986) coincides with the peak of resin flow (September) in *P. taeda* under generally mild soil water deficit (Lorio *et al.*, 1990). This suggests that these two defense systems respond in accordance with the growth and differentiation principles as discussed in Chapter 5 and Lorio *et al.* (1990). These two defense systems may act in concert, rather than one after the other, in containing invasion.

Qualitative and quantitative variations in the chemistry of this response have been reported (Shain, 1967; Hadwiger and Schwochan, 1969; Kosuge, 1969; Shain and Hillis, 1971; Russell and Berryman, 1976; Hemingway *et al.*, 1977; Shrimpton, 1978; Ryan, 1979; Cates and Alexander, 1982; Raffa and Berryman, 1983; Gambliel *et al.*, 1985; Paine *et al.*, 1987). Comparing monoterpenes of induced lesion and sapwood resin from *P. taeda* inoculated with *O. minus*, Paine *et al.* (1987) showed significant differences in the quantitative and qualitative composition of these two defensive systems. The lesion resin is qualitatively richer in monoterpenes than is the sapwood resin. Two unknown monoterpenes were detected in the lesion but not in the sapwood resin. Quantitative characterization of the secondary chemistry of this induced response in *P. taeda* revealed considerable variation among individual trees (Gambliel *et al.*, 1985) which suggests an influence of individual tree vigor in the effectiveness of the induced system for infection containment. Although Gambliel *et al.* (1985) found similar monoterpenes, resin acids, and condensed tannins in both sterile and fungal inoculations, the latter caused greater than 40-fold increase in the accumulation of monoterpenes and a 20-fold increase in resin acids over those found in the sterile inoculations. Levels of condensed tannins, a reflection of protein-complexing capacity, decreased slightly relative to the controls. Gambliel *et al.* (1985) suggested that condensed tannins do not play a defensive role against *O. minus*. In fact, Hemingway *et al.* (1977) found that the condensed tannins, catechin, and flavonols were degraded by *O. minus*. The stilbenes pinosylvin, pinosylvin monomethyl ether, and resveratrol also were found to decline in *P. taeda* phloem and xylem inoculated with *O. minus*, during the later stages of incubation, suggesting the detoxification of these compounds by *O. minus*. Also isolated from *O. minus*-inoculated *P. taeda* phloem in significant quantity was 4-allyl anisole, a phenylpropanoid previously unreported in *P. taeda* (Gambliel *et al.*, 1985). Interestingly, this compound recently has been demonstrated to be highly inhibitory to the three symbiotic fungi associated with *D. frontalis* (Bridges, 1987). Whether this compound is a newly evolved chemical, as a result of selection pressure, or a pre-existing compound missed in previous investigations, is an interesting but challenging question worth investigating.

Although our understanding of the chemistry of this induced defense system in pines, as discussed above, is limited, it appears that many of the phenolic (polyphenolic) compounds isolated and identified from the hypersensitive lesion of *P. taeda* play very little or no role in host defense against the fungi associated with *D. frontalis*. Effects of individual compounds or groups of compounds from the hypersensitive lesions of specific hosts on known specific invading organisms should be assessed experimentally. This would improve understanding of the defensive strategies of conifers against the invasion by bark beetle-fungal associations and ultimately produce a clearer picture of the host-bark beetle-microorganism interactions.

8.3 TREE DECLINE

8.3.1 Decline process

After bark beetles have successfully overcome the tree's preformed and induced defensive system by mass attack, colonization begins. This involves two major events: (1) introduction of associated microorganisms into the phloem and xylem tissues and (2) the physical construction of the egg gallery.

Tree-killing bark beetle species introduce a number of fungal organisms. The amount of fungal inoculum introduced into the xylem generally is accepted to be important with respect to eventual death of the tree (Wood, 1972, 1892; Hodges *et al.*, 1985). The fungi penetrate and kill live tissue surrounding bark beetle galleries and enhance host colonization by the beetles (Safranyik, 1988).

Successful attack by most bark beetles usually results in the death of the tree. Anderson (1960) suggested several mechanisms by which the associated fungi could cause rapid host death. These include: (1) toxin production, (2) mycelial plugging of the tracheids, (3) release of gas bubbles into the tracheids, and (4) production of particles that block the pit openings by causing tori aspiration. Shepard and Watson (1959) suggested that *Ophiostoma* spp. reduce stored food in the paranchyma cells and restrict water conduction by destroying the ray parenchyma cells that partially control water movement.

There are, however, strip attacks in which the beetles successfully kill vertical strips of phloem, but the tree survives (Miller and Keen, 1960). When resinosis is extensive, the invading beetles are pitched out or heavy brood mortality results; *Ophiostoma* becomes entombed in the resinous tissue, and the tree survives. However, in successful *D. rufipennis* attacks on fully or strip-attacked trees, windfall, and logs, neither resin-producing tissues nor wound periderm is produced to any significant extent (Safranyik, 1988).

We put forth a hypothesis (Hodges *et al.*, 1985) that has served as a basis for investigating the role of the microorganisms in the death of the tree in the southeastern US. The success of beetle attack, and thus the death of the tree, is directly related to the complete circumferential introduction of fungal inoculum into the living tree. The subsequent production of toxic substances by the pathogen in turn alters physiological processes (e.g. water relations) in the tree, and makes the tree far more susceptible to subsequent attacks. Relative resistance is, therefore, a function of the ease with which inoculum can be introduced and successfully established.

An important consequence of beetle attack is the disruption of normal plant water relations that eventually leads to severe water stress (DeAngelis *et al.*, 1986). Secondary symptoms often attributed to water stress include: (1) rapid drop in turgor pressure throughout living cells of the bole, as evidenced by reduction in oleoresin exudation pressure shortly after successful beetle invasion; (2) blockage or aspiration of tracheids and concomitant reduction in the volume of water transported to the crown (Basham, 1970); and (3) drying of the outer bole tissues as water is withheld from below and within (Nelson, 1934; Basham, 1970). Cause and effect relationships have not been elucidated clearly. Symptoms may result from blockage of xylem tracheids by fungal hyphae, by toxic fungal metabolites, or by aspiration of individual tracheids when tracheid walls are penetrated by growing hyphae (DeAngelis *et al.*, 1986). Any or all of the above may occur after fungal inoculation, but none has been proven responsible for tree death, nor has the possible involvement of other mechanisms been eliminated (Hodges *et al.*, 1989).

Accumulating evidence suggests the involvement of toxic fungal metabolites in overcoming tree resistance and causing the death of the tree. Hemingway *et al.* (1977) and McGraw and Hemingway (1977) isolated three phenolic acids from liquid cultures of *O. minus* and from infected tissues of *P. taeda*. The compounds isolated were 6,8-dihydroxy-3-hydroxymethyl isocoumarin (Compound I), 6,8-dihydroxy-3-methyl isocoumarin (Compound II) and 3,6,8-trihydroxy-2-tetralone (Compound III). Compound I was in greatest abundance followed by II. Compound III was isolated only from the liquid culture and not from infected *P. taeda* tissue. DeAngelis *et al.* (1986) verified the production of Compounds I and II and demonstrated that purified extracts from liquid cultures greatly increased the rate of water loss from seedlings when applied to cut stems.

Concerning the effects of these compounds on various physiological processes, Hodges *et al.* (1989) found that at least one of the phenolic compounds (Compound I) appears to increase transpiration rates in *P. taeda*. Apparently, this occurs as a result of decreased stomatal control of water loss in infected trees. Of special interest is the slowness with which moisture stress develops even though transpiration has increased. This suggested that water transport in the stem is not markedly affected, at least in the early stages of beetle attack. Large reserves of water in the bole could ameliorate the effects of increased transpiration. Parmeter *et al.* (1989, 1992) reported limited occlusion of sapwood by *O. minus* and *Leptographium terebrantis* in 17–32 cm DBH (diameter at breast height) *P. ponderosa*; water moved readily through uncolonized sapwood. The rate of sapwood occlusion was considered inadequate *per se* to account for moisture stress in colonized trees. Even though moisture stress does not develop rapidly, the increased water loss over time is likely to be a major factor in desiccation of the foliage and eventual death of the tree.

The mechanism(s) responsible for the decrease in oleoresin flow and pressure in trees attacked by bark beetles has not been identified. However, Blanche *et al.* (1983) concluded that moisture stress does not increase rapidly in infested trees as a result of excessive water loss, introduction of air bubbles, or blockage of the transport system, as suggested by Anderson (1960). Thus, the decrease in resin flow is not the result of a rapid increase in moisture stress. This factor may contribute later to the death of the tree but is not responsible for the initial decrease in resistance. One possible explanation for the rapid decrease in flow is a decrease in turgor pressure of the epithelial cells surrounding the resin ducts. This could result from toxic effects of compounds produced by *O. minus* or other fungi (Hodges *et al.*, 1989).

Mechanical wounding, to simulate beetle attacks, causes a rapid decrease in resin flow. Thus, reduced resin flow may result simply from rapid draining of the oleoresin reservoir through the hundreds of wounds caused by attacking beetles. This does not rule out the effects of associated fungi on tree water relations and the eventual death of the tree (DeAngelis *et al.*, 1986).

8.3.2 Visual symptoms

Boring dust and pitch tubes are usually the first signs of bark beetle attack. The initial boring dust is the color of the bark. This dust accumulates on the upper surfaces of the understory and in spider webs at or near the base of the tree. The initial resin flow is clear, but turns pink as boring dust becomes mixed with it. The resin then begins to turn white as it reacts with the air and hardens. The size of the pitch tube is proportional to the size of the attacking beetle and amount of resin available to defend against invasion at that point in time. Although the amount of available resin depends on a number of factors, pitch tubes may not be formed during drought periods, and may be larger than normal during periods of excess moisture.

Foliar color changes occur as a result of bark beetle/fungal colonization and subsequent death of the tree. The foliage gradually changes from green to light brown ("fader") to red ("red top"), and finally falls to the ground. The rate of color change varies with season of the year, geographic location, and tree-species (Coulson and Witter, 1984). An interesting observation is that the foliage of the host tree always exhibits a deep red color when killed by *D. frontalis* but not by other agents (personal observation).

The point at which tree death has occurred has been debated for a long time. We suggest that a tree is dead when it no longer can recover sufficiently for resin flow to resume. For example, when a tree is struck by lightning the resin flow ceases but resumes after a period of time if the tree is not colonized by bark beetles or killed by the strike (Blanche *et al.*, 1985a).

8.3.3 Fungal growth and wood staining

Fares *et al.* (1980a) described basic tree anatomy and general fungal growth patterns. The hyphae of *O. minus* are initiated from spores deposited by beetles in their egg galleries in the phloem-cambium tissues. The hyphae enter the wood primarily through the wood rays. Thus, their initial growth is radially inward. Fares *et al.* (1980b) estimated fungal growth rate at 2.5 mm day^{-1} under favorable conditions, using data presented by Lagerberg *et al.* (1927). Paine and Stephen (1987a) observed that *O. minus* associated with *D. frontalis* grew 20 mm in the phloem within 48 hours of inoculation. At 15-30°C the rates of spread were 10–15 mm week^{-1} and 50 mm week^{-1} in the radial and longitudinal directions, respectively, in artificially inoculated samples of *P. sylvestris* wood (Hennington and Lundstrom, 1974). Eventually, the three-dimensional pattern of *O. minus* spread varies 1:2:7 to 1:4:15 in the tangential, radial, and longitudinal directions, respectively (Fares *et al.*, 1980a). Goldhammer *et al.* (1989) investigated the radial growth rate and chlamydospore production of *O. minus*, *O. minus* var. *barrasii* and SJB 122 on three different media. They concluded, as have others (Parmeter *et al.*, 1992), that growth rate depends on tree condition or nutritional status.

Parmeter *et al.* (1989) measured growth of single and paired isolates of *O. minus* and *O. nigrocarpum* from *D. brevicomis*, and *O. ips* and *Letographium terebrantis* from *D. valens* inoculated into small (10–25 cm DBH) *P. ponderosa*. Rates of spread in the radial direction varied from 2 mm week^{-1} for *O. nigricarpum* to 6 mm week^{-1} for *L. terebrantis* and in the longitudinal direction from 0.3 mm week^{-1} for *O. nigricarpum* to 20 mm week^{-1} for *L. tere-*

brantis. Radial penetration of *L. terebrantis* alone was significantly greater than penetration when this pathogen was coinoculated with *O. ips*.

A generally observed phenomenon in bark beetle-colonized trees is the blue coloration of the sapwood. This staining has been attributed to *Ophiostoma* introduced by the bark beetles during the colonization process. This sapwood discoloration is believed to be a result of chelation of iron by iron chelating agents called siderophores. Metabolites formed by *O. clavigerum*, *O. ips*, and *O. huntii* in still culture include 2,3-dihydroxybenzoic acid, a known iron-chelating agent (Ayer *et al*., 19.86). Ayer *et al*. (1986) were able to induce blue staining in *P. taeda* sapwood shavings using a methanol solution of 2,3-dihydroxybenzoic acid. In our tests of cell membrane impairment by metabolites produced by *Ophiostoma*, we also observed blue staining of *P. taeda* callus when treated with aqueous solution of the same compound (Blanche, Nebeker, and Hodges, unpublished data).

Ophiostoma minus in still cultures forms a different iron-chelating agent identified as ceratenolone (Ayer *et al*., 1987). Although 2,3-dihydroxybenzoic acid and ceratenolone are capable of forming ferric complexes, these were not detected in the blue-stained wood of diseased *P. taeda* (Ayer and Migaj, 1989). Thus, the iron chelates of these compounds may not be responsible for the blue coloration, although the iron chelates could have been lost in the extraction and partitioning process or, alternatively, other iron chelating agents may remain to be identified.

The role of staining in host colonization is not entirely clear but early investigations demonstrated that stained portions of the sapwood were generally drier than unstained portions, and that water conduction was impaired in stained portions (Nelson, 1934; Caird, 1935; Bramble and Holst, 1940; Mathre, 1964). Hence, the implication of the blue-staining fungi in the rapid death of the host has attracted considerable research attention. However, observations of pines successfully colonized by bark beetles without *Ophiostoma* (Hetrick, 1949; Whitney and Cobb, 1972; Bridges *et al*., 1985) challenged the importance of the fungi in host mortality. In fact, we have observed *P. taeda*, attacked by *D. frontalis*, that remained alive despite extensive blue staining of the sapwood (Blanche and Nebeker, unpublished data). Based on their studies on the relationship between *Ophiostoma* and bark beetle activity, Bridges *et al*. (1985) suggested that the absence of blue stain in *D. frontalis*-infested trees can be an indicator of *D. frontalis* outbreaks (see Chapter 7). In light of these observations, efforts to elucidate the roles of other bark beetle-associated microorganisms (non-staining) in the death of the host must be intensified to gain a full understanding of the cause of host mortality.

8.4 CONCLUSIONS

The preformed and induced systems have been assumed to be responsible for host resistance to successful beetle attack. The induced system may be particularly important to tree survival, by preventing growth or spread of the pathogenic fungi associated with the beetles. However, the importance of these systems in preventing death of the tree cannot be evaluated adequately until we know the exact role of the fungi in the death of the tree.

The preformed (constitutive) oleoresin system of pines is considered to be the primary defense against bark beetle attack. This assumption appears justified, but has not been demonstrated conclusively, largely because oleoresin flow in individual trees cannot be manipulated without seriously altering other processes in the tree. Also, the decrease in oleoresin flow after beetle attack has been assumed to result from moisture stress, but our recent work indicates other factors. We suggest that the rapid reduction in oleoresin flow following bark beetle attack results from mass wounding and a depletion of the oleoresin reservoir in the tree, not from moisture stress. We further suggest that the fungi, or compounds produced by the associated fungi, are not responsible for the initial decrease in oleoresin flow but do have an effect on tree water relations and are likely to be involved in the eventual death of the tree.

When bark beetles attack a tree, death of the tree occurs as a result of progressive events initiated by both the beetle and its associated fungi. Steps in the decline process involve suppression of initial tree resistance, inoculation of fungi in beetle galleries, production of fungal metabolites, alterations of host physiological functions by the metabolites, and subsequent death of the tree .

REFERENCES

Anderson, R.F. (1960). "Forest and Shade Tree Entomology" John Wiley and Sons, New York, 428 pp.

Anderson, N.H. and Anderson, D.B. (1968). *Ips* bark beetle attacks and brood development on a lightning-struck pine in relation to its physiological decline. *Florida Entomol.* **51(1)**,23-30.

Ayer, W.A. and Migaj, B.S. (1989). Acids from blue-stain diseased lodgepole pine. *Can. J. Bot.* **67**,1426-1428.

Ayer, W.A., Browne, L.M., Feng, M.C., Orszanska, H. and Saeedi-Ghomi, H. (1986). The chemistry of the blue-stain fungi. I. Some metabolites of *Ceratocystis* species associated with mountain pine beetle infected lodgepole pine. *Can. J. Chem.* **64**, 904-909.

Ayer, W.A., Attah-Poku, S.K., Browne, L.M. and Orszanska, H. (1987). The chemistry of the blue-stain fungi. III. Some metabolites of *Ceratocystis minor* (Hedgc.) Hunt. *Can. J. Chem.* **65**, 765-769.

Barras, S.J. and Hodges, J.D. (1969). Carbohydrates of inner bark of *Pinus taeda* as affected by *Dendroctonus frontalis* and associated microorganisms. *Can. Entomol.* **101**, 489-493.

Basham, H.G. (1970). Wilt of loblolly pine inoculated with blue-stain fungi of the genus *Ceratocystis*. *Phytopathology* **60**, 750-754.

Berryman, A.A. (1969). Response of *Abies qrandis* to attack by *Scolytus ventralis* (Coleoptera: Scolytidae). *Can. Entomol.* **101**, 1033-1041.

Berryman, A.A. (1972). Resistance of conifers to invasion by bark beetle–fungus association. *BioScience* **22**, 598-602.

Blanche, C.A., Hodges, J.D., Nebeker, T.E. and Moehring, D.M. (1983). Southern pine beetle: the host dimension. Mississippi Agric. For. Exp. Stn. Bull. 917, 29 pp.

Blanche, C.A., Hodges, J.D. and Nebeker, T.E. (1985a). Changes in bark beetle susceptibility indicators in a lightning-struck loblolly pine. *Can. J. For. Res.* **15**, 397-399.

Blanche, C.A., Nebeker, T.E., Hodges, J.D., Karr, B.L. and Schmitt, J.J. (1985b). Effect of thinning damage on bark beetle susceptibility indicators in loblolly pine. *In* "Proc. 3rd Biennial Southern Silvicultural Res. Conf." (E. Shoulders, ed). pp. 471-479. USDA Forest Serv. Gen. Tech. Rep. S0-54. USDA Forest Serv., Southern Forest Exp. Stn., New Orleans, LA.

Bordasch, R.P. and Berryman, A.A. (1977). Host resistance to the fir engraver beetle, *Scolytus ventralis* (Coleoptera: Scolytidae) 2. Repellency of *Abies qrandis* resins and some monoterpenes. *Can. Entomol.* **109**, 95-100.

Bramble, W.C. and Holst, E.C. (1940). Fungi associated with *Dendroctonus* in killing shortleaf pines and their effect on conduction. *Phytopathology* **30**, 881-899.

Bridges, J.R. (1987). Effects of terpenoid compounds on growth of symbiotic fungi associated with the southern pine beetle. *Phytopathology* **77**, 83-85.

Bridges, J.R., Netteleton, W.A. and Conner, M.D. (1985). Southern pine beetle (Coleoptera: Scolytidae) infestations without the blue-stain fungus, *Ceratocystis minus*. *J. Econ. Entomol.* **78**, 325-327.

Caird, R.W. (1935). Physiology of pines infected with bark beetles. *Bot. Gaz.* **96**, 709-733.

Callaham, R.Z. (1966). Nature of resistance of pines to bark beetles. *In* "Breeding Pest-resistant Trees"

(H.D. Gerhold, R.E. McDermott, E.J. Schreiner and J.A. Winieski, eds), pp. 197-206. Pergamon Press, Oxford.

Cates, R.G. and Alexander, H. (1982). Host resistance and susceptibility. *In*: "Bark Beetles in North American Conifers: Ecology and Evolution." (J.B. Mitton and K.B. Sturgeon, eds), pp. 212-263. University of Texas Press, Austin, TX.

Christiansen, E. and Horntvedt, R. (1983). Combined *Ips/Ceratocystis* attack on Norway spruce, and defensive mechanisms of the trees. *Z. Ang. Entomol.* **96**, 110-118.

Christiansen, E., Waring, R.H. and Berryman, A.A. (1987). Resistance of conifers to bark beetle attack: searching for general relationships. *For. Ecol. Manage.* **22**, 89-106.

Cook, S.P. and Hain, F.P. (1987). Four parameters of the wound response of loblolly and slash pines to inoculation with the blue-staining fungus associated with the southern pine beetle. *Can. J. Bot.* **65**, 2403–2409.

Cook, S.P., Hain, F.P. and Nappen, P.B. (1986). Seasonality of the hypersensitive response by loblolly and shortleaf pine to inoculation with fungal associates of the southern pine beetle (Coleoptera: Scolytidae). *J. Entomol. Sci.* **21(3)**, 283-285.

Coulson, R.N. and Witter, J.A. (1984). "Forest Entomology: Ecology and Management?" John Wiley & Sons, Inc., New York. 669 pp.

Coulson, R.N., Hennier, P.B., Flamm, R.O., Rykiel, E.J., Hu, L.C. and Payne, T.L. (1983). The role of lightning in the epidemiology of the southern pine beetle. *Z. Ang. Entomol.* **96**, 182-193.

Coulson, R.N., Saunders, M.C., Payne, T.L., Flamm, R.O., Wagner, T.L., Hennier, P.B. and Rykiel, E.J. (1985). A conceptual model of the role of lightning in the epidemiology of the southern pine beetle. *In* "The Role of the Host in Population Dynamics of Forest Insects" (L. Safranyik, ed.), pp. 136-146. Canadian Forestry Serv., Pacific Forestry Centre, Victoria, BC.

Coyne, J.F. and Lott, L.H. (1976). Toxicity of substances in pine oleoresin to southern pine beetles. *J. Ga. Entomol. Soc.* **11(4)**, 297-301.

DeAngelis, J.D., Nebeker, T.E. and Hodges, J.D. (1986). Influence of tree age and growth rate on the radial resin duct system in loblolly pine (*Pinus taeda*). *Can. J. Bot.* **64**, 1046-1049.

Elkinton, J.S. and Wood, D.L. (1980). Feeding and boring behavior of the bark beetle *Ips paraconfusus* (Coleoptera: Scolytidae) on the bark of a host and non-host tree species. *Can. Entomol.* **112**, 797-809.

Fares, Y., Goeschl, J.D. and Sharpe, P.J.H. (1980a). Dynamics of bark beetle–fungus symbiosis: I. pine tree anatomy and fungus growth pattern. *In* "Modeling Southern Pine Beetle Populations" (F.M. Stephen, J.L. Searcy and G.D. Hertel, eds), pp. 54-60. USDA Forest Serv. Tech. Bull. No. 1630. USDA Forest Serv., Washington, DC.

Fares, Y., Magnuson, C.E., Doraiswamy, P.C. and Sharpe, P.J.H. (1980b). Dynamics of bark beetle–fungus symbiosis: II. pine tree drying model. *In* "Modeling Southern Pine Beetle Populations" (F.M. Stephen, J.L. Searcy and G.D. Hertel, eds), pp. 61-74. USDA Forest Serv. Tech. Bull. No. 1630. USDA Forest Serv., Washington, DC.

Gambliel, H.A., Cates, R.G., Caffey-Maquin, M.K. and Paine, T.D. (1985). Variation in the chemistry of loblolly pine in relation to infection by the blue-stain fungus. *In* "Integrated Pest Management Research Symposium: The Proceedings" (S.J. Branham and R.C. Thatcher, eds), pp. 177-184. USDA Forest Serv. Gen. Tech. Rep. SO-56. USDA Forest Serv., Southern Forest Exp. Stn., New Orleans, LA.

Goldhammer, D.S., Stephen, F.M. and Paine, T.D. (1989). Average radial growth and chlamyhdospore production of *Ceratocvstis minus*, *Ceratocystis minus* var. *barrasii* and SJB 122 in culture. *Can. J. Bot.* **67**, 3498–3505.

Hadwiger, L. and Schwochan, M. (1969). Host resistance responses — an induction hypothesis. *Phytopathology* **7**, 13-22.

Hain, F.P., Mawby, W.D., Cook, S.P. and Arthur, F.H. (1983). Host conifer reaction to stem invasion. *Z. Ang. Entomol.* **96**, 247-256.

Hemingway, R.W., McGraw, G.W. and Barras, S. (1977). Polyphenols in *Ceratocystis minus* infected *Pinus taeda*: fungal metabolites, phloem and xylem phenols. *Agric. Food Chem.* **25**, 717-722.

Hennington, B. and Lundstrom, H. (1974). "The Growth of Insect-borne Blue-stain, and Effects caused by Immersion of the Wood; Some Laboratory Tests" Rapporter, Institutionen for Virkeslara No. R 92, 20 pp.

Hetrick, L.A. (1949). Some overlooked relationships of southern pine beetle. *J. Econ. Entomol.* **42**, 466-469.

Hodges, J.D. and Pickard, L.S. (1971). Lightning in the ecology of the southern pine beetle,

Dendroctonus frontalis (Coleoptera: Scolytidae). *Can. Entomol.* **103**, 44-51.

Hodges, J.D., Elam, W.W. and Watson, W.F. (1977). Physical properties of the oleoresin system of the four major southern pines. *Can. J. For. Res.* **7**, 520-525.

Hodges, J.D., Elam, W.W., Watson, W.F. and Nebeker, T.E. (1979). Oleoresin characteristics and susceptibility of four southern pines to southern pine beetle (Coleoptera: Scolytidae) attacks. *Can. Entomol.* **111**, 889-896.

Hodges, J.D., Nebeker, T.E., DeAngelis, J.D., Karr, B.L. and Blanche, C.A. (1985). Host resistance and mortality: a hypothesis based on the southern pine beetle–microorganism–host interactions. *Bull. Entomol. Soc. Am.* **31(1)**, 31-35.

Hodges, J.D., Nebeker, T.E., Blanche, C.A., Honea, C.R., Fisher, T.H. and Schultz, T.P. (1989). Southern pine beetle–microorganisms–host interactions: influence of compounds produced by *Ceratocystis minor*. *In* "Proceedings 5th Silvicultural Research Conference" (J.H. Miller, ed.). pp. 567-572. USDA Forest Serv. Gen. Tech. Rep. SO-74. USDA Forest Serv., Southern Forest Exp. Stn., New Orleans, LA.

Howe, V.K., Oberle, A.D., Keeth, T.G. and Gordon, W.J. (1971). The role of microorganisms in the attractiveness of lightning-struck pines to southern pine beetles. Western Illinois University Bull. Vol. L, No. 3. (Series in Biological Sciences 9), 44 pp.

Johnson, P.C. (1966). Attractiveness of lightning-struck ponderosa pine trees to *Dendroctonus brevicomis* (Coleoptera: Scolytidae). *Ann. Entomol. Soc. Am.* **59**, 615.

Kosuge, T. (1969). The role of phenolics in host response to infection. *Annu. Rev. Phytopathol.* **7**, 195-222.

Lagerberg, T., Lundberg, G. and Melin, E. (1927). Biological and practical researches into blueing in pine and spruce. Sveska Skogsvardsfor. *Tidsk Haft* **II**, 145-272.

Lorio, P.L., Jr., Sommers, R.A., Blanche, C.A., Hodges, J.D. and Nebeker, T.E. (1990). Modeling pine resistance to bark beetles based on growth and differentiation balance principles. *In* "Forest Growth: Process Modeling of Responses to Environmental Stress" (R.K. Dixon, R.S. Meldahl, G.A. Ruark, and W.G. Warren, eds), pp. 402-409. Timber Press, Portland, OR.

Mathre, D.E. (1964). Effect of *Ceratocystis ips* and *C. minus* on the free sugar pool in ponderosa pine sapwood. *Contrib. Boyce Thompson Inst.* **22**, 509-512.

Mattson, W.J., and Haack, R.A. (1987). The role of drought in outbreaks of plant-eating insects. *BioScience* **37**, 110-118.

McGraw, G.W. and Hemingway, R.W. (1977). 6,8-Dihydroxy-3hydroxymethyl isocoumarin, and other phenolic metabolites of *Ceratocystis minus*. *Phytochemistry* **16**, 1315-1316.

Miller, J.M. and Keen, F.P. (1960). Biology and Control of the Western Pine Beetle, USDA Forest Serv. Misc. Publ. 800, USDA Forest Serv., Washington, DC, 381 pp.

Miller, R.H. and Berryman, A.A. (1985). Energetics of conifer defense against bark beetles and associated fungi. *In* "The Role of the Host in the Population Dynamics of Forest Insects." (L. Safranyik ed.), pp. 13–23. Canadian Forestry Serv., Pacific Forestry Centre, Victoria, BC.

Nebeker, T.E. and Hodges, J.D. (1983). Influence of forestry practices on host-susceptibility to bark beetles. *Z. Ang. Entomol.* **96**, 194-208.

Nebeker, T.E., Hodges, J.D., Honea, C.R. and Blanche, C.A. (1988). Preformed defensive system in loblolly pine: variability and impact on management practices. *In* "Integrated Control of Scolytid Bark Beetles" (T.L. Payne and H. Saarenmaa, eds), pp. 147-162. Virginia Polytechnic Institute Press, Blacksburg, VA.

Nelson, R.M. (1934). Effect of blue-stain fungi on southern pine attacked by bark beetles. *Phytopathology* **7**, 327–353.

Owen, D.R., Lindahl, K.Q., Jr., Wood, D.L. and Parmeter, J.R., Jr. (1987). Pathogenicity of fungi isolated from *Dendroctonus valens, D. brevicomis*, and *D. ponderosae* to ponderosa pine seedlings. *Phytopathology* **77**, 631-636.

Paine, T.D., Stephen, F.M. and Cates, R.G. (1985). Induced defenses against *Dendroctonus frontalis* and associated fungi: variation in loblolly pine resistance. *In* "Integrated Pest Management Research Symposium: the Proceedings" (Branham, S.J. and R.C. Thatcher, eds), pp. 169-176. USDA Forest Service Gen. Tech. Rep. SO-56. USDA Forest Serv. Southern Forest Exp. Stn., New Orleans, LA.

Paine, T.D. (1984). Seasonal response of ponderosa pine to inoculation of the mycangial fungi from the western pine beetle. *Can. J. Bot.* **62**, 551-555.

Paine, T. D. and Stephen, F. M. (1987a). Fungi associated with southern pine beetle: avoidance of induced defense response in loblolly pine. *Oecologia* **74**, 377-379.

Paine, T.D. and Stephen, F.M. (1987b). The relationship of tree height and crown class to the induced plant defenses of loblolly pine. *Can. J. Bot.* **65**, 2090-2092.

Paine, T.D., Blanche, C.A., Nebeker, T.E. and Stephen, F.M. (1987). Composition of loblolly pine resin defenses: comparison of monoterpenes from induced lesion and sapwood resin. *Can. J. For. Res.* **17**, 1202-1206.

Parmeter, J.R., Jr., Slaughter, G.W., Chen, M.-M., Wood, D.L. and Stubbs, H.A. (1989). Single and mixed inoculations of ponderosa pine with fungal associates of *Dendroctonus* spp. *Phytopathology* **79**, 768-772.

Parmeter, J.R., Jr., Slaughter, G.W., Chen, M.-M. and Wood, D.L. (1992). Rate and depth of sapwood occlusion following inoculation of pines with bluestain fungi. *Forest Sci.* **38**, 34-44.

Popp, M.P., Johnson, J.D. and Massey, T.L. (1991). Stimulation of resin flow in slash and loblolly pine by bark beetle vectored fungi. *Can. J. For. Res.* **21**, 1124–1126.

Raffa, K.F. and Berryman, A.A. (1982a). Gustatory cues in the orientation of *Dendroctonus ponderosae* (Coleoptera: Scolytidae) to host trees. *Can. Entomol.* **114**, 97-104.

Raffa, K.F. and Berryman, A.A. (1982b). Physiological differences between lodgepole pines resistant and suceptible to the mountain pine beetle and associated microorganisms. *Environ. Entomol.* **11**, 486-492.

Raffa, K.F. and Berryman, A.A. (1983). Physiological aspects of lodgepole pine wound responses to a fungal symbiont of the mountain pine beetle, *Dendroctonus ponderosae* (Coleoptera: Scolytidae). *Can. Entomol.* **115**, 724-734.

Raffa, K.F. and Smalley, E.B. (1988). Seasonal and long-term responses of host trees to microbial associates of the pine engraver, *Ips pini*. *Can. J. For. Res.* **18**, 1624–1634.

Reid, R.W., Whitney, H.S. and Watson, J.A. (1967). Reactions of lodgepole pine to attack by *Dendroctonus ponderosae* Hopkins and blue-stain fungi. *Can. J. Bot.* **45**, 1115–1126.

Richmond, J.A., Mills, C. and Clark, E.W. (1970). Chemical changes in loblolly pine, *Pinus taeda* L., inner bark caused by blue-stain fungus, *Ceratocystis minus* (Hedg.) Hunt. *J. Elisha Mitchell Sci. Soc.* **86**, 171

Rudinsky, J.A. (1966). Host selection and invasion by the Douglas-fir beetle, *Dendroctonus pseudotsugae* Hopkins, in coastal Douglas-fir forests. *Can. Entomol.* **98**, 98–lll.

Russell, C.E. and Berryman, A.A. (1976). Host resistance to the fir engraver beetle. I. Monterpene composition of *Abies grandis* pitch blisters and fungus-infected wounds. *Can. J. Bot.* **54**, 14-18.

Ryan, C.A. (1979). Proteinase inhibitors. *In* "Herbivores: Their Interaction With Secondary Plant Metabolites" (G. Rosenthal and D. Janzen, eds), pp. 599-618. Academic Press, New York.

Safranyik, L. (1988). The population biology of the spruce beetle in western Canada and implications for management. *In* "Integrated Control of Scolytid Bark Beetles" (T.L. Payne and H. Saarenmaa, eds), pp. 3-23. Virginia Polytechnic Institute Press, Blacksburg, VA.

Shain, L. (1967). Resistance of sapwood in stems of loblolly pine to infection by *Fomes annosus*. *Phytopathology* **57**, 1034-1045.

Shain, L. and Hillis, W.E. (1971). Phenolic extractives in Norway spruce and their effects on *Fomes annosus*. *Phytopathology* **61**, 841-845.

Shepherd, R.F. and Watson, J.A. (1959). Blue-stain fungi associated with the mountain pine beetle. Canadian Dept. Agric. For. Biol. Div. Prog. Rep. 15(3), pp. 2-3.

Shrimpton, D.M. (1973). Extractives associated with the wound response of lodgepole pine attacked by the mountain pine beetle and associated microorganisms. *Can. J. Bot.* **51**, 527-534.

Shrimpton, D.M. (1978). Resistance of lodgepole pine to mountain pine beetle infestations. *In* "Theory and Practice of Mountain Pine Beetle Management in Lodgepole Pine Forests — a Symposium." (A.A. Berryman, G.D. Amman, R.W. Stark and D.L. Kibbee, eds), pp. 6476. College of Forest Resources, Univ. of Idaho, Moscow.

Smith, R. H. (1966). Resin quality as a factor in the resistance of pines to bark beetles. *In* "Breeding Pest-resistant Trees" (Gerhold, H.D., R.E. McDermott, E.J. Schreiner and J.A. Winieski, eds), pp. 189-196. Pergamon Press, Oxford.

Smith, R.H. (1975). Formula for describing effect of insect and host tree factors on resistance to western pine beetle attack. *J. Econ. Entomol.* **68**, 841-844.

Stephen, F.M. and Paine, T.D. (1985). Seasonal patterns of host tree resistance to fungal associates of the southern pine beetle. *Z. Ang. Entomol.* **99**, 113-122.

Taylor, A.R. (1974). Ecological aspects of lightning in forests. Ann. Proc. Tall Timbers Fire Ecol. Conf. No. 13. Tallahassee, FL, pp. 455-496.

Thomas, H.A., Richmond, J.A. and Bradley, E.L. (1981). Bioassay of pine bark extracts as biting stimulants for the southern pine beetle. USDA Forest Serv. Res. Note SE-302. USDA Forest Serv. Southeastern Forest Exp. Stn. Asheville, NC, 5 pp.

Vité, J.P. (1961). The influence of water supply on oleoresin exudation pressure and resistance to bark beetle attack in *Pinus ponderosa*. *Contrib. Boyce Thompson Inst.* **21**, 37-66.

White, J.D. (1981). A bioassay for tunneling responses of southern pine beetle to host extractives. *J. Ga. Entomol. Soc.* **16(4)**, 484-492.

Whitney, H.S. and Cobb, F.W., Jr. (1972). Non-staining fungi associated with the bark beetle *Dendroctonus brevicomis* (Coleoptera: Scolytidae) on *Pinus ponderosa*. *Can. J. Bot.* **50**, 1943-1945.

Wood, D.L. (1972). Selection and colonization of ponderosa pine by bark beetles. *In* "Insect/Plant Relationships" (H.F. van Emden, ed.), pp. 101-117. Blackwell Scientific Publications, Oxford.

Wood, D.L. (1982). The role of pheromones, kairomones, and allomones in the host selection and colonization behavior of bark-beetles. *Annu. Rev. Entomol.* **27**, 411–446.

–9–

Effects of Pathogens and Bark Beetles on Forests

D. J. GOHEEN[1] and E. M. HANSEN[2]
[1]*USDA Forest Service, Pacific Northwest Region, Portland, OR, USA*
[2]*Department of Botany and Plant Pathology, Oregon State University, Corvallis, OR, USA*

9.1 INTRODUCTION

Pathogenic fungi and bark beetles are important components of most coniferous forest ecosystems. Their relationships to each other and to the trees that dominate these plant communities have been moderated by millions of years of evolution into a dynamic equilibrium. Both pathogenic fungi and bark beetles have successful life history strategies for exploiting scattered stressed hosts and, under natural conditions, neither threatens the long-term productivity of the forests on which they depend.

Pathogenic fungi and herbivorous and saprophagous insects play important roles as consumers and decomposers in the energy flow and nutrient cycles of the forest. Humans are also important components of most forest ecosystems, but in evolutionary terms, they have arrived on the scene very recently. Changes in the forest environment resulting from human activity have, in many cases, increased populations and activity of root disease fungi and bark beetles. Humans also brought the concept of economic value to the forest. From this perspective, root pathogens and bark beetles are among the most destructive of the many threats to forest productivity.

This chapter addresses the varied roles that root pathogens and bark beetles play in western coniferous forests as (1) regulators of ecological structure and processes, (2) arbiters of management success and (3) agents of significant economic loss. Pathologists, entomologists, and forest managers often speak of the "impact" of fungal and insect "pests" on forest values. This terminology carries connotations of death and destruction that reflect only part of the role that these organisms play in the forest. The death of a tree may represent the loss of many cubic meters of timber, but at the same time, may improve soil fertility and grow-

BEETLE–PATHOGEN INTERACTIONS IN CONIFER FORESTS
ISBN 0-12-628970-0

ing conditions for surrounding trees and increase the non-economic diversity value of the forest by promoting non-host vegetation and creating new habitat for cavity-nesting birds and mammals. In most situations, it is impossible to compute a net "impact" because of the mixture of economic and non-economic values and ecological processes involved. We have chosen to avoid this problem by referring to the "effects" of root pathogens and bark beetles on various values and processes. Our goal, in part, is to provide a compilation of both economic and ecological effects of these organisms.

9.2 EFFECTS OF BARK BEETLE–PATHOGEN INTERACTIONS

The ultimate effect of pathogens and bark beetles is tree mortality. The effect of tree death on forests depends on the management context in which the effect is judged (Leuschner and Berck, 1985). The economic and ecological consequences of tree mortality are often quite different in forests where tree harvest is a significant objective compared to forests managed for other values. Both management scenarios are important in North America today, and a full accounting of the effects of pathogens and insects on forests must consider both.

The root disease fungi and bark beetles in North American forests are diverse groups both taxonomically and behaviorially (see Chapter 2, 3 and 6). It should not be surprising that their effects on forests are at least as varied. The forests themselves are dramatically different, ranging from vast natural and planted monocultures to the most diverse assemblages of conifers in the world. The consequences of infection by the root pathogen *Heterobasidion annosum*, for example, are very different in coastal *Tsuga heterophylla* forests from those in interior *Abies* stands. The result of infection in *Tsuga* forests is butt rot with only a slightly increased chance of tree breakage, while in *Abies* forests, trees often are killed in gradually expanding infection centers. Insects appear to play no significant role when *T. heterophylla* is infected, but *Scolytus ventralis* regularly attacks and hastens the death of *H. annosum*-infected *Abies*. In recognition of these limitations, we have focused our review on specific organisms acting in specific ecological situations.

Five major and widespread interactions are developed in detail: (1) *Phellinus weirii* and *Dendroctonus pseudotsugae* on *Pseudotsuga menziesii* west of the Cascade Mountains in the Pacific Northwest, (2) *Leptographium wageneri* var. *pseudotsugae* and associated beetles on *P. menziesii*, (3) *L. wageneri* var. *ponderosum* and associated beetles on *Pinus ponderosa*, (4) several species of root pathogens and *Dendroctonus ponderosae* on *P. contorta*, and (5) several species of root pathogens and *Scolytus ventralis* on *A. concolor* and *A. grandis*. Interactions among *Dendroctonus frontalis* and root pathogens in *Pinus* forests in the southern US and interactions among several root-feeding beetles and root pathogens in *P. resinosa* forests in the north central states also will be described.

We will explore the nature and extent of the relationships between these root pathogens and their bark beetle associates, summarize available information on the economic losses to timber production that accrue as a result of their activity, and summarize their effects on forest community structure, dynamics, and ecosystem processes.

9.2.1 Interactions in *Pseudotsuga* forests: *Phellinus* and *Dendroctonus*

9.2.1.1 Nature of relationship

In western Oregon and Washington, USA, and British Columbia, Canada, endemic populations of *D. pseudotsugae* primarily infest scattered windthrown, injured, and diseased

Pseudotsuga menziesii (Furniss and Carolin, 1977). *Phellinus weirii* is a major cause of root disease in Pacific Northwest forests, and *D. pseudotsugae* is especially likely to be found attacking trees infected by this fungus (Wright and Lauterback, 1958; Hadfield, 1985; Hadfield *et al.*, 1986). Rudinsky (1966) demonstrated that *D. pseudotsugae* is able to detect differences between volatile oleoresin terpenes in healthy and physiologically stressed *Pseudotsuga menziesii* and is thus able to locate low-vigor hosts over substantial distances (see Chapter 6). *Dendroctonus pseudotsugae* outbreaks do develop periodically, and numerous healthy trees are infested. Usually, such outbreaks follow major windthrow events (Furniss and Carolin, 1977).

Beetle-brood production is optimal in previously healthy, windthrown *P. menziesii*, especially if the fallen material is shaded. The large populations of beetles emerging from substantial numbers of windthrown *P. menziesii* may attack and kill nearby standing trees (Wright and Lauterbach, 1958; Furniss and Carolin, 1977). Outbreaks collapse rapidly, because brood success is poor in such vigorous hosts. Large populations cannot be sustained unless there are additional fallen trees. *Phellinus weirii* plays an important role in the dynamics of *D. pseudotsugae* by providing stressed hosts that maintain low-level beetle populations between disturbance events that can trigger outbreaks.

Dendroctonus pseudotsugae may enhance long-term survival of *P. weirii* by attacking and killing some declining *Pseudotsuga menziesii* in root disease centers before the trees are windthrown. Beetle-killed trees are less prone to be windthrown than live trees because they lack foliage to catch the wind. The fungus survives in dead roots that remain in the soil. When trees are windthrown and infected roots are pulled out of the ground, most, if not all, inoculum is prevented from spreading.

The fungus survives for up to 50 years in infected stumps and snags, and infects susceptible trees that are regenerated on the site as growing roots contact infected material (Buckland and Wallis, 1956; Hansen, 1979; Tkacz and Hansen, 1982). The pathogen spreads little, if at all, by windborne spores or any means other than mycelial growth on or within roots. It usually takes 12–15 years for contacts to develop between roots of stumps and adjacent saplings and for the fungus to grow along the roots and up to the root collars of the young trees. Significant amounts of mortality in the new stand generally appear at this age, initially involving scattered trees adjacent to the old stumps. Disease centers develop as the pathogen subsequently grows from tree to tree across roots at a rate of about 30 cm $year^{-1}$ (Nelson and Hartman, 1975; McCauley and Cook, 1980).

9.2.1.2 Extent of relationship

Though many investigators have noted the common association between *Phellinus weirii* and *D. pseudotsugae*, few have attempted to quantify the degree of relationship. Goheen and Schmitt (unpublished data) surveyed on the ground a randomly selected sample of aerially detected, endemic *D. pseudotsugae* infestations in western Oregon (e.g. Fig. 9.1). They found that 77% of the areas where trees had been infested by bark beetles were infected with *P. weirii*. The remaining areas with infested trees were associated with undiseased, windthrown trees on steep slopes. Goheen *et al.* (unpublished data) found that over 90% of dead *P. weirii*-infected *Pseudotsuga menziesii* 20 cm or greater diameter at breast height (DBH) exhibited evidence of *D. pseudotsugae* galleries in a 7290 hectare area surveyed in northwest Oregon. Both standing dead and windthrown diseased *P. menziesii* exhibited evidence of infestation. However, beetle larval gallery development appeared to be more extensive on the windthrown trees. *Pseudotsuga menziesii* smaller than 20 cm DBH frequently

Fig.9.1. *Phellinus weirii* infection center in a mature *Pseudotsuga menziesii* stand in the Oregon Coast Range Mountains. *P. menziesii* have been killed progressively by the fungus and associated *Dendroctonus pseudotsugae*, a large opening has developed in the formerly closed stand, and the open area has been colonized by *Alnus rubra*, a hardwood species immune to the pathogen.

showed evidence of infestation by other bark beetles, primarily *Scolytus unispinosis* and *Pseudohylesinus nebulosus*.

9.2.1.3 Timber losses

Phellinus weirii causes annual losses estimated at 900 000 m^3 of wood in Oregon and Washington forests (Childs and Shea, 1967) and 1 050 000 m^3 in British Columbia (Wallis, 1967). Much of this loss is the result of *Pseudotsuga menziesii* mortality that also involves D. *pseudotsugae* infestation. Timber loss might better be attributed to the root disease–bark beetle association (Fig. 9.1). These estimates do not result from actual surveys but rather are based on projections that at least 5% of the area occupied by the *P. menziesii* forest type is affected, and losses within infected areas approach 50% over a rotation. No regional root disease surveys have been done, but the few local surveys suggest that perhaps even larger proportions of this forest type may be affected (Table 9.1). Within *Phellinus weirii* centers, reduction in *Pseudotsuga menziesii* volume has been measured at 10% (Bloomburg and Reynolds, 1985) to 55% (Goheen *et al.*, unpublished data).

Phellinus weirii also increases the chances of windthrow during storms. Root-rotted trees obviously are prone to windthrow, but healthy trees exposed to high winds at the edge of root rot openings also are more vulnerable. Windthrow resulting from winter storms, in February 1990, on Mary's Peak in the Oregon Coast Range was almost exclusively related to root rot or to recent clear-cut margins, even in thinned stands (Hansen, unpublished data). Losses attributed to *D. pseudotsugae*, alone, during outbreaks have not been well documented but can be considerable. In outbreaks associated with windthrow, one green tree commonly is infested for every four windthrown trees (D. Bridgewater, USDA Forest Serv., unpublished data). Wright and Lauterbach (1958) reported that 2.8 billion cubic meters of timber were windthrown in major storms during the winters of 1949–1950 and 1951 in

Table 9.1. Local area surveys for incidence of *Phellinus weirii* in *Pseudotsuga menziesii* forests in western Oregon, USA

Investigators	Total area surveys	% of area affected
Goheen (1979)	Mapleton Ranger District (50 000 ha)	4.7
Lawson et al. (1983)	Black Rock Forest (195 ha)	4.9
Kanaskie (1985)	Columbia and Clatsop Co. (40 000 ha)	5.3
Goheen and Goheen (unpublished)	Alsea Ranger District (38 000 ha)	16.0
Goheen et al. (unpublished)	Scappoose Block (7290 ha)	10.9
Kastner and Kral (unpublished)	Tillamook Resource Area (30 000 ha)	7.4
Hansen (1978)	Sweet Home Ranger District (3840 ha)	11.0

western Oregon and Washington. By 1953, when populations collapsed, *D. pseudotsugae* had killed an additional 0.8 billion cubic meters of standing *Pseudotsuga menziesii* in the vicinity of the blowdown.

9.2.1.4 Ecological effects

Most of the forest area west of the Cascade Mountains in Oregon, Washington, and British Columbia is dominated by the coastal variety of Douglas-fir (*P. menziesii* var. *menziesii*). This tree develops in relatively pure stands following major disturbances such as fires, wind storms, bark beetle outbreaks, and clearcut harvest. *Pseudotsuga menziesii* grows rapidly, and by about age 30 often forms a closed canopy. It continues to dominate the site until the next major disturbance, sometimes for 500 years or more. On all but the driest sites, however, *T. heterophylla* is considered the climax species. This is the "*Tsuga heterophylla* zone" of Franklin and Dyrness (1969). While *T. heterophylla* can establish itself and grow in the shade beneath a *P. menziesii* canopy, it will not assume dominance until the canopy is opened, through death of *P. menziesii*. *Phellinus weirii* and associated bark beetles are the principal agents of mortality in the *Pseudotsuga menziesii* forest, allowing establishment of the *T. heterophylla* climax (Holah, 1991).

In typical *P. menziesii* stands, *Phellinus weirii* and endemic populations of *D. pseudotsugae* act together, preferentially killing *Pseudotsuga menziesii* and creating gradually expanding openings in the canopy where non-host tree species and shrubs are favored. Only conifers are susceptible to *Phellinus weirii*, and *Pseudotsuga menziesii* is the most susceptible of those commonly encountered in the *Tsuga heterophylla* zone. *Tsuga heterophylla*, *Pinus* spp., and *Thuja plicata* may be infected but are seldom killed. The fungus is usually confined as a butt rot in these species. *Dendroctonus pseudotsugae* attacks only *Pseudotsuga menziesii*.

Succession is either advanced or reset to more seral stages by *Phellinus weirii*, depending on the proximity of the infection center to seed sources or vegetative propagules of climax or seral species. *Tsuga heterophylla*, *Picea sitchensis*, *Thuja plicata*, *Calocedrus decurrens*, *Taxus brevifolia* and *Pinus monticola*, in addition to such hardwoods as *Alnus rubra*, *Acer macrophyllum*, and *A. circinatus* commonly proliferate in disease centers (Fig. 9.1). The effects of the *Phellinus weirii*/*D. pseudotsugae* complex on community composition and structure change with stand age.

Plant community development after catastrophic disturbance such as wildfire or clearcut harvesting is affected by *P. weirii* almost from the beginning. Woody shrubs and hardwood trees, established in root disease openings in the previous stand, sprout back and may quickly dominate the former infection center. The rodent *Aplodontia rufa* also may be especially abundant in old infection centers after disturbance. In mature stands, their colonies are largely confined to areas with herbaceous vegetation, such as root disease centers. The combination of early shrub competition and animal damage may delay the establishment of conifers on the site. Susceptible conifers growing through the seral vegetation are increasingly likely to contact old *P. weirii*-infected roots.

Understory vegetation generally is shaded out as the conifer canopy closes. Root disease centers, in contrast, represent gradually expanding islands of light in the dark forest. Mining of dead *Pseudotsuga menziesii* by *D. pseudotsugae* accelerates decomposition and nutrient turnover from wood (Edmonds and Eglitis, 1989; Schowalter *et al.*, 1992). Nitrogen availability is increased as killed trees decompose (Waring *et al.*, 1987). Temperature and humidity fluctuate more widely. Herbaceous plant diversity is higher than in the closed-canopy forest. There are corresponding differences in populations of soil microbes (Hutchins and Rose, 1984) and herbivorous animals. The regular accretion of snags and fallen trees provides habitat for a different assemblage of bird, small mammal, and amphibian species.

In most plant associations of the *Tsuga heterophylla* zone, *Pseudotsuga menziesii* cannot regenerate successfully in brush-filled root disease centers, but the disease-tolerant *T. heterophylla* and *Thuja plicata* can. As a consequence, infection centers may assume quite different compositions as old-growth stands and as young forests. Vegetational response to root pathogen infection has been examined in an old-growth forest of the central Oregon Cascades (Holah, 1991). In these stands, *P. menziesii* killed by *Phellinus weirii* and *D. pseudotsugae* is replaced by *Tsuga heterophylla* as the pathogen slowly spreads through the forest. The resulting *T. heterophylla* climax, in turn, forms a denser canopy than the *Pseudotsuga* canopy, with the result that there may be less understory vegetation in the old infection centers than outside the areas of infection.

On particularly dry sites (*Pseudotsuga/Holodiscus* association; Franklin and Dyrness, 1969), stands are more open, and *P. menziesii* is the climax species. Accumulating evidence suggests that *Phellinus weirii* is both more common and more extensive on these sites where *Pseudotsuga menziesii* regenerates successfully in its own shade (Bloomberg and Beale, 1985; W. Kastner, unpublished data).

9.2.2 Interactions in *Pseudotsuga* forests: *Leptographium* and bark beetles

9.2.2.1 Nature of relationship

Leptographium wageneri var. *pseudotsugae* affects *P. menziesii* in California, Oregon, Washington, Idaho, Montana, and British Columbia, but it is most widely distributed and severe in 10- to 30-year-old plantations and young natural stands on disturbed sites in northwest California and southwest Oregon (Fig. 9.2). Two species of curculionids, *Pissodes fasciatus* and *Steremnius carinatus*, and one scolytid, *Hylastes nigrinus*, have been implicated as vectors of *L. wageneri* var. *pseudotsugae* (Witcosky and Hansen, 1985; Harrington *et al.*, 1985; Witcosky *et al.*, 1986a,b). *Pissodes fasciatus*, *S. carinatus*, and *H. nigrinus* prefer or are most successful in roots of injured, stressed, or dying *P. menziesii* (Blackman, 1941; Chamberlain, 1958; Zethner-Moller and Rudinsky, 1967; Furniss and Carolin, 1977), and *H. nigrinus* has been shown to be attracted to low molecular weight volatiles produced by stressed hosts (Rudinsky and Zethner-Moller, 1967), including *L. wageneri*-infected roots (Witcosky *et al.*, 1987).

Unlike most root pathogens, *L. wageneri* does not cause a root decay but rather a vascular wilt-type disease. The fungus grows in the xylem tracheids and physically blocks water conduction by the host. Hosts decline rapidly and are infested commonly by stem-attacking

Fig. 9.2. Tree decline and mortality caused by *Leptographium wageneri* var. *pseudotsugae* in a 20-year-old *Pseudotsuga menziesii* plantation in western Oregon. Fifty-nine per cent of the trees in this plantation were infected at this time. The fungus probably was vectored into the plantation by the root-feeding bark beetles, *Hylastes nigrinus*.

bark beetles in addition to the root-feeding insects. Large trees (20 cm DBH or greater) may be attacked by *D. pseudotsugae*, while smaller trees usually are infested by *Scolytus unispinosis*, *Pseudohylesinus nebulosis*, *Dryocoetes autographus*, buprestids, and cerambicids (Goheen and Hansen, 1978).

9.2.2.2 Extent of relationship

Pissodes fasciatus, *Steremnius carinatus*, and *H. nigrinus* commonly breed in root systems of *L. wageneri*-infected *Pseudotsuga menziesii*, and several generations of beetles may develop before diseased trees die (Harrington *et al.*, 1985; Witcosky and Hansen, 1985). *Leptographium wageneri* sporulates in beetle galleries. Conidiospores are borne in sticky slime droplets and adhere to the bodies of beetles as they emerge. Adult beetles disperse and locate new hosts in spring. *Hylastes nigrinus* and *Pissoides fasciatus* can fly considerable distances. *Steremnius carinatus* is flightless but can walk up to 100 m in 7 weeks, based on mark–recapture data (J. Witcosky, unpublished data). All three species burrow through soil to roots and engage in maturation feeding, sometimes visiting and wounding roots of several trees before constructing galleries. *Leptographium wageneri* requires wounds penetrating to the xylem for infection to occur. Witcosky *et al.* (1986b) isolated *L. wageneri* from 2.3% of *H. nigrinus*, 0.5% of *S. carinatus*, and 0.5% of *P. fasciatus* in a sample of beetles emerging from diseased *Pseudotsuga menziesii*. The vector relationship was confirmed by Harrington *et al.* (1985) and Witcosky *et al.* (1986b) when it was demonstrated that both artificially and naturally infested beetles introduced *L. wageneri* into living *Pseudotsuga menziesii* seedlings. Successful infection was associated with 1–5% of seedlings caged with naturally infested beetles, suggesting that the vectors are highly efficient at inoculating this fungus. This is borne out by the large numbers and widespread distribution of new *L. wageneri* centers observed in most years in areas where site and stand conditions are favorable for the disease.

Occurrence of *L. wageneri* in *P. menziesii* is strongly associated with a history of site disturbance. Disease centers are most likely to be found along roads, in areas where drainage patterns have been changed and especially on or near old tractor trails and landings (used to drag and stack harvested timber for transport), where soils have been compacted or topsoil removed (Goheen and Hansen, 1978; Hansen, 1978; Lawson and Cobb, 1986; Hansen *et al.*, 1988; Morrison and Hunt, 1988). Occurrence of infection centers in areas where low tree vigor has resulted from adverse site conditions undoubtedly reflects vector preference for stressed hosts.

Leptographium wageneri also is much more likely to be found in *P. menziesii* plantations that have been precommercially thinned than in unthinned plantations (Harrington *et al.*, 1983; Witcosky *et al.*, 1986a; Hansen *et al.*, 1988). Insect vector activity increases greatly in thinned stands, especially if green slash is created or still present early in the year when insect dispersal occurs. Witcosky *et al.* (1986a) found particularly dramatic insect population increases in plantations thinned after August and before June. They also found considerable evidence of new *L. wageneri* infections associated with vector attacks in *P. menziesii* crop-trees and stumps.

The effect of stem-attacking beetles on *L. wageneri*-infected *P. menziesii* has received less attention than the vector relationship. Nevertheless, it appears that these insects play an important part in administering the *coup de grace* to disease-weakened trees. In a sample of 231 recently dead *L. wageneri*-infected *P. menziesii* in western Oregon and Washington, Goheen and Hansen (1978) found that 52% were infested by stem-attacking beetles. Twenty-four percent of the trees exhibited galleries of *Pseudohylesinus nebulosis*. Other beetles were also present, including *Scolytus unispinosis* (14%), buprestids (14%), cerambycids (4%), *Dryocoetes autographus* (3%), and *Dendroctonus pseudotsugae* (1%). Typical of *L. wageneri*-affected trees in the region, this sample was composed mainly of trees smaller than 20 cm DBH. *Dendroctonus pseudotsugae* is more abundant when larger trees are affected.

In southwestern Oregon where *L. wageneri* var. *pseudotsugae* is of considerable concern to forest managers, 10- to 25-year-old *Pseudotsuga menziesii* plantations on about 400 000 hectares were surveyed in three separate evaluations (Goheen *et al.*, unpublished data; Hessburg *et al.*, USDA Forest Serv., unpublished data; Table 9.2). The surveys indicated that *L. wageneri* was common and widely distributed throughout *P. menziesii* plantations in the areas evaluated, but levels of associated mortality were quite variable. On the Siskiyou National Forest and the Medford District, 90% or more of the affected plantations exhibited low levels of mortality, averaging less than 2% of the trees. However, in the Tioga Resource Area, half of all units examined showed more than 2% mortality, and 10% suffered more than 10% mortality. Some plantations in this area exhibited extremely severe losses, as much as 50% of all *P. menziesii* (Fig. 9.2). Losses were especially severe considering that this mortality represented crop trees after precommercial thinning. Greater damage in the

Table 9.2. Frequency of black stain root disease (BSRD) in southwest Oregon, USA

Survey area	Plantations surveyed		Percent of trees infected in infected plantations			
	Number	Percent with BSRD	<0.1%	0.1–2.0%	2.1–10%	>10
Siskiyou N.F.	100	25	45	45	10	0
Tioga R.A.	100	29	40	10	40	10
Medford Dist.	500	19	93	3	4	1

Tioga Resource Area, compared to the other areas surveyed, probably resulted from greater site disturbance associated with previous tractor logging.

In southwest Oregon, *L. wageneri* appears in plantations at age 10–15, spreads at a rate of 0.4–0.9 m year^{-1} in radially expanding infection centers, and slows dramatically at age 30–40 (Hansen and Goheen, 1988a). The disease seldom is observed affecting older trees. In Mendocino County, California, however, high levels of infection have been detected in 60- to 80-year-old trees (Jackman and Hunt, 1975; Lawson and Cobb, 1986).

In parts of western North America other than southwest Oregon and northwest California, *L. wageneri* on *P. menziesii* generally is much less widely distributed and damaging (Goheen and Hansen, 1978; Byler *et al.*, 1983; Morrison and Hunt, 1988). When found, *L. wageneri* usually affects small numbers of young trees in plantations or scattered older *P. menziesii* that have been weakened by other root diseases, adverse site factors, or defoliation by *Choristoneura occidentalis* (Goheen and Hansen, personal observation).

9.2.2.3 Ecological effects

L. wageneri usually occurs on *P. menziesii* in the early stages of forest succession and is strongly associated with soil disturbance, usually the result of human activity. It is most common in relatively pure young *P. menziesii* stands that have been planted on sites where the former stand was harvested by clearcutting. It also is found in naturally regenerated *P. menziesii* on sites where drastic disturbance such as road building or fire has destroyed the previous forest.

Three host specific variants of *L. wageneri* (Harrington and Cobb, 1984, 1986; Chapter 3) are known. *Leptographium wageneri* var. *pseudotsugae* affects only *P. menziesii* in nature, although there is crossover to other species in artificial inoculations. Non-host tree species, shrubs, and herbaceous plants increase in frequency and cover as *L. wageneri* kills *P. menziesii* in disease centers. Although the insect vectors usually introduce *L. wageneri* into stressed or injured *P. menziesii*, the pathogen subsequently spreads to adjacent trees via root contact, regardless of vigor. *Pseudotsuga menziesii* is not eliminated from the infection centers since some individuals escape or tolerate infection (Hansen and Goheen, 1988a; Morrison and Hunt, 1988).

In southwest Oregon, the typical *L. wageneri* infection center contains a much higher proportion of *Tsuga heterophylla, Thuja plicata, Calocedrus decurrens, Pinus monticola* and/or *P. lambertiana* than nearby unaffected portions of the same stand. However, this depends on seed availability. In northwestern California, *Sequoia sempervirons* also is favored by death of *Pseudotsuga menziesii*. Hardwood trees and shrub species increase dramatically. Herbaceous plant species richness is higher in the diseased areas, and many wildlife species appear to prefer the habitat in disease centers to that in healthy *P. menziesii* stands, especially as canopies close and light becomes limiting (Goheen, personal observation). Because disease spread often diminishes after stands reach ages of 30–40 years, the disease-caused openings gradually fill with shade-tolerant tree species.

9.2.3 Interactions in *Pinus ponderosa* forests

9.2.3.1 Nature of relationship

Leptographium wageneri var. *ponderosum* occurs at scattered locations throughout western North America, causing a vascular wilt disease of hard pines (Wagener and Mielke, 1961; Smith, 1967; Filip and Goheen 1982). The disease is especially common and damaging on

P. ponderosa in northeastern California and the central Sierra Nevada. The causal fungus is believed to be insect vectored, probably by the root-feeding bark beetle *Hylastes macer* (Goheen and Cobb, 1978). The biology of *H. macer* is imperfectly known, though the insect has been shown to attack the roots of weakened pines, appears to be attracted to wounds, and is most active when soils are moist (Blackman, 1941; Chamberlain, 1958; Bright and Stark, 1973).

Leptographium wageneri on *P. ponderosa* has been demonstrated to play a highly significant role as a predisposing agent for stem-infesting bark beetles (Wagener and Mielke, 1961; Cobb *et al.*, 1973; Goheen and Cobb, 1980; Cobb, 1988b). Results of some studies indicate that several species of tree-attacking bark beetles actually are attracted to *L. wageneri*-infected trees (Goheen *et al.*, 1985; Witcosky *et al.*, 1987). Other beetles may land on trees at random and simply have a better chance of successfully infesting diseased trees due to their weakened condition (Wood, 1972; Moeck *et al.*, 1981; Owen, 1985; Chapter 6).

9.2.3.2 Extent of relationship

An insect vector for *L. wageneri* var. *ponderosum* had been suspected for some time, based on (1) the often widely separated distribution of *L. wageneri* infection centers in *P. ponderosa* stands, (2) the apparent association of the disease with weakened and injured trees in early stages of infection center development, and (3) the morphology of the causal fungus, which is particularly well adapted for insect dispersal. Evidence to support this hypothesis was provided when both the perfect and imperfect states of the fungus were found in galleries of *H. macer*, and excavation studies showed a close association between *H. macer* galleries and discrete areas of xylem staining by *L. wageneri* far down the roots of some newly infected pines (Goheen and Cobb, 1978). These findings were made in 1975, a year with an unusually moist spring. *Hylastes macer* galleries with *L. wageneri* fruiting were not found in subsequent attempts during drier years.

Cobb *et al.* (1982) studied sequential aerial photographs of *P. ponderosa* stands on the Georgetown Divide in the Sierra Nevada of California, where *L. wageneri* is common and widely distributed, and found that the average rate of generation of new infection centers was quite low, only one per 1000 ha per year. Although insect vectoring likely does occur, efficiency of disease transmission apparently is not as great in the *P. ponderosa* system as in the *Pseudotsuga menziesii* system. Furthermore, vector success may be dependent on uncommonly high moisture conditions. Both *H. macer* and *L. wageneri* var. *ponderosum* are known to be favored by high soil moisture (Blackman, 1941; Goheen *et al.*, 1978).

In areas where *L. wageneri* var. *ponderosum* occurs, the primary (tree-killing) bark beetles of *P. ponderosa*, *Dendroctonus brevicomis*, and *D. ponderosae*, are found much more commonly infesting diseased than healthy trees (Cobb *et al.*, 1973). Cobb (1988a) showed substantially reduced oleoresin yield and oleoresin exudation pressure in severely diseased trees, suggesting that these trees might be more susceptible to successful colonization by stem-attacking bark beetles than are uninfected trees. In the central Sierra Nevada, Goheen and Cobb (1980) monitored a sample of 256 pines with predetermined levels of *L. wageneri* var. *ponderosum* infection for 4 years. They found that 73% of the severely diseased trees (those with 50% or more of the root collar stained) were infested by *D. brevicomis* and D. *ponderosae* during the study, while 29% of the moderately diseased trees (those with 1–50% of the root collar stained) and only 2% of the apparently healthy trees were infested. Several other species of beetles, including *D. valens*, *Ips* spp., *Spondylis upiformis*, and *Melanophila* spp., also more commonly constructed galleries on *L. wageneri*-infected trees.

9.2.3.3 Timber losses

Leptographium wageneri var. *ponderosum* and associated bark beetles cause rapidly expanding mortality centers in *P. ponderosa* stands of any age, although 60- to 100-year-old sawtimber size trees are affected most commonly (Fig. 9.3). Once established in a suitable stand, the pathogen spreads to surrounding hosts at a rate of 1 m $year^{-1}$ across closely associated root systems (Cobb *et al.*, 1982). Losses due to the disease can be locally great but have not been documented often. Byler *et al.* (1979) found that *L. wageneri* occurred on 65 of 11 560 ha on the Georgetown Divide, California. The disease and associated bark beetles had killed trees containing 110 000 m^3 of wood.

9.2.3.4 Ecological effects

In nature, *L. wageneri* var. *ponderosum* is specific to the hard pines (Harrington and Cobb, 1984, 1986). It rarely affects other tree species in *P. ponderosa* infection centers although on occasion, *P. contorta, P. monticola,* and *P. lambertiana* growing in association with *P.*

Fig. 9.3. Mortality caused by *Leptographium wageneri* var. *ponderosum* and associated *Dendroctonus brevicomis* and *D. ponderosae* in a 60-year-old *Pinus ponderosa* stand in the central Sierra Nevada of California. *Abies concolor* and *Calocedrus decurrens* are replacing *P. ponderosa* as the latter die.

ponderosa are damaged. Generally, *L. wageneri* var. *ponderosum* promotes other conifer species, as well as hardwood trees and shrubs.

Leptographium wageneri and associated bark beetles tend to affect *P. ponderosa* in pure, heavily stocked stands. In the Central Sierra Nevada of California, *L. wageneri* occurs in 60- to 80-year-old essentially pure pine stands having basal areas of >50 m^2 ha^{-1} (Byler *et al.*, 1979). Such patches of *P. ponderosa* are scattered within a mixed-conifer forest composed of *P. lambertiana, Pseudotsuga menziesii, Calocedrus decurrens, Abies concolor*, and *Quercus kelloggii*, as well as *Pinus ponderosa.* The pure *P. ponderosa* stands exist in areas where logging disturbance at the turn of the century was especially great. Many of the sites were burned with high intensity fires or had soil removed during skidding of logs to transport sites and railroad operations. *Leptographium wageneri* and associated bark beetles speed up succession in such locations by killing most of the *P. ponderosa.* Late successional (shade-tolerant) tree species, especially *C. decurrens* and *A. concolor*, which are immune to the disease, regenerate readily in the openings (Fig. 9.3).

In Oregon, as well, *L. wageneri* and associated bark beetles often are found in dense pine stands on poor or disturbed sites. However, when old-growth pines are affected, other species may not respond as readily, particularly on the drier sites of the "*Pinus ponderosa* zone" (Franklin and Dyrness, 1969). Rather, infection centers remain as gaps in the conifer canopy that progressively enlarge. *Leptographium wageneri* and beetles kill trees as they stand, often with extensive resinosus in the butt and roots. These snags are attractive to cavity nesting birds and may stand longer than snags created by other agents.

9.2.4 Interactions in *Pinus contorta* forests

9.2.4.1 Nature of relationship

In interior western North America, *D. ponderosae* is the primary bark beetle species killing *P. contorta*. Epidemic beetle populations periodically infest *P. contorta* over extensive areas. Large diameter (16 cm DBH or greater), mature (80 years old or more) *P. contorta* in dense stands are the most likely to be infested (Cole and Amman, 1973; Safranyik *et al.*, 1974; Burnell, 1977). The bark beetles appear to be most successful on trees that are under stress from competition (Berryman, 1978; Mitchell *et al.*, 1983). Apparent associations between *D. ponderosae* and the root pathogens, *Phaeolus schweinitzii, Leptographium wageneri*, and *Armillaria* sp., also have been reported by Geiszler *et al.* (1980), Hunt and Morrison (1986), and Tkacz and Schmitz (1986), respectively. Root diseases may play a role in providing suitable host material that maintains endemic *D. ponderosae* populations (Cobb *et al.*, 1973). They also may trigger population increases at the beginning of outbreaks. Some evidence indicates that endemic *D. ponderosae* populations are preferentially attracted to root-diseased trees (Gara *et al.*, 1984).

9.2.4.2 Extent of relationship

Hunt and Morrison (1986) reported that in surveys of *P. contorta* stands severely infected by *L. wageneri* var. *ponderosum* in the Kootenay Region of British Columbia, Canada, *D. ponderosae* infestations were observed infrequently. However, when bark beetle infestations were encountered, they almost always were associated with *L. wageneri* infection. Other bark beetles, especially *Ips latidens* and *I. mexicanus*, also were found attacking *L. wageneri*-infected trees. In *P. contorta* stands with endemic *D. ponderosae* populations in

Utah, Tkacz and Schmitz (1986) found *D. ponderosae* infesting 90% of a sample of pines with *Armillaria* sp. infections. Only 14% of the trees without detectable Armillaria root disease were infested. In central Oregon, Geiszler *et al.* (1980) found that 59% of a sample of *D. ponderosae*-killed *P. contorta* were infected by *Phaeolus schweinitzii*, while no infection was found in a sample of uninfested live trees from the same area.

9.2.4.3 Timber losses

Timber losses due to *D. ponderosae* outbreaks in *P. contorta* have been enormous. When populations of this aggressive bark beetle reach large size, they are capable of killing living apparently healthy trees (see Chapter 6). It is estimated that *D. ponderosae* has killed trees containing about 12 000 000 m^3 of wood annually since 1895 (Wood, 1963). In 1970 alone, *D. ponderosae* killed trees containing 13 400 000 m^3 of wood in the Rocky Mountain states (Cole and Amman, 1980). Losses in *P. contorta* stands in eastern Oregon have been correspondingly great. Root disease is not directly involved in causing these large timber losses. Rather, the pathogens probably enhance long-term survival of *D. ponderosae* populations when stand conditions are suboptimal for the insects. Where they occur, root diseases certainly predispose *P. contorta* to beetle infestation.

9.2.4.4 Ecological effects

Pinus contorta is a seral species in several forest zones in western North America. It has an ability to grow on a wide range of forest sites and is also the climax species on many poorly drained or frost-prone sites. *Pinus contorta* produces both serotinus and open cones, regenerates readily, and develops very dense stands. Trees grow rapidly when young, but stands have a tendency to become stagnant at relatively early ages. In the past, periodic intense wildfires burned large areas of *P. contorta* forests and essentially regulated stand development (Brown, 1975). In the absence of fire, *D. ponderosae* also can act as a stand regulating agent by killing trees over substantial areas. The concentration of dead trees in turn increases the probability of subsequent wildfire (Leuschner and Berck, 1985). Where *P. contorta* is a climax species, trees killed by *D. ponderosae* are replaced by new *P. contorta* stands. Where *P. contorta* is seral, however, beetle-killed stands may succeed to more shade-tolerant species such as *A. lasiocarpa*, *A. grandis*, and *Picea engelmanii*.

9.2.5 Interactions in *Abies* forests

9.2.5.1 Nature of relationship

Scolytus ventralis is considered to be the major bark beetle species killing *Abies* spp. throughout western North America (Strubel, 1957; Furniss and Carolin, 1977). It is especially common on *A. concolor* and A. *grandis* in interior mixed-conifer stands. In addition to infesting entire trees, *S. ventralis* also commonly attacks and kills tops or portions of trees. *Scolytus ventralis* rarely infests vigorous *Abies*. Rather, it usually is found attacking injured, overmature, drought-stressed, insect-defoliated, or diseased trees (Fig. 9.4). The bark beetle may be attracted to unhealthy hosts (Ferrell, 1971). Root diseases are especially likely to be associated with *S. ventralis* infestation of *A. concolor* and *A. grandis* (Partridge and Miller, 1972; Cobb *et al.*, 1973; Miller and Partridge, 1974; Hertert *et al.*, 1975; Ferrell and Smith, 1976; Lane and Goheen, 1979: James and Goheen, 1981; Filip and Goheen, 1982; James *et*

Fig. 9.4. Mortality center in a mixed conifer stand in eastern Oregon. *Scolytus ventralis* are killing *Abies grandis* that are infected with *Phellinus weirii* and *Armillaria* sp. *Pinus ponderosa* are not being affected.

al., 1984; Hadfield *et al.*, 1986; Cobb, 1988a; Hagel and Goheen, 1988; Goheen and Goheen 1990). Root pathogens commonly involved are *Phellinus weirii*, *Armillaria* spp., and *Heterobasidion annosum* occurring individually or together as disease complexes (Goheen and Filip, 1980, Hadfield *et al.*, 1986, Hansen and Goheen, 1988b).

9.2.5.2 Extent of relationship

The degree of relationship between root diseases and *S. ventralis* on *A. concolor* has been investigated by examining the root systems of large samples of *S. ventralis*-infested trees and determining the incidence and degree of infection by pathogenic fungi (Table 9.3). Most *Abies* in these samples exhibited extensive root system colonization by the pathogens. The trees undoubtedly had been infected for considerable periods of time before being infested by *S. ventralis*. Cobb *et al.* (1973) investigated a sample of *A. concolor* with tops killed by *S. ventralis* and determined that 40% of the live, beetle-damaged trees were root-diseased. It is evident that root diseases predispose *A. concolor* to *S. ventralis* infestation. Under normal circumstances, they appear to be critical to maintaining *S. ventralis* populations when at low densities. Many uninfected trees are infested, however, when there are other weakening fac-

Table 9.3. Associations between root pathogens and *Scolytus ventralis* in the western United States

Investigators	Location	% beetle-infested trees with disease	Root pathogens
Miller & Partridge (1974)	Idaho	98	*Armillaria, Phellinus*
Hertert *et al.* (1975)	Idaho	96	*Armillaria, Phellinus*
Cobb *et al.* (1973)	California	92	*Armillaria, Heterobasidion*
Lane and Goheen (1979)	Oregon, Washington	86	*Armillaria, Phellinus, Heterobasidion*
James and Goheen (1981	Colorado	81	*Armillaria, Heterobasidion*

tors such as major droughts or severe defoliation by insects such as *Choristoneura occidentalis* and *Orgyia pseudotsugata* (Berryman and Wright, 1978; Chapter 4).

9.2.5.3 Timber losses

Chronic mortality and growth losses due to root diseases and associated *S. ventralis* are common throughout western North America. Few large-scale damage assessments have been done, but it is believed that losses are great in a high proportion of mixed conifer stands. In surveys of several individual stands in Oregon and Washington, mortality losses of 4–55% of the trees and 7–33% of the volume have been reported (Filip and Goheen, 1984). Numerous unpublished surveys by USDA Forest Service, Forest Pest Management, Pacific Northwest Region, show similar or even higher losses in eastern Oregon and Washington stands.

Damage is particularly substantial in stands where *A. concolor* comprises 10% or more of the overstory and where one or more selective harvest entries have occurred (Schmitt *et al.*, 1984). In surveys of 192 randomly selected stands on the Ochoco and Fremont National Forests in eastern Oregon, Schmitt *et al.* (1984) found that 13% of *A. concolor* in all stands had been killed by *P. weirii, Armillaria* sp., *H. annosum,* and *S. ventralis*. Survey plots that contained trees affected by the root disease–beetle association showed a 45% reduction in basal area compared to unaffected plots.

In the northern Rocky Mountains, surveys show that about 1% of the commercial forest land is composed of large active root disease pockets and another 13% contains numerous small scattered infection centers (James *et al.*, 1984). Root disease centers are most common in mixed-conifer types, and the *A. grandis*–root disease–*S. ventralis* association is the cause of a substantial amount of the mortality.

9.2.5.4 Ecological effects

In this century, fire exclusion policies have favored development of late successional communities in much of the *Abies grandis* zone (Franklin and Dyrness, 1969) and comparable mixed-conifer forest types at mid-elevation in interior western North America. *Abies concolor* and *A. grandis* establish themselves readily on all but the coldest and driest sites and grow rapidly. *Abies* spp., however, are very prone to pest problems. In addition to root diseases and *S. ventralis,* they suffer significant predisposing damage from stem decays and defoliating insects, especially *Choristoneura occidentalis* and *Orgyia pseudotsugata.*

Root disease and *S. ventralis,* acting in concert, selectively kill *A. concolor* and *A. grandis*. If *Phellinus weirii* or *Armillaria* sp. are the root pathogens involved, *Pseudotsuga men-*

ziesii also may be killed. Seral tree species, such as *Pinus* spp. and *Larix occidentalis*, are rarely damaged, and mature individuals may be favored in mixed-conifer stands with root disease (Fig. 9.4). It is uncommon, however, for *Pinus* spp. or *L. occidentalis* to establish themselves from seed in mortality centers. These species typically germinate on bare soil in open areas following fire. In the absence of fire or other catastrophic disturbance, *A. concolor* and *A. grandis* usually regenerate in disease centers, ensuring a continuous progression of regeneration, reinfection, and mortality within diseased stands. The accumulation of dead wood increases the probability of stand replacement wildfire and the opportunity for recolonization of the site by seral tree species. Disease centers may provide superior wildlife habitat because of the response of understory vegetation to the stand opening. *Abies concolor* and *A. grandis* snags provide only short-term benefit to cavity-using birds and mammals, however, because of their rapid deterioration.

9.2.6 Interactions in eastern forests

Dendroctonus frontalis is the most devastating forest insect in *Pinus* forests of the southeastern U.S.; populations also are found in Arizona, Mexico and Central America (Payne, 1980). These forests occupy portions of the eastern deciduous forest biome where fire frequency was sufficient to prevent replacement of *Pinus* spp. by hardwoods (Schowalter *et al.*, 1981a). *Dendroctonus frontalis* causes extensive mortality to *Pinus* spp., especially in dense stands composed primarily of *P. taeda* and *P. echinata* (Payne, 1980).

Endemic populations of *D. frontalis* and *Ips* spp. may be maintained in *Leptographium procerum* and *Heterobasidion annosum* infection centers in *Pinus* forests in the southern US (Hicks, 1980; Alexander *et al.*, 1981; Skelly *et al.*, 1981). However, *D. frontalis* is a very aggressive species. Large populations colonize all *Pinus*, regardless of condition or size, within gradually expanding infestations, much like the infection centers of root pathogens (Schowalter *et al.*, 1981b). Schowalter *et al.* (1981b) reported that *D. frontalis* populations become self-sustaining after reaching threshold sizes of about 100 000 beetles by June. Smaller populations may be unable to reach sufficient attack densities to cause continued tree mortality during the hot summer.

Thinning, harvest, and salvage operations (designed to control *D. frontalis* infestations) facilitate spread of *H. annosum* (Hodges, 1969). These organisms interacting in southern forests have caused widespread and severe timber losses (Payne, 1980; Leuschner and Berck, 1985). Mortality of *Pinus* in mixed pine–hardwood forests also affects forest composition and availability of resources for other species (Leuschner and Berck, 1985).

Reduced canopy cover and transpiration may affect hydrologic processes, but effects of *D. frontalis* on water yield and quality probably are minimal (Leuschner and Berck, 1985). However, altered canopy coverage and species composition likely affect nutrient cycling (Schowalter *et al.*, 1981a), and extensive tree mortality may fuel subsequent wildfire (Coulson *et al.*, 1983, 1985).

Widespread decline of *Pinus resinosa* plantations has become prevalent in the Great Lakes region of eastern North America (Klepzig *et al.*, 1991). This decline is related to interaction among *Leptographium procerum*, *L. terebrantis* and several bark beetle and root weevil vectors. The sequence of pathogen transmission, development of disease symptoms, and colonization by *Ips pini* and its fungal associate, *Ophiostoma ips*, associated with tree death (Klepzig *et al.*, 1991) closely resembles that described above for *Leptographium*–beetle interactions in *Pseudotsuga* and *Pinus ponderosa* forests.

9.3 CONCLUSIONS

Root disease fungi and bark beetles occur together in predictable associations by host and geographic region throughout the coniferous forests of North America. The insects and fungi emphasized in this chapter are all native species acting in relationships that have evolved with the conifer forest habitat. Tree killing is an essential part of the life history strategy of these organisms.

Despite the close association between fungus and insect in most of our examples, the relationship is primarily opportunistic, not symbiotic. Both beetles and root pathogens lead independent lives, each without any requirement for the other. In most cases, the root pathogens weaken trees, which are then vulnerable to attack by bark beetles. These relationships are especially important when beetle populations are low. During outbreaks, the beetles can attack and kill apparently healthy trees. Beetles also may favor root pathogens by killing root-diseased trees before they are windthrown and thereby maintaining inoculum in the soil. Spores of *Leptographium wageneri* are transported on emerging adult root-feeding beetles to infection courts on healthy trees.

Comprehensive figures on timber loss are not available, and there is danger of significant overestimation if losses attributed to individual pathogens and beetles simply are added for a total. However, we estimate that 6 000 000 m^3 of timber volume are lost annually to root disease/bark beetle associations in western North America. This excludes the losses from bark beetles acting alone during outbreaks. There are additional significant losses resulting from death of immature trees, loss of productive land to non-commercial species, and growth loss of trees damaged but not killed.

Root diseases and bark beetles have profound impacts on forest community structure, composition, and succession. Specific effects depend on whether trees are killed and remain standing or fall over, on the pattern of mortality in the forest, on the rate of spread, and on the forest type. In general, these organisms create discontinuities in canopy structure and function, with gaps of altered light availability, temperature and humidity, evapotranspiration, and interception of precipitation.

Plants and animals respond to the altered environment resulting from tree death. Increased light on the forest floor is probably the most immediate change, but temperature and humidity fluctuation also increases. Nutrient availability increases as nutrients stored in wood are released during decomposition, which is stimulated by bark beetles and other wood borers mining the wood and inoculating it with saprophytic microorganisms. Soil fertility may be enhanced over the long term. In most situations, the diversity and cover of understory vegetation increases in mortality centers, at least until a new overstory canopy is established. However, tree mortality also may increase the probability of subsequent stand-replacement wildfire.

REFERENCES

Alexander, S.A., Skelly, J.M. and Webb, R.S. (1981). Effects of *Heterobasidion annosum* on radial growth in southern pine beetle infested loblolly pine. *Phytopathology* **71**, 479–481.

Berryman, A.A. (1978) A synoptic model of the lodgepole pine/mountain pine beetle interaction and its potential application in forest management. *In* "Theory and Practice of Mountain Pine Beetle Management in Lodgepole Pine Forests" (A.A. Berryman, G.D. Amman, and R.W. Stark, eds) p. 98–105. University of Idaho, Moscow.

Berryman, A.A. and Wright, L.C. (1978). Defoliation, tree condition, and bark beetles. *In* "The Douglas-fir Tussock Moth: a Synthesis" (M.H. Brookes, R.W. Stark and R.W. Campbell, eds), pp. 81–88. USDA Forest Serv. Tech. Bull. 1585. USDA Forest Serv., Washington, DC.

Blackman, M.W. (1941). Bark beetles of the genus *Hylastes* Erickson in North America. USDA Misc. Publ. 417. USDA, Washington, DC, 27 pp.

Bloomberg, W.J. and Beale, J.D. (1985). Relationship of ecosystem to *Phellinus weirii* root rot on southern Vancouver Island. *In* "Proc. 33rd Annual Western International Forest Disease Work Conference," pp. 20–28. Oregon State University Press, Corvallis.

Bloomberg, W.J. and Reynolds, G. (1985). Growth loss and mortality in laminated root rot infection centers in second-growth Douglas-fir on Vancouver Island. *Forest Sci.* **31**, 497–508.

Bright, D.E., Jr., and Stark, R.W. (1973). The bark and ambrosia beetles of California (Coleoptera: Scolytidae and Platypodidae). Bull. Calif. Insect Surv. Vol. 16. University of California Press, Berkeley, 169 pp.

Brown, J.K. (1975). Fire cycles and community dynamics in lodgepole pine forests. *In* "Management of Lodgepole Pine Ecosystems" (D.M. Baumgartner, ed.), pp. 427–456. Washington State University Press, Pullman.

Buckland, D.C., and Wallis, G.W. (1956). The control of yellow laminated root rot of Douglas-fir. *Forestry Chron.* **32**, 14–16.

Burnell, D.G. (1977). A dispersal-aggregation model for mountain pine beetle in lodgepole pine stands. *Res. Pop. Ecol.* **19**, 99–106.

Byler, J.W., Cobb, F.W., Jr. and Rowney, D.L. (1979). An evaluation of black stain root disease on the Georgetown Divide, El Dorado County. USDA Forest Serv. Forest Insect and Disease Management Rpt. 79-2. USDA Forest Serv. Region 5, San Francisco, CA, 15 pp.

Byler, J.W., Harrington, T.C., James, R.L., and Haglund, S. (1983). Black stain root disease in Douglas-fir in western Montana. *Plant Dis.* **67**, 1037–1038.

Chamberlain, W.J. (1958). The Scolytidae of the Northwest: Oregon, Washington, Idaho, and British Columbia. Oregon State Monogr. Studies in Entomology, No. 2. Oregon State University, Corvallis, 205 pp.

Childs, T.W., and Shea, K.R. (1967). Annual losses from diseases in Pacific Northwest forests. USDA Forest Service Resource Bull. PNW-20. USDA Forest Serv. Pacific Northwest Res. Stn., Portland, OR, 19 pp.

Cobb, F.W., Jr. (1988a). Interactions among root disease pathogens and bark beetles in coniferous forests. *In* "Proceedings of the Seventh International Conference on Root and Butt Rots" (D.J. Morrison, ed.), pp. 142–148. Forestry Canada, Pacific Forestry Centre, Victoria, BC.

Cobb. F.W., Jr. (1988b). *Leptographium wageneri*, cause of black-stain root disease: a review of its discovery, occurrence and biology with emphasis on pinyon and ponderosa pine. *In* "Leptographium Root Diseases on Conifers" (T.C. Harrington and F.W. Cobb, Jr., eds), pp. 41–62. American Phytopathological Society Press, St. Paul, MN.

Cobb, F.W., Jr., Parmeter, J.R., Jr., Wood, D.L., and Stark, R.W. (1973). Root pathogens as agents predisposing ponderosa pine and white fir to bark beetles. *In* "Proceedings of the Fourth International Conference on *Fomes annosus*" (E.G. Kuhlman, ed.), pp. 8–15. Athens, Georgia.

Cobb, F.W., Jr., Slaughter, G.W., Rowney, D.L. and Demars, C.J. (1982). Rate of spread of *Ceratocystis wageneri* in ponderosa pine stands in the central Sierra Nevada. *Phytopathology* **72**, 1359–1362.

Cole, W.E. and Amman, G.D. (1973). Mountain pine beetle infestations in relation to lodgepole pine diameters. USDA Forest Serv. Res. Note INT-95, USDA Forest Serv. Intermountain Res. Stn., Ogden, UT, 7 pp.

Cole, W.E. and Amman, G.D. (1980). Mountain pine beetle dynamics in lodgepole pine forests part I: course of an infestation. USDA Forest Service, Gen. Tech. Rpt. INT-89, USDA Forest Serv. Intermountain Forest Res. Stn., Ogden, UT, 56 pp.

Coulson, R.N., Hennier, P.B., Flamm, R.O., Rykiel, E.J., Hu, L.C. and Payne, T.L. (1983). The role of lightning in the epidemiology of the southern pine beetle. *Z. Ang. Entomol.* **96**, 182–193.

Coulson, R.N., Saunders, M.C., Payne, T.L., Flamm, R.O., Wagner, T.L., Hennier, P.B. and Rykiel, E.J. (1985). A conceptual model of the role of lightning in the epidemiology of the southern pine beetle. *In* "The Role of the Host in Population Dynamics of Forest Insects" (L. Safranyik, ed.), pp. 136–146. Canadian Forestry Serv., Pacific Forestry Centre, Victoria, BC.

Edmonds, R.L. and Eglitis, A. (1989). The role of the Douglas-fir beetle and wood borers in the decomposition and nutrient release from Douglas-fir logs. *Can. J. For. Res.* **19**, 853–859.

Ferrell, G.T. (1971) Host selection by the fir engraver, *Scolytus ventralis* (Coleoptera: Scolytidae): preliminary field studies. *Can. Entomol.* **103**, 1717–1725.

Ferrell, G.T., and Smith, R.S., Jr., (1976). Indicators of *Fomes annosus* root decay and bark beetle susceptibility in sapling white fir. *Forest Sci.* **22**, 365–369.

Filip, G.M., and Goheen, D.J. (1982). Tree mortality caused by root pathogen complex in Deschutes National Forest, Oregon. *Plant Dis.* **66**, 240–243.

Filip, G.M., and Goheen, D.J. (1984). Root diseases cause severe mortality in white and grand fir stands in the Pacific Northwest. *Forest Sci.* **30**, 138–142.

Franklin, J.F., and Dyrness, C.T. (1969). "Vegetation of Oregon and Washington." USDA Forest Service, Research Paper PNW-80. USDA Forest Serv. Pacific Northwest Res. Stn., Portland, OR.

Furniss, R.L. and Carolin, V.M. (1977). "Western Forest Insects." USDA Forest Service, Misc. Publication No. 1339. USDA Forest Serv., Washington, DC, 654 pp.

Gara, R.I., Geiszler, D.R. and Littke, W.R. (1984). Primary attraction of the mountain pine beetle to lodgepole pine in Oregon. *Ann. Entomol. Soc. Am.* **77**, 333–334.

Geiszler, D.R., Gara, R.I., Driver, C.H., Gallucci, V.F. and Martin, R.E. (1980). Fire, fungi, and beetle influences on a lodgepole pine ecosystem of south-central Oregon. *Oecologia (Berl.)* **46**, 239–243.

Goheen D.J. (1979). Survey for laminated root rot on the Mapleton District of the Siuslaw National Forest. *In* "Proc. 27th Annual Western International Forest Disease Work Conference," pp. 74–75. USDA Forest Serv., Rocky Mountain Forest & Range Exp. Stn., Fort Collins, CO.

Goheen, D.J. and Cobb, F.W., Jr. (1978). Occurrence of *Verticicladiella wageneri* and its perfect state, *Ceratocystis wageneri* sp. nov., in insect galleries. *Phytopathology* **68**, 1192–1195.

Goheen, D.J. and Cobb, F.W., Jr. (1980). Infestation of *Ceratocystis wageneri*-infected ponderosa pines by bark beetles (Coleoptera: Scolytidae) in the central Sierra Nevada. *Can. Entomol.* **112**, 725–730.

Goheen, D.J. and Filip, G.M. (1980). Root pathogen complexes in Pacific Northwest forests. *Plant Dis.* **64**, 793–794.

Goheen, D.J. and Hansen, E.M. (1978). Black stain root disease in Oregon and Washington. *Plant Dis. Rep.* **62**, 1098–1102.

Goheen, D.J., Cobb, F.W., Jr. and McKibbin, G.N. (1978). Influence of soil moisture on infection of ponderosa pine by *Verticicladiella wagenerii. Phytopathology* **68**, 913–916.

Goheen, D.J., Cobb, F.W., Jr., Wood, D.L. and Rowney, D.L. (1985). Visitation frequencies of some insect species on *Ceratocystis wageneri* infected and apparently healthy ponderosa pines. *Can. Entomol.* **117**, 1535–1543.

Goheen, E.M. and Goheen, D.J. (1990). Losses caused by annosus root disease in Pacific Northwest forests. *In* "Symposium on Annosus Root Disease" (W.J. Otrosina and R.F. Scharpf, eds), pp. 66–69. USDA Forest Serv. Gen. Tech. Rpt. PSW-116. USDA Forest Serv. Pacific Southwest Forest and Range Experiment Station, Berkeley, CA.

Hadfield, J.S. (1985). Laminated root rot, a guide for reducing and preventing losses in Oregon and Washington forests. USDA Forest Service Pacific Northwest Region, Portland, OR, 13 pp.

Hadfield, J.S., Goheen, D.J., Filip, G.M., Schmitt, C.L. and Harvey, R.D. (1986). Root diseases in Oregon and Washington conifers. USDA Forest Serv. Region 6 Report R6-FPM-250-86. USDA Forest Service Region 6, Portland, OR, 27 pp.

Hagel, S.K. and Goheen, D.J. (1988). Root disease response to stand culture. *In* "Proc. Future Forests of the Mountain West: a Stand Culture Symposium" (W.C. Schmidt, ed.), pp. 303–309. USDA Forest Serv. Gen. Tech. Rpt. INT-243. USDA Forest Serv. Intermountain Res. Stn., Ogden, UT.

Hansen, E.M. (1978). Incidence of *Verticicladiella wagenerii* and *Phellinus weirii* in Douglas-fir adjacent to and away from roads in western Oregon. *Plant Dis. Rep.* **62**, 179–181.

Hansen, E.M. (1979). Survival of *Phellinus weirii* in Douglas-fir stumps after logging. *Can. J. For. Res.* **9**, 484–488.

Hansen, E.M. and Goheen, D.J. (1988a). Rate of increase of black stain root disease in Douglas-fir plantations in Oregon and Washington. *Can J. For. Res.* **18**, 942–946.

Hansen, E.M. and Goheen, D.J. (1988b). Root disease complexes in the Pacific Northwest. *In* "Proc. Seventh International Conference on Root and Butt Rots" (D.J. Morrison, ed.), pp. 129–141. Forestry Canada, Pacific Forestry Centre, Victoria, BC.

Hansen, E.M., Goheen, D.J., Hessburg, P.F., Witcosky, J.J. and Schowalter, T.D. (1988). Biology and management of black-stain root disease in Douglas-fir. *In* "Leptographium Root Diseases on

Conifers" (T.C. Harrington and F.W. Cobb, Jr., eds), pp. 63–80. American Phytopathological Society Press, St. Paul, MN.

Harrington, T.C. and Cobb, F.W., Jr. (1984). Host specialization of three morphological variants of *Verticicladiella wageneri*. *Phytopathology* **74**, 286–290.

Harrington, T.C. and Cobb, F.W., Jr. (1986). Varieties of *Verticicladiella wageneri*. *Mycologia* **78**, 562–567.

Harrington, T.C., Reinhart, C., Thornburgh, D.A. and Cobb, F.W., Jr. (1983). Association of black-stain root disease with precommercial thinning of Douglas-fir. *Forest Sci.* **29**, 12–14.

Harrington, T.C., Cobb, F.W., Jr. and Lownsbery, J.W. (1985). Activity of *Hylastes nigrinus*, a vector of *Verticicladiella wageneri*, in thinned stands of Douglas-fir. *Can. J. For. Res.* **15**, 579–523.

Hertert, H.D., Miller, D.L. and Partridge, A.D. (1975). Interactions of bark beetles (Coleoptera: Scolytidae) and root rot pathogens in northern Idaho. *Can. Entomol.* **107**, 899–904.

Hicks, R.R., Jr. (1980). Climate, site, and stand factors. *In* "The Southern Pine Beetle" (R.C. Thatcher, J.L. Searcy, J.E. Coster and G.D. Hertel, eds), pp. 55–68. USDA Forest Serv. Tech. Bull. 1631. USDA Forest Serv., Washington, DC.

Hodges, C.S. (1969). Modes of infection and spread of *Fomes annosus*. *Annu. Rev. Phytopathol.* **7**, 247–266.

Holah, J.C. (1991). Effects of *Phellinus weirii* on plant community composition and succession of mature and old-growth Douglas-fir forests. M.S. Thesis, Oregon State University, Corvallis, 107 pp.

Hunt, R.S. and Morrison, D.J. (1986). Black stain root disease on lodgepole pine in British Columbia. *Can. J. For. Res.* **16**, 996–999.

Hutchins, A.S. and Rose, S.L. (1984). The variation in antagonistic *Streptomyces* populations in soils from different vegetation types in western Oregon. *Northwest Sci.* **58**, 249–255.

Jackman, R.E. and Hunt, R. (1975). Black-stain root disease in Douglas-fir on Jackson State Forest. California State For. Note No. 58:3-4. California State Department of Forestry, Sacramento.

James, R.L. and Goheen, D.J. (1981). Conifer mortality associated with root disease and insects in Colorado. *Plant Dis.* **65**, 506–507.

James, R.L., Stewart, C.A. and Williams, R.E. (1984). Estimating root disease losses in northern Rocky Mountain national forests. *Can. J. For. Res.* **14**, 652–655.

Kanaskie, A. (1985). Root disease assessment on non-Federal lands in western Oregon: methodology and preliminary results. *In* "Proc. 33rd Annual Western International Forest Disease Work Conference," pp. 89–92. Oregon State University Press, Corvallis.

Klepzig, K.D., Raffa, K.F. and Smalley, E.B. (1991). Association of an insect–fungal complex with red pine decline in Wisconsin. *Forest Sci.* **37**, 1119–1139.

Lane, B.B. and Goheen, D.J. (1979). Incidence of root disease in bark beetle-infested eastern Oregon and Washington true firs. *Plant Dis. Rep.* **63**, 262–266.

Lawson, T.T. and Cobb., F.W., Jr. (1986). Stand and site conditions associated with mortality by *Verticicladiella wageneri* in Mendocino County, California. *Phytopathology* **76**, 1058.

Lawson, T.T., Berg, A.B. and Hansen, E.M. (1983). Damage from laminated root rot at the Black Rock Forest Management Research Area in western Oregon. Forest Research Laboratory Res. Note 75. Oregon State University, Corvallis, 7 pp.

Leuschner, W.A. and Berck, P. (1985). Impacts on forest uses and values. *In* "Integrated Pest Management in Pine–Bark Beetle Ecosystems" (W.E. Waters, R.W. Stark and D.L. Wood, eds), pp. 105–120. John Wiley & Sons, New York.

McCauley, K.J. and Cook, S.A. (1980). *Phellinus weirii* infestation of two mountain hemlock forests in the Oregon Cascades. *Forest Sci.* **26**, 23–29.

Miller, D.L. and Partridge, A.D. (1974). Root rot indicators in grand fir. *Plant Dis. Rep.* **58**, 275–276.

Mitchell, R.G., Waring, R.H. and Pitman, G.B. (1983). Thinning lodgepole pine increases tree vigor and resistance to mountain pine beetle. *Forest Sci.* **29**, 204–211.

Moeck, H.A., Wood, D.L. and Lindahl, K.Q., Jr. (1981). Host selection behavior of bark beetles (Coleoptera: Scolytidae) attacking *Pinus ponderosa*, with special emphasis on the western pine beetle, *Dendroctonus brevicomis*. *J. Chem. Ecol.* **7**, 49–83.

Morrison, D.J., and Hunt, R. S. (1988). *Leptographium* species associated with root disease of conifers in British Columbia. *In* "*Leptographium* Root Diseases of Conifers" (T.C. Harrington and F.W. Cobb, Jr. eds), pp. 81–95. American Phytopathological Society Press, St. Paul, MN.

Nelson, E.E. and Hartman, T. (1975). Estimating spread of *Poria weirii* in a high elevation mixed conifer stand. *J. Forestry* **73**, 141–142.

Owen, D.R. (1985). The role of *Dendroctonus valens* and its vectored fungi in the mortality of ponderosa pine. Ph.D. Dissertation, University of California, Berkeley, 64 pp.

Partridge, A.D. and Miller, D.L. (1972). Bark beetles and root rots related in Idaho conifers. *Plant Dis. Rep.* **56**, 498–500.

Payne, T.L. (1980). Life history and habits. *In* "The Southern Pine Beetle" (R.C. Thatcher, J.L. Searcy, J.E. Coster and G.D. Hertel, eds), pp. 7–28. USDA Forest Serv. Tech. Bull. 1631. USDA Forest Serv., Washington, DC, 266 pp.

Rudinsky, J.A. (1966). Host selection and invasion by the Douglas-fir beetle, *Dendroctonus pseudotsugae* Hopkins, in coastal Douglas-fir forests. *Can. Entomol.* **98**, 98–111.

Rudinsky, J.A. and Zethner-Møller, O. (1967). Olfactory responses of *Hylastes nigrinus* to various host materials. *Can. Entomol.* **99**, 911–938.

Safranyik, L., Shrimpton, D.M. and Whitney, H.S. (1974). Management of lodgepole pine to reduce losses from mountain pine beetle. Canadian Forestry Serv. Forestry Tech. Rep. 1. Forestry Canada, Pacific Forestry Research Centre, Victoria, BC, 24 pp.

Schmitt, C.L., Goheen, D.J., Goheen, E.M. and Frankel, S.J. (1984). Effects of management activities and dominant species type on pest-caused mortality losses in true fir on the Fremont and Ochoco National Forests. USDA Forest Service, Pacific Northwest Region, Forest Pest Management, Portland, OR, 34 pp.

Schowalter, T.D., Coulson, R.N. and Crossley, D.A., Jr. (1981a). Role of southern pine beetle and fire in maintenance of structure and function of the southeastern coniferous forest. *Environ. Entomol.* **10**, 821–825.

Schowalter, T.D., Pope, D.N., Coulson, R.N. and Fargo, W.S. (1981b). Patterns of southern pine beetle (*Dendroctonus frontalis*) infestation enlargement. *Forest Sci.* **27**, 837–849.

Schowalter, T.D., Caldwell, B.A., Carpenter, S.E., Griffiths, R.P., Harmon, M.E., Ingham, E.R., Kelsey R.G., Lattin, J.D. and Moldenke, A.R. (1992). Decomposition of fallen trees: effects of initial conditions and heterotroph colonization rates. *In* "Tropical Ecosystems: Ecology and Management" (K.P. Singh and J.S. Singh, eds), pp. 373–383. Wiley Eastern Ltd., New Delhi.

Skelly, J.M., Alexander, S.A. and Webb, R.S. (1981). Association of annosus root rot with southern pine beetle attacks. *In* "Site, Stand, and Host Characteristics of Southern Pine Beetle Infestations" (J.E. Coster and J.L. Searcy, eds), pp. 50–67. USDA Forest Serv. Tech. Bull. 1612. USDA Forest Serv., Washington, DC.

Smith, R.S., Jr. (1967). *Verticicladiella* root disease of pines. *Phytopathology* **57**, 935–938.

Strubel, G.T. (1957). The fir engraver a serious enemy of western true firs. USDA Forest Serv. Production Research Report No. 11. USDA Forest Service, California Forest and Range Experiment Station, Berkeley, CA 18 pp.

Tkacz, B.M. and Hansen, E.M. (1982). Damage by laminated root rot in two succeeding stands of Douglas-fir. *J. Forestry* **80**, 788–791.

Tkacz, B.M. and Schmitz, R.F. (1986). Association of an endemic mountain pine beetle population with lodgepole pine infected by Armillaria root disease in Utah. USDA Forest Service Res. Note INT-353. USDA Forest Serv. Intermountain Res. Stn., Ogden, UT, 7 pp.

Wagener, W.W. and Mielke, J.L. (1961). A staining-fungus root disease of ponderosa, Jeffrey, and pinyon pines. *Plant Dis. Rep.* **45**, 831–836.

Wallis, G.W. (1967). *Poria weirii* root rot. *In* "Important Forest Insects and Diseases of Mutual Concern to Canada, the United States, and Mexico." Canadian Dept. Forest. and Rural Devel. Publ. 1180. Canadian Department of Forestry, Ottawa, Ontario.

Waring, R.H., Cromack, K., Matson, P.A., Boone, R.D. and Stafford, S.G. (1987). Responses to pathogen-induced disturbance: decomposition, nutrient availability, and tree vigour. *J. Forestry* **60**, 219–227.

Witcosky, J.J. and Hansen, E.M. (1985). Root-colonizing insects recovered from Douglas-fir in various stages of decline due to black-stain root disease. *Phytopathology* **75**, 399–402.

Witcosky, J.J., Schowalter, T.D. and Hansen, E.M. (1986a). The influence of time of thinning on the colonization of Douglas-fir by three species of root-colonizing insects. *Can. J. For. Res.* **16**, 745–749.

Witcosky, J.J., Schowalter, T.D. and Hansen, E.M. (1986b). *Hylastes nigrinus* (Coleoptera: Scolytidae), *Pissodes fasciatus*, and *Steremnius carinatus* (Coleoptera: Curculionidae) as vectors of black-stain root disease of Douglas-fir. *Environ. Entomol.* **15**, 1090–1095.

Witcosky, J.J., Schowalter, T.D. and Hansen, E.M. (1987). Host-derived attractants for the beetles *Hylastes nigrinus* (Coleoptera:Scolytidae) and *Steremnius carinatus* (Coleoptera: Curculionidae). *Environ. Entomol.* **16**, 1310–1313.

Wood, D.L. (1972). Selection and colonization of ponderosa pine by bark beetles. *In* "Insect/Plant Relationships" (H.F. van Emden, ed.). Symposia of the Royal Entomological Society of London, Vol. 6, pp. 101–117.
Wood, S.L. (1963). A revision of the bark beetle genus *Dendroctonus* Erickson (Coleoptera: Scolytidae). *Great Basin Nat.* **23**, 1–117.
Wright, K.W. and Lauterbach, P.G. (1958). A 10-year study of mortality in a Douglas-fir sawtimber stand in Coos and Douglas Counties, Oregon. USDA Forest Serv. Res. Paper 27. USDA Forest Serv. Pacific Northwest Res. Stn., Portland, OR, 29 pp.
Zethner-Møller, O. and Rudinsky, J.A. (1967). On the biology of *Hylastes nigrinus* in western Oregon. *Can. Entomol.* **99**, 897–911.

PART V

Management of Interactions

–10–

Modeling Interactions

C.G. SHAW III and B.B. EAV
USDA Forest Service, Fort Collins, CO, USA

10.1 INTRODUCTION

Information on the various interactions among bark beetles, pathogens, and conifers comprises the major portion of this book. Integration of this material, particularly as it relates to predicting pest outbreaks and quantifying their impacts, would markedly enhance the information's utility for both forest managers and scientists (e.g. Stephen *et al.*, 1980). Computer models can meet this need by capturing in a set of mathematical equations the available biological data along with the conceptual models used by forest managers dealing with these pests and scientists studying them (Stephen *et al.*, 1980; Lynch, 1990). In combination with existing stand growth and yield models, e.g PROGNOSIS (Stage, 1973; Wykoff *et al.*, 1982), a bark beetle–pathogen–conifer model (Fig. 10.1) could be used by forest managers to evaluate potential outcomes of various silvicultural prescriptions and direct pest control options.

Research scientists should be actively involved in defining the biological assumptions necessary to develop such models. If they are, then the resulting product becomes not only a tool of immediate use to managers, but also a quantitative description of a series of hypotheses about the dynamics, behavior, and impact of these various interactions. As such, it can aid scientists in identifying serious data gaps and thus help to define research priorities. An important benefit of the modeling process is a highlighting of our still inadequate understanding of many aspects of these host/pest interactions.

10.2 DEVELOPMENT AND USE OF PEST MODELS

The primary model discussed in this chapter was designed to assist foresters develop, explore, and evaluate alternative management prescriptions in stands where pest impacts involve interactions among bark beetles and fungi that cause root diseases. To be useful,

BEETLE–PATHOGEN INTERACTIONS IN CONIFER FORESTS
ISBN 0-12-628970-0

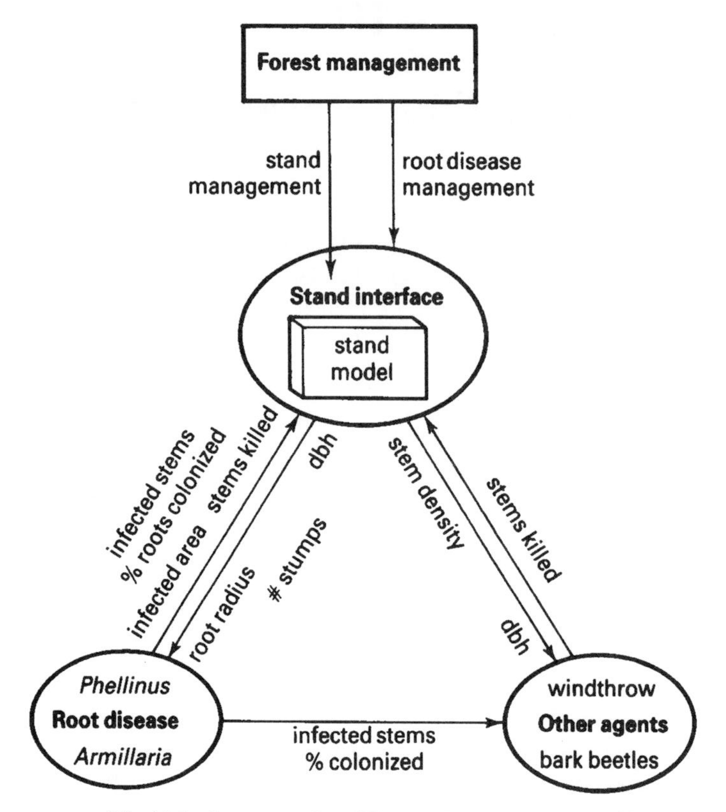

Fig.10.1. Structure of the Western Root Disease Model.

such models must enable forest managers to project probable changes in forest stands under various management scenarios, such as (1) the stand is allowed to develop naturally without management intervention, (2) silvicultural prescriptions are applied to maximize production of certain resources (i.e. timber), (3) silvicultural prescriptions are applied to reduce pest impacts on current and future stand growth and development, and (4) direct pest control actions (i.e. pesticide spraying) are applied with or without other silvicultural treatments.

The preferred method for developing a simulation model depends on the nature of the system being modeled and the state of knowledge about that system. In pest modeling, knowledge about the system is usually inadequate. When this is the case, modeling through the protocols of "adaptive environmental assessment" (Holling, 1978) can be successful, as it was to development of the Western Root Disease Model (Shaw *et al.*, 1985; Eav and Shaw, 1987; Stage *et al.*, 1990).

In this procedure, various experts in insect and disease recognition, biology, impact, and management, as well as potential users of the model, met for several short periods of intense interaction. Through the direction and assistance of model coordinators, they developed a conceptual model of the problem and possible management alternatives (Brookes, 1985). The time between workshops was devoted to synthesizing ideas, computer programming, behavior testing, and refining the model. At each successive workshop, participants reviewed model behavior and provided the modeling team with ideas for further improvements. The process itself was not new, but creatively extended the use of the scientific method from the individual investigator to a corporate surrogate (Walters, 1986).

A recognized strength of this procedure is that it gave ownership of the process, and thus a desire to create a quality product in a timely manner, to all involved. Also, because both scientists and managers participated in model building, scientists could appreciate the need to provide managers with the best current understanding of the situation, and managers could recognize the critical uncertainties in our knowledge of interactions among root pathogens and insects and the need for further research.

Cooperation between scientists and managers is necessary to accomplish an appropriate balance between model precision and utility. Model performance is sensitive to the quality of data provided for initialization. For example, Schowalter *et al.* (1982) evaluated the cost, in terms of accuracy (difference between sample and target means) and precision (replicability of results), of estimating *Dendroctonus frontalis* populations from different subsets of data collected during intensive research on infested trees. Data options ranged from all data on infested surface area and beetle life stage and density to only number of infested trees (data collected during conventional surveys). Accuracy and precision of estimation were compromised by using smaller subsets of data. The magnitude of error increased during beetle development as the relationship between beetle density and tree characteristics became subject to other factors. Predicted tree mortality, using the TAMBEETLE population dynamics model (Turnbow *et al.*, 1982), was underestimated by 40% when the model was given only number and infested surface areas of attacked trees, compared to predicted tree mortality when all data were provided (Schowalter *et al.*, 1982). Clearly, the reliability of prediction is determined by the quality of data used to initialize the model.

10.3 THE WESTERN ROOT DISEASE MODEL

10.3.1 Overview

At present, the Western Root Disease Model is the only pest model in North America that deals with interactions among root pathogens, bark beetles, and conifers. This model is both a pest dynamics model and a pest impact model. It provides a dynamic representation of the spatial and temporal epidemiology of pathogenic species of *Armillaria* or *Phellinus weirii* (McNamee *et al.*, 1989; Stage *et al.*, 1990; Shaw *et al.*, 1991). The model simulates spatial relationships among locations of infected stumps and infected and uninfected live trees to predict the spread of disease in the stand. It also simulates the persistence or legacy of disease from a harvested stand in succeeding stands.

To use the model, an inventory of stand conditions must be available and the extent of pest problems defined. For example, the compartment examination procedure described by Stage and Alley (1972) and USDA Forest Service Handbook for Region 1 (USDA Forest Service 1986) can supply the necessary stand data if it is augmented with information on stumps infected with root pathogens and stems harboring bark beetles. With this information available, an array of silvicultural prescriptions can be formulated and translated into model control commands that, along with the stand and pest inventory data, are required to begin a simulation. The projected changes over time in stand conditions, resource outputs, and pest conditions from each test prescription then can be compared as an aid in determining the best (depending on management objectives) treatment option for the stand.

The major relationships captured in the model are: the susceptibility of trees to infection by root pathogens; the vulnerability of trees to death once infected; disease-related reductions in stem growth; disease spread to previously uninfected areas; inoculum dynamics in infected dead trees and stumps; and the effects of windthrow and attack by bark beetles.

Physical and biological agents that may influence root disease incidence and severity can be included optionally in the simulation.

Default values in the model reflect current understanding, expressed primarily as the judgements and experiences of knowledgeable scientists and managers, of the epidemiology of pathogenic species of *Armillaria* and *P. weirii* in the interior western United States (Stage *et al.*, 1990). These parameters can be adjusted through a keyword mechanism to reflect more accurately particular pathogen attributes and stand or site conditions.

10.3.2 Model organization

In the Western Root Disease Model, root disease dynamics are modeled in six major parts:

(1) size and distribution of root disease areas (centers) at the start of a simulation;
(2) the dynamics of tree infection inside root disease centers;
(3) tree mortality and growth inside root disease centers;
(4) spread and enlargement of root disease centers;
(5) the persistence of root disease following tree harvest, through a newly regenerated stand;
(6) the interaction of "other agents" (windthrow and four types of bark beetles) with root disease.

In this chapter we concentrate on describing the portions of the Western Root Disease Model that deal with root disease/bark beetle interactions. Additional information on other model components appear elsewhere (Eav and Shaw, 1987; McNamee *et al.*, 1989; Shaw *et al.*, 1985, 1989, 1991; Stage *et al.*, 1990).

10.3.3 Simulating interactions

The Other Agents portion of the Western Root Disease Model is actually a collection of separate models, each of which simulates the action of one type of agent, such as bark beetles, on trees in the stand. These models require the user to specify the time when such an event can first occur and its severity.

Through the keyword mechanism users can "turn on" one or more of these agents during any desired projection cycle. From this time on, the agent remains potentially active. If the required stand conditions develop, then the specified event occurs and the switch is "turned off." Thus, a user could specify a cycle year of 2001, but not have the event occur until 2031 because conditions from 2001 to 2031 were not suitable for the event to occur.

Four types of bark beetle interactions and their effects on tree mortality can be simulated. In the first three types, bark beetles interact with root disease fungi primarily by changing inoculum levels and potential for disease spread. For example, trees killed by bark beetles can become *Armillaria* inoculum only if they were infected by the fungus prior to death. Thus, beetles that kill uninfected trees near disease centers could potentially slow disease spread because the model assumes that these root systems will not become inoculum sources. However, the model assumes that *Armillaria* quickly colonizes the entire root system of infected trees after tree death. In the fourth type, the probability of tree death depends on its distance from the root disease center. In all of these bark beetle events, infected trees that are killed by bark beetles do act as inoculum sources.

10.3.3.1 Type 1 bark beetles

An outbreak of Type 1 bark beetles occurs when the density of susceptible trees of a given species and minimum diameter exceeds a user-defined minimum. An attack of

Dendroctonus ponderosae on *Pinus contorta* is used as the prototype interaction from which parameters for Type 1 interactions are determined (see Chapter 9). *Dendroctonus ponderosae* or *D. brevicomis* attacks on *P. ponderosa* are also examples of Type 1 interactions. Default values for this type of an attack to occur are at least 10 *P. contorta* ≥ 20 cm DBH. When these minimum conditions are present, then 85% of the stems that meet these criteria will be killed.

Extensions to the Western Root Disease Model include effects of root disease caused by *Heterobasidion annosum* (Shaw *et al.*, 1989; McNamee *et al.*, 1991). These provide some additional options for Type 1 bark beetles. For example, a Type 1 event can occur more than once in a rotation and be triggered by a relative index of stand density rather than an actual minimum number of trees per hectare. In addition, users will have the option, regardless which measure of stand density they select, of calculating density for the entire stand, or only for areas outside of root disease centers. As reconfigured, this bark beetle type will effectively define stand density above which bark beetle induced tree mortality will occur.

10.3.3.2 Type 2 bark beetles

A Type 2 beetle infestation depends on the number of trees windthrown in the current time step. A time step is a period of time, usually 10 years, that the PROGNOSIS model of stand development uses to compute tree growth and mortality. An outbreak occurs if the number of windfallen stems of a suitable size for beetle breeding exceeds a user-specified minimum. Standing live trees are also attacked in proportion to their density and the density of windthrown trees that are attacked. An attack of *D. pseudotsugae* on *Pseudotsuga menziesii* is used as the prototype for a Type 2 interaction from which default parameters are set (see Chapter 9).

The number of live trees killed by the bark beetle is only slightly fewer than the number of host trees windthrown. For example, if 12 *P. menziesii* of the size specified were windthrown and attacked, then 11 more live trees would be attacked and killed by the bark beetle, if this many eligible stems remained after windthrow. Default values for this type of an attack to occur are at least three standing *P. menziesii* ≥ 25 cm DBH and 10 windthrown stems. When these minimum conditions are present, then 88% of the standing, live trees that meet these criteria will be killed by a Type 2 bark beetle event.

The windthrow submodel simulates major blowdown events in root disease centers. Dominant and codominant trees (defined as the largest 20% of the stems in the stand) are windthrown if the density of eligible stems exceeds a user-defined minimum (default = 0). For infected trees that are windthrown, the model determines the number of stems that tip over, removing their root systems and inoculum from the soil, as opposed to snapping off and thus leaving the root system and inoculum in the ground (see Chapter 9). This decision is based on the proportion of the root system of each tree that is infected by root pathogens. Infected, windthrown stems that tip over and expose their entire root system do not contribute further to root disease spread, but can harbor bark beetles. Infected trees that are snapped off become inoculum sources for root disease as well as a breeding ground for bark beetles. To simulate a Type 2 bark beetle event effectively, the user must change the windthrow default for the proportion of eligible stems to be windthrown from 0 to a value that will ensure that at least 25 trees per hectare are windthrown.

10.3.3.3 Type 3 bark beetles

An outbreak of Type 3 bark beetles occurs if the density of a given tree species with sufficient size and proportion of root system infected by root pathogens exceeds a user-defined minimum. That is, this type of bark beetle event is dependent on root infection, tree species, and tree size and density. An attack by *Scolytus ventralis* on *Abies grandis* is used as the prototype Type 3 interaction from which parameters are estimated (see Chapter 9). Default values for this type of an attack to occur are at least 10 trees > 25 cm DBH, each of which has at least 30% of its root system infected by root pathogens. When these minimum conditions are present, then 88% of the stems that meet these criteria will be killed.

10.3.3.4 Type 4 bark beetles

Like a Type 3 event, an outbreak of Type 4 bark beetles occurs if the density of a given tree species with sufficient size and proportion of root system infected by root pathogens exceeds a user-defined minimum density. The difference lies in the impact of the event, because the probability of a tree dying depends on its location relative to that of root disease centers. Three types of tree conditions and locations are recognized: (I) uninfected trees at a defined distance from root disease centers; (II) uninfected trees inside root disease centers or within "close" proximity to root disease centers; and (III) infected trees inside root disease centers.

The criteria for closeness to a root disease center are determined by a keyword (Stage *et al.*, 1990). This keyword originally was designed to extend the radius of a root disease center after a clearcut harvest by the mean radius of all trees in the stand to account for non-symptomatic trees on the edge of the root disease center that actually were infected by root pathogens. Trees that would fall within this distance, as calculated at the time the event occurs, are counted as being close to the center and thus fall into the above-mentioned category II. For the Type 4 bark beetle, these trees are considered vulnerable and subject to attack, as are trees within the infection center. Suggested default values for the probability of death for trees in categories I, II, and III are, respectively, 0.001, 0.4, and 0.88.

At present, trees killed by root disease or other factors within a time step are not counted in the density requirements that must be met for the bark beetle event to occur. In the Annosus extension to the Western Root Disease Model, trees that died within the simulation time step will be eligible for counting to meet the density requirements for the event to occur. Part of the logic behind this extension is that weakened trees destined to die within the time step are prime candidates for attack by bark beetles.

10.4. EXAMPLE SIMULATION

The following example illustrates interactions between *Armillaria* root disease and a Type 3 bark beetle (i.e. *Scolytus*) on *Abies concolor* as simulated by the Western Root Disease Model. The example stand, known as the Strawberry Stand, was about 100 years old when inventoried in 1990. The stand is on a 6–16% slope with northwest orientation. The stand is composed mainly of *Abies concolor* (84%), *Pinus ponderosa* (9%) and *Calocedrus decurrens* (6%). Tree DBH ranges from 18 to 177 cm. The average tree has a DBH of 36.6 cm and a height of 25.3 m. Approximately 83% of the 4 ha stand is infected with root disease. Within its 10 root disease centers, 29–30 of the 311 trees ha^{-1} are infected. Three scenarios are simulated for 120 years using 1990 inventory data from this stand: (1) simulation of stand development assuming no losses from root disease or bark beetles; (2) simulation of stand development with root disease alone, and (3) simulation of stand development with root disease and a bark beetle outbreak in the year 2000.

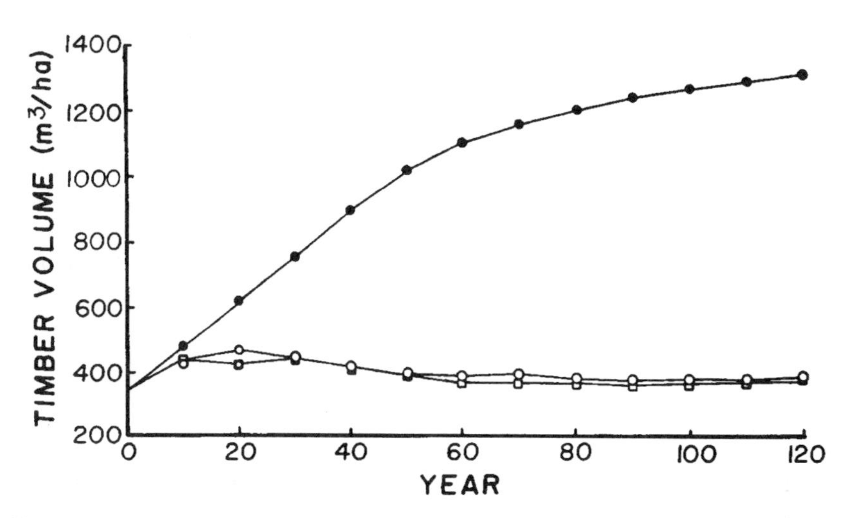

Fig. 10.2. Projected timber volumes in Strawberry Stand with no root disease or bark beetles (closed circles), root disease only (open circles), and root disease with bark beetles (squares).

Fig. 10.2 illustrates the gross overestimation (250% at year 2110) of timber "available" for harvest when the stand is assumed to develop with neither root disease or bark beetles present. The model's projection of stand development when root disease and bark beetles are considered is far more realistic than the uninfected stand assumption. Simply put, if a stand model alone is used to project volume for forest resource planning purposes, then erroneous conclusions are likely to be reached in stands affected by these pests — a not uncommon situation in conifer forests of western North America.

In the long run, the model projects that the additional impact of bark beetles over that of root disease alone is minimal. This result is not unexpected since the model allows bark beetles to kill only trees already infected by root pathogens. These infected trees eventually would die from root disease alone. This situation also is depicted in Fig. 10.3, which shows the number of trees killed per ha within root disease centers through time. The bark beetle

Fig. 10.3. Timber mortality as a result of root disease alone (open circles) or root disease and associated bark beetles (closed circles).

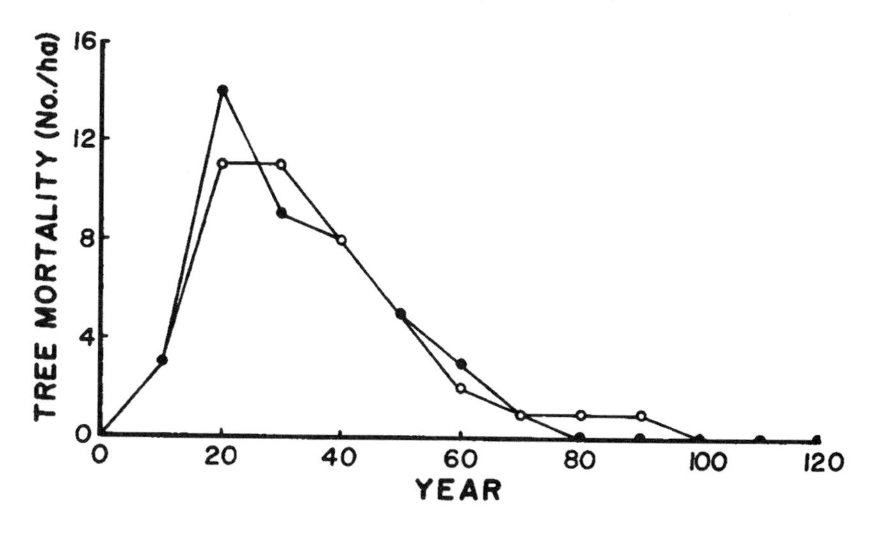

outbreak in year 2000 is reflected in the increased number of dead trees tallied in year 2010. However, the removal of a number of infected trees (from the available pool of live-infected trees to be killed by root disease) by bark beetles is reflected in the lower mortality rates during the next two projection cycles. The bark beetle simply killed these trees before they succumbed to root disease. In contrast to these scenarios, we would expect some additional impact from a Type 4 bark beetle attack because some trees uninfected by root disease also would be killed.

As shown in Fig. 10.4, this Type 3 bark beetle outbreak did, however, increase the stand area affected by root disease. This phenomenon results from the model assumption that, after tree death, *Armillaria* quickly colonizes the entire root system of affected trees.

10.5 CONCLUSIONS

Interactions among bark beetles and root pathogens are important to the dynamics of both pests in the forest, as well as to overall ecosystem structure, function, development, and productivity. Consequently, development of models that describe and predict bark beetle and pathogen epidemiologies and tree mortality has become a major focus of attention.

The Western Root Disease Model is the only pest model in North America that currently deals with these interactions. Although far from complete, the current model should continue to improve as a result of validation studies and new information. In particular, the various effects of tree mortality on timber value and ecosystem processes, as discussed in Chapter 9, must be addressed. The model is being analyzed for its sensitivity to changes in the various parameters that control it, thus testing the assumptions and hypotheses under which it was developed. We believe that the parameters to which the model is most sensitive should be the focus of future research to improve model performance.

This model and the process used to develop it provide a framework for transferring technology on the dynamics and impacts of pest interactions to forest managers for use both in short term, site specific management decisions and in long term planning. To help ensure their utility, such models should be developed so that they can function in concert with

Fig. 10.4. Stand area occupied by root disease over time with associated bark beetles present (closed circles) or absent (open circles).

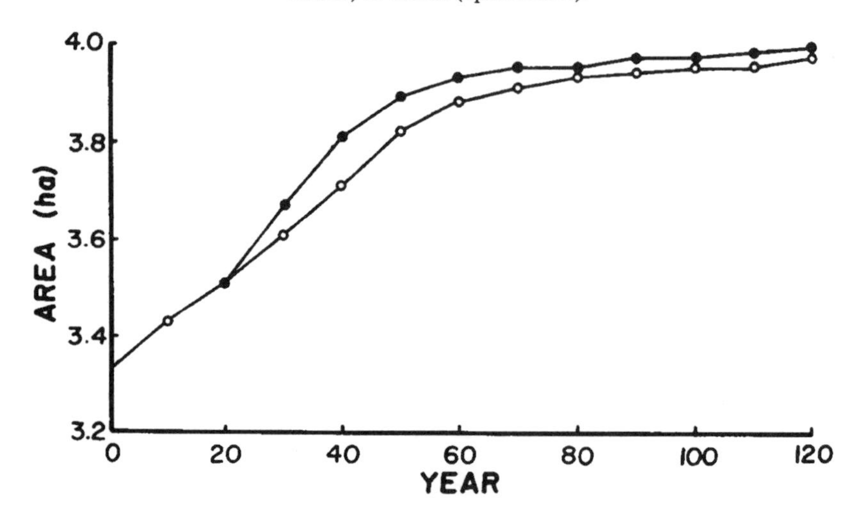

existing models that predict other stand attributes (i.e, overall growth and development). The process of model development is a valuable tool for scientists to clarify the current state of knowledge and to focus management-oriented research needs.

REFERENCES

Brookes, M. (1985). Scientists, managers, and modelers work on root diseases. Forest Research West (August), pp. 12–15.

Eav, B.B. and Shaw, C.G. III. (1987) The Western Root Disease Model: A status report. *In* "Proc. 35th Western International Forest Disease Work Conference" (G.A. Denitto, ed.), pp. 84–92. USDA Forest Serv., Pacific Southwest Res. Stn., Berkeley, CA.

Holling, C.S. (ed). (1978) "Adaptive Environmental Assessment and Management." John Wiley, New York, 377 pp.

Lynch, A.M. (1990) Modeling pest systems for sampling multiple damaging agents. *In* "Proc. 1990 Society of American Foresters National Convention", pp. 160–165. Society of American Foresters, Washington, DC.

McNamee, P.J., Sutherland, G., Shaw, C.G. III *et al.* (1989) Description of a multi-species root disease model developed for silvicultural planning and management in coniferous forests of western North America. *In* "Proc 7th IUFRO Conference on Root and Butt Rot in Forest Trees" (D.J. Morrison, ed.), pp. 320–335. Forestry Canada, Pacific Forest Research Centre, Victoria, BC.

McNamee, P.J., Kurz, W.A., Daniel, C.J., Robinson, D.C.E. and Deering, M.G. (1991). Description of the annosus root disease model. Environmental and Social Systems Analysis, Ltd. Vancouver, BC. Report prepared for USDA Forest Serv., 70 pp.

Schowalter, T.D., Coulson, R.N., Turnbow, R.H. and Fargo, W.S. (1982). Accuracy and precision of procedures for estimating populations of the southern pine beetle (Coleoptera: Scolytidae) by using host tree correlates. *Environ. Entomol.* **75**, 1009–1016.

Shaw, C.G. III, Stage, A.R., and Webb, T.M. (1985) Development of a root disease subroutine for use with stand growth models of western forests. *In* "Proc. 33rd Western International Forest Disease Work Conference" (W.G. Thies, ed.), pp. 48–54. USDA Forest Serv., Pacific Northwest Res. Stn., Portland, OR.

Shaw, C.G. III, Goheen, D.J. and Eav, B.B. (1989) Simulation of impacts of annosus root disease with the Western Root Disease Model. *In* "Proc. Symposium on Research and Management of Annosus Root Disease (*Heterobasidion annosum*) in Western North America," pp. 129–139. USDA Forest Serv. Gen. Tech. Rep. PSW-116. USDA Forest Service, Pacific Southwest Forest and Range Exp. Stn., Berkeley, CA.

Shaw, C.G., III, Stage, A.R. and McNamee, P. (1991) Modeling the dynamics, behavior, and impacts of Armillaria root disease. *In* "Armillaria Root Disease" (G.G. Shaw III and G.A. Kile, eds), pp. 150–156. USDA Forest Serv., Agric. Handbook 691, Washington, DC.

Stage, A.R. (1973) Prognosis model for stand development. USDA Forest Serv. Res. Paper INT-137. USDA Forest Serv., Intermountain Forest and Range Exp. Stn., Ogden, UT. 32 pp.

Stage, A.R. and Alley, J.R. (1972) An inventory design using stand examinations for planning and programing timber management. USDA Forest Serv. Res. Paper INT-126. USDA Forest Serv., Intermountain Forest and Range Exp. Stn., Ogden, UT, 17 pp.

Stage, A.R., Shaw, C.G. III, Marsden, M.A., *et al.* (1990) "Users' Manual for Western Root Disease Model." USDA Forest Serv. Gen. Tech. Rep. INT. No. 267. USDA Forest Serv., Intermountain Research Station, Ogden, UT, 49 pp

Stephen, F.M., Searcy, J.L. and Hertel, G.D., eds (1980). "Modeling Southern Pine Beetle Populations: Symposium Proceedings." USDA Forest Serv. Tech. Bull. 1630. USDA Forest Serv., Washington, DC, 174 pp.

Turnbow, R.H., Coulson, R.N., Hu, L. and Billings, R.F. (1982). "Procedural Guide for Using the Interactive Version of the TAMBEETLE Model of Southern Pine Beetle Population and Spot Dynamics." Texas Agric. Exp. Stn. Publ. MP-1518, Texas A&M University, College Station, TX.

USDA Forest Service. (1986) Field instructions: stand exam: Region 1. Timber management data handbook FSH2409.21h, Rl chapter 400. USDA Forest Serv., Northern Region R-l, Missoula, MT, 442 pp.

Walters, C.J. (1986) "Adaptive Management of Renewable Resources." MacMillan Press, New York, 374 pp.

Wykoff, W.R., Crookston, N.L. and Stage, A.R. (1982) User's Guide to the Stand Prognosis Model. USDA Forest Serv. Gen. Tech. Rep. INT-133. USDA Forest Serv., Intermountain Forest and Range Exp. Stn., Ogden, UT, 112 pp.

–11–

Managing Root Disease and Bark Beetles

S. HAGLE[1] and R. SCHMITZ[2]
[1] *USDA Forest Service, Northern Region, Missoula, MT, USA*
[2] *USDA Forest Service, Intermountain Research Station, Forestry Sciences Lab., Ogden, UT,*

11.1 INTRODUCTION

Root pathogens and bark beetles are natural components of conifer forest ecosystems (Chapter 9). In most cases endemic populations of these organisms function to remove weak individual trees, leaving the stand healthier. Hence, widespread epidemics of these organisms should be relatively rare events in healthy natural forests and in stands managed according to ecologically sound tenets. This has been a premise of forest pest management for some decades, but over the past 20 years severe bark beetle and root pathogen epidemics have been common in western and southern North America. Where have we gone wrong?

We are still in the process of trying to understand what the ecologically sound tenets should be. For example, autecological research indicated that young stands of *Pseudotsuga menziesii* that were carefully planted from seed of superior-appearing parents, tended by thinning at proper times to avoid excessive competition, and harvested as they reached maturity should develop into healthy forests. We realize now that this may be a partial answer on some sites but certainly not on all, or even most, sites. We have come to understand that stands with apparently ideal growing conditions, produced from local, presumably locally adapted, seed sources, thinned or even fertilized to maintain optimum growth, i e. stands that should be vigorous, still are beset with root pathogen and bark beetle epidemics. We now are looking beyond particular tree species and stands to the community and site, specifically to the natural succession of communities characteristic of the site. The roles bark beetles, pathogens and associated species play in forests may vary considerably among successional stages on any one site.

BEETLE–PATHOGEN INTERACTIONS IN CONIFER FORESTS
ISBN 0-12-628970-0

Epidemics of insects and pathogens induced by inappropriate management practices are not easily rectified. Pathogenic fungi, once established tend to endure in stumps and other residues for many decades (Chapter 9). Prevention provides the surest means to manage root diseases, as well as other potential pests. However, many common management practices create conditions that may affect forest insects and pathogens in unknown ways.

Changes are required to make forest management consistent with the ecological principles that govern forest health and productivity, as we understand them. This does not mean that we necessarily should return stands to the conditions that existed prior to Euro-American settlement, although a management emphasis on maintaining healthy forests and preventing pest outbreaks probably would result in a significant shift toward those conditions. Forest management practices during the past century have caused changes in species composition, stand age, and disturbance patterns (e.g. Barrett and Arno, 1982; Gruell, 1985; Schowalter *et al.*, 1981). The effects of these management practices on bark beetles and root pathogens and options for averting pest activity in the future are the focus of this chapter.

11.2 EFFECTS OF PAST AND CURRENT MANAGEMENT PRACTICES

11.2.1 Species composition

Populations of pathogens and insects are dependent largely on host abundance and distribution. The increased abundance of *Pseudotsuga menziesii* and *Abies* spp. in western North American forests and increased abundance of *Pinus taeda* in southern forests have been accompanied by increased incidence of bark beetle and pathogen outbreaks. The ease with which *Phellinus weirii* finds a suitable host greatly influences its survival on a site. The same can be said for *Dendroctonus pseudotsugae* and other species of insects and root pathogens.

Light ground fires once swept through southern and inland western forests at frequencies of only a decade or two. Many of these fires were ignited by lightning strikes, others by Native Americans for a variety of purposes (Barrett and Arno, 1982; Gruell, 1985). Whatever the source, fires would burn to the limits of available fuel or favorable weather. Euro-American settlement in North America rapidly changed fire frequencies. By the turn of the century, fire frequencies in pine forests throughout North America were reduced considerably (Arno, 1976; Stokes and Dieterich, 1980; Martin, 1982; Steele *et al.*, 1986). The change in fire frequencies can be explained entirely by human activity. The resulting changes in forest stand species composition, age and diameter distributions, and subsequent status of some particularly troublesome disease and insect problems can be linked directly to these human activities (Schowalter *et al.*, 1981).

At about the same time fire intervals were being altered, selective logging was creating additional changes in species composition in North American forests. Commercial cutting of *Pinus monticola, P. ponderosa* and *Thuja plicata* in interior western forests left the inferior quality, mostly understory *Pseudotsuga menziesii* and *Abies* spp. (Figs 11.1 and 11.2). *Pinus monticola* abundance and distribution were affected further by mortality to *Dendroctonus ponderosae* and the pathogen *Cronartium ribicola*, that was introduced accidentally from Europe shortly after the turn of the century in western Canada. The disease spread rapidly through vast stands of aging *P. monticola* in western North America, stimulating additional logging to salvage the threatened pine. Most *P. monticola* seedlings suc-

Fig. 11.1. Mature *Pinus monticola* stand, 1938. (Photo courtesy of H. Miller Cowling, USDA Agric. Res. Serv., Bureau of Entomology and Plant Quarantine, Division of Plant Disease Control.)

Fig. 11.2 Selective removal of *Pinus monticola* from stands radically altered species composition. *Pseudotsuga menziesii* and *Abies grandis* became the predominant seed sources for regenerating this stand. (Photo 1953, courtesy of H. Miller Cowling, USDA Agric. Res. Serv., Bureau of Entomology and Plant Quarantine, Division of Plant Disease Control.)

cumb to the disease within their first two decades, effectively limiting natural regeneration of this species.

Planting programs, that became increasingly important to regenerating harvested land and land burned by wildfires from the 1920s to 1980s, generally were aimed at producing monocultures of commercially valuable tree species. For example, *Pseudotsuga menziesii* has been widely planted as single-species plantations on inland sites that previously had supported mixed-species seral stands with major components of *Pinus monticola*, *P. contorta* and *Larix occidentalis* or climax stands of *T. plicata* or *Tsuga heterophylla*. Although this practice continues in many areas, there has been a major shift in recent years toward promoting mixed-species stands through artificial and natural regeneration.

Pacific coastal forests are more mesic and consequently experience stand replacement fires less frequently. *The Pseudotsuga menziesii* sere of coastal forests can survive for 500 or more years and typically is followed by an extended *Tsuga heterophylla* and *Thuja plicata* climax of 1000 years or more (Barrows, 1984; Hermann, 1985). *Phellinus weirii* occurs in discreet patches in coastal *Pseudotsuga menziesii* forests, killing all or most of the trees within patches (Chapter 9). Dying *P. menziesii* are replaced by *Thuja plicata* and *Tsuga heterophylla* (Hadfield and Johnson, 1977) or immune hardwoods, such as *Acer circinatum* and *Alnus rubra* (Chapter 9). Disease inoculum likely would decline during long periods of non-host domination of infected sites in premanaged forests, perhaps lingering in surviving infected trees. Clearcut harvesting and replanting with *Pseudotsuga menziesii* changed this situation and promoted continued disease spread in young forests. Consequently, *Phellinus weirii* in western Oregon is thought to be more severe in second generation *Pseudotsuga menziesii* forests than it was in the first at a comparable age (Tkacz and Hansen, 1982; Chapter 9).

Similarly, forests on the southeastern coastal plain were dominated by *Pinus palustris* at the time of European settlement. This species is relatively resistant to *D. frontalis* (Hodges *et al.*, 1979) and tolerant of drought and fire that occurred frequently in this region (Schowalter *et al.*, 1981; Coulson *et al.*, 1983, 1985). However, *P. palustris* is relatively slow-growing, especially for the first 3–6 years ("grass stage") during which development of the extensive root system necessary for drought tolerance precludes shoot growth (Harlow and Harrar, 1950). Consequently, the more shade-tolerant and rapidly growing *P. taeda* has been favored for timber production, but this species is more susceptible to *D. frontalis* (Hodges *et al.*, 1979; Schowalter *et al.*, 1981).

11.2.2 Stand age and density

Fire exclusion also has resulted in development of older forests. Stand replacement fires were a particularly important feature of interior *Pinus contorta* forests. Fires burning through stands of *P. contorta* killed trees in patches and left other patches unburned or only lightly burned. The result was a mosaic of stands of various ages with varying susceptibility to bark beetles and pathogens (Gara *et al.*, 1985). Fire exclusion has produced uniformly old stands of *P. contorta* that provide ideal conditions, e.g. adequate phloem thickness and closely spaced hosts, for continued survival and spread of *D. ponderosae* (Schmidt, 1989).

During the past 20 years, bark beetle outbreaks have been particularly severe in the western United States, especially within the Northern Rocky Mountains, in areas last burned in the early 1900s. The regeneration that followed these extensive wildfires resulted in vast areas of pure, even-aged stands that subsequently were protected from wildfire. The exclusion of fire allowed many stands to reach maturity and overmaturity with little change in species composition and age structure, contributing to severe and prolonged bark beetle infestations. The most destructive of these infestations involves *D. ponderosae* in *P. contorta* forests.

Pseudotsuga menziesii stands suffer increasing volume loss to *Phaeolus schweinitzii* as they mature. Root infections that began when trees were young slowly develop in roots and, eventually, in butts of trees. Direct killing of immature trees by this fungus generally is limited to especially harsh sites. While this effect is locally important, it accounts for only a minor part of the overall impact of this disease. Butt rot, windthrow and bark beetle attack account for most of the volume losses attributable, in part, to infection by this root pathogen.

Old *Pseudotsuga menziesii* throughout its range typically is affected by *Phaeolus schweinitzii*. This probably was not an unusual condition in pre-Euro-American times. Old-growth *P. menziesii* stands are still in evidence, but these trees are now the overstory of dense *P. menziesii* and *Abies* stands. Historically, these overstory components were protected from competition by frequent cool fires that killed much of the regeneration that developed following each fire. Today, the dense understory vegetation competes for water and nutrients with the old overstory trees. *Arceuthobium* spp. readily colonize hosts in the understory, and ground fire no longer removes small, infected trees. Aggressive root pathogens now spread easily from tree to tree where root contact among closely spaced trees provides access to adjacent hosts. In the past, fungus inoculum for such epidemics may have languished for years in isolated root systems of killed trees that were too far from live roots of suitable hosts to facilitate spread. Change in forest structure and condition has changed the relative importances of various interactions. Where a relatively mild pathogen, *Phaeolus schweinitzii*, once posed the greatest threat, aggressive *Armillaria ostoyae* and *Phellinus weirii* are now responsible for increasing mortality.

Pseudotsuga menziesii are reaching diameters suitable for *D. pseudotsugae* in stands of high density and species purity. These conditions are ideal for *D. pseudotsugae* colonization and reproduction (Furniss *et al.*, 1981). Relatively pure, dense stands of *Pinus* in the southern US are susceptible to *D. frontalis* (Lorio, 1980). Pure, dense stands of *Abies grandis* are highly susceptible to *Scolytus ventralis* attack.

Endemic populations of *D. pseudotsugae* and *S. ventralis* are found in association with patches of root disease (Chapter 9). Major windthrow events, drought, defoliation, or abundant logging debris generally can be identified as triggers for epidemics of these beetles in dense, nearly pure stands of host trees, but annual losses associated with endemic populations of these beetles far exceed those of sporadic epidemics (Chapter 9). Therefore, managing these beetles requires management of factors that maintain endemic populations (i.e. root diseases).

11.2.3 Site or stand disturbance

Root disease fungi and some bark beetles capitalize on stand disturbances that increase availability of stumps and wounded, live trees. *Armillaria, Phellinus* and *Heterobasidion* use stumps and roots of dead trees (Fig. 11.3) as food bases from which to spread to live trees (Greig and Pratt, 1976; Filip, 1979; Tkacz and Hansen, 1982). Spores of *H. annosum* infect basal wounds on live trees (Hodges, 1969) and freshly cut stump surfaces. *Leptographium wageneri* is carried into stands by insects attracted by stand disturbance (Harrington *et al.*, 1985; Witcosky *et al.*, 1986; Chapter 9). Once established, root pathogens move aggressively outward into other susceptible residual trees or regeneration.

Repeated partial harvest, such as that resulting from economic selection cutting, sanitation/salvage cutting, or uneven-age management, increases both the frequency and severity

Fig. 11.3. *Pinus ponderosa* saplings killed by *Armillaria ostoyae* that originated from an infected residual *P. ponderosa* stump. (Photo by G.M. Filip.)

of root disease in stands. Even one harvest entry in stands has been found to increase the frequency of root disease greatly, relative to stands that have not been entered for tree harvest. Surveys of the Fremont and Ochoco National Forests of Oregon revealed a high infection rate of *Abies concolor* and *A. grandis* in stands with at least one harvest entry compared to those that had no harvest history (Schmitt *et al.*, 1984; Goheen and Goheen, 1989). *Heterobasidion annosum* was found affecting *Abies* spp. in 0% and 12% of the surveyed stands with no harvest history on the Ochoco and Fremont National forests, respectively. With a single harvest entry, 50% and 40% of the stands were affected by the disease, and with multiple entries, 33% and 100% were affected. Disease severity was correspondingly greater in the harvested stands than those with no harvest history: 0.5% and 1.7% of the *Abies* spp were killed in the once-entered stands, and 2.9% and 20.8% in the stands with multiple harvest entries, compared to 0% and 0.04% killed in the stands with no harvesting.

Byler *et al* (1987) reported a doubling of root disease frequency in stands on the Lolo National Forest in Montana with at least one harvest entry compared to those with no history of tree harvest. *Armillaria ostoyae* and *P. weirii* were the most damaging root pathogens in these stands. On the Winema National Forest in Oregon, 37% of the timber volume in a twice-entered stand was affected by *Armillaria ostoyae*, that killed primarily *Abies concolor* (Filip, 1977). *Pinus ponderosa* had been selectively logged. More than half of the infected trees already had died. Commercial thinning in a 55-year-old *Pseudotsuga menziesii* stand on the Flathead Indian Reservation in Montana was followed by a steadily increasing annual mortality rate resulting in death of 60% of the stand after 10 years (Hagle and Goheen, 1988). At the present rate, the stand will be eliminated by root disease long before it reaches the planned rotation age of 90 years.

Factors that influence the amount of mortality in a stand after partial harvest include, but are not limited to, inoculum density prior to tree harvest, amount and severity of tree wounding, tree species composition before and after harvest, tree diameter distribution, and pathogen and insect species. Consequently, we are unable to accurately predict mortality rates following partial stand harvest.

Few root disease problems have been reported to result from thinning stands of trees that were not of commercial size (20 cm DBH). A notable exception is *L. wageneri*. This fungus is transmitted by root-feeding insects that are attracted to disturbed sites and damaged trees (Harrington *et al.*, 1985; Witcosky *et al.*, 1986; Cobb, 1988; Hansen *et al.*, 1988; Chapter 9). Precommercial thinning of stands of *Pinus ponderosa* in central Oregon did not significantly influence rates of mortality caused by *A. ostoyae*, but did increase volume production (Filip *et al.*, 1989).

High use areas such as campgrounds, picnic areas, administrative and home sites have special problems resulting in root disease (Filip and Goheen, 1982a) and bark beetle outbreaks. Site development nearly always involves some selective tree cutting. Unprotected stump surfaces have provided infection courts for *H. annosum* in both *P. ponderosa* and mixed-conifer stands (Morrison and Johnson, 1978). Additionally, selection of sites for development seldom has included provision for avoiding stands in which root pathogens were already epidemic. The result often has been discovery of root disease problems shortly after completion of a development project.

11.3 MANAGEMENT STRATEGIES FOR ECOLOGICAL BALANCE

Recognition of detrimental effects of human activity on bark beetle and pathogen populations has led to efforts to reverse human-caused changes. While our understanding of interactions between bark beetles and root diseases is far from complete, existing knowledge is sufficient to devise alternative management strategies that will help maintain pest populations at endemic levels. At the same time, we recognize that bark beetle epidemics also can result from uncontrollable factors such as wind, fire, and prolonged drought that cause rapid increases in the numbers of susceptible trees and favor the development of population outbreaks (Chapter 4). In such situations, non-silvicultural mitigation may be required to bring bark beetle or pathogen populations under control.

11.3.1 Maintaining disease resistant or tolerant species

Identifying species that can resist or tolerate root pathogens or bark beetles is based on observation of tree species mortality rates in affected stands. *Pinus* spp. and *Larix occidentalis* often are more resistant to root pathogens than are *Pseudotsuga menziesii* and *Abies* spp. (Hadfield *et al.*, 1986; Hagle and Goheen, 1988). Byler *et al.* (1990) were able to rate stands for risk of root disease occurrence based on site factors. *Abies grandis* and *T. heterophylla* habitat types (Pfister *et al.*, 1977) and *P. menziesii* habitat types on southern aspects were at highest risk of *Armillaria* and *Phellinus* occurrence. Species composition on these site types now is quite different from that occurring prior to fire exclusion practices (Byler, 1984). *Abies grandis* and *T. heterophylla* habitat types originally were dominated by seral *Pinus monticola, L. occidentalis* and *P. contorta.* South-facing *Pseudotsuga menziesii* habitat types generally were occupied by low-density *Pinus ponderosa* stands maintained by frequent low-intensity fires. On these sites, *P. ponderosa, P. monticola, L. occidentalis* and *P. contorta* often are resistant or tolerant to root disease while *Pseudotsuga menziesii* and *Abies* spp. are highly susceptible. Seral species seem to do well on these sites and, indeed, natural monocultures of *Pinus ponderosa* were common consequences of natural fire (and perhaps root pathogen) cycles on *Pseudotsuga menziesii* habitat types.

Artificial regeneration often is the only means available to re-establish seral stands because seed sources were eliminated during previous timber harvest. Clearcuts are generally very favorable for establishment of shade-intolerant seral species. Seed sources must be matched carefully to the site when using artificial regeneration methods. Most conifer species have genetically distinct, locally adapted populations. Trees that are planted on sites to which they are not adapted may be considerably more susceptible to disease than is characteristic of the species on native sites (Hagle and Shaw, 1991). Seed tree (individual trees left as seed sources) and, on some sites, shelterwood methods also are effective for regenerating seral species. While spread of inoculum from stumps of harvested trees may result in a slightly elevated mortality rate for the disease-resistant seed tree or shelterwood residual trees, most of the trees can be expected to live long enough to serve their purpose of producing seed or shelter through the regeneration phase.

Other methods are being employed on sites that have few seral-species trees available to leave as seed trees but on which clearcutting is undesirable for silvicultural or non-silvicultural reasons. Leaving widely spaced *P. menziesii* or *Abies* spp. on such sites and planting with *Pinus* or *L. occidentalis* is a method that allows establishment of a root disease-resistant stand without clearcutting. Most of the overstory is killed by root disease and bark beetle attack within a few decades but will have served its purpose of providing shelter, visual diversity, watershed protection, etc.

Intermediate levels of resistance or tolerance to root diseases are recognized in some tree species. *Abies grandis* is killed by *Armillaria ostoyae, Phellinus weirii*, and *H. annosum* at about half the rate found in *Pseudotsuga menziesii* in Idaho and Montana, but the reverse may be true in eastern Oregon and southeastern Washington (Filip and Schmitt, 1979; Hagle and Goheen, 1988). *Pseudotsuga menziesii, Abies grandis* and *A. concolor* are more susceptible than are *A. lasiocarpa* and other *Abies* species (Hadfield *et al.*, 1986; Hagle, 1985). The relative proportions of species with intermediate levels of resistance may influence the impact of disease in infected stands greatly. Relative resistance cannot be uniformly applied throughout North American and should not be used as a replacement for monitoring regional performance of species.

Mortality rates resulting from *A. ostoyae* changes in some tree species at about 30 years of age. *Pinus* spp., *L. occidentalis* and coastal *Pseudotsuga menziesii* are readily killed by *A. ostoyae* as seedlings and saplings (Byler *et al.*, 1985; Filip *et al.*, 1989). Those trees that survive these stages have decreasing probabilities of death from *A. ostoyae* (Morrison, 1981; Filip and Goheen, 1982b; Hagle and Goheen, 1988). The mechanism accounting for this change is not known. Resistance or tolerance to infection may increase in these species or, perhaps, *A. ostoyae* fails or ceases to spread from tree to tree in older stands of these tree species.

Managing for seral species often is the answer to root diseases caused by *A. ostoyae* or *Phellinus weirii* where *Pseudotsuga menziesii* or *Abies* spp. are the most susceptible hosts and *Pinus* or *L. occidentalis* would have been the major stand components under natural fire regimes. However, there are some sites that, for unknown reasons, display the opposite pattern of species resistance to root disease caused by *A. ostoyae* (Shaw *et al.*, 1976). In this case, *P. ponderosa* is the most susceptible tree species, and *Pseudotsuga menziesii* (the coastal variety that becomes somewhat resistant to *Armillaria* after age 30) is managed as the preferred resistant tree species. Also, other pests, notably *Pinus* (P) strains of *H. annosum* and *Leptographium wageneri*, and *D. brevicomis* and *D. ponderosae*, cause heavy damage in seral and climax *Pinus ponderosa* stands where tree cutting has created unnatural conditions (Bega and Smith, 1966). Other management strategies are needed to combat these problems. Host specificity in *H. annosum* provides an advantage to the forest manager. The two intersterility groups in the western United States (Chase, 1989; Chapter 3) show important differences in host specificity. The P-type of *H. annosum* causes a disease of *Pinus* spp. to which *Abies* spp. are resistant, whereas the S-type causes a disease of *Abies* spp. to which *Pinus* spp. are resistant (Worrall *et al.*, 1983; Chapter 3). Rapid tests are being developed for distinguishing these biological species, but observed mortality by tree species also will indicate which fungus is present in a stand.

11.3.2 Maintaining species diversity

Stands with a high degree of diversity can withstand the loss of one tree species during a bark beetle or pathogen epidemic better than can stands with low diversity. Unattacked trees would continue to provide at least partial shade and other resources, reduce erosion, etc. A combination of host and non-host species may be sufficient in many cases to avert pathogen or insect epidemics. Efforts to minimize tree mortality to bark beetles by altering species composition must take into account the susceptibility of all tree species to shared bark beetles and pathogens, as well as other potential pests.

Root pathogens and bark beetles are favored in stands that have adequate densities of preferred hosts. On the Lolo National Forest of Montana, stands in which *Pseudotsuga menziesii* and *A. grandis* comprised more than 40% of the basal area had a significantly higher percentage of plots with trees killed by *Armillaria* spp. or *Phellinus weirii* than stands with lower proportions of *Pseudotsuga menziesii* and *Abies grandis* (Byler *et al.*, 1987). The proportion of *P. menziesii* is an important factor in rating stands for susceptibility to *D. pseudotsugae* attack as well (Furniss *et al.*, 1981). Stands in which *P. menziesii* comprises less than one-third of the basal area have a low probability of *D. pseudotsugae* epidemics. *Heterobasidion* infection of *Pinus ponderosa* was less prevalent in mixed-conifer stands (4% stump infection) compared to nearly pure *P. ponderosa* stands (18% stump infection) on the McCloud Ranger District in northern California (DeNitto, USDA Forest Serv., unpublished data). Some species of bark beetles, such as *D. frontalis* and *D. ponderosae*, attack more than one tree species. Where *P. contorta* and *P. ponderosa* grow intermingled, *P. contorta* often is killed by *D. ponderosae* at a greater initial rate than is *P. ponderosa*. However, given sufficient numbers, *D. ponderosae* also will kill *P. ponderosa*. Similarly, *D. frontalis* typically attacks *P. taeda* and *P. echinata*, but at sufficient densities will kill other *Pinus* and some *Picea* spp. as well (Payne, 1980).

The flight capability of beetles also requires that managers consider the threat posed by infestations developing in nearby areas not managed to minimize bark beetle populations. Natural processes or human activities that create extensive areas of nearly pure mature stands favor severe and prolonged bark beetle outbreaks. Epidemic populations of *D. ponderosae* may find and kill any *P. contorta* until most suitable trees, generally over vast areas, are dead (Amman and Baker, 1972; Wickman, 1990). Similarly, epidemic populations of *D. frontalis* in the southeastern US discovered and colonized scattered *P. rigida* persisting in late successional hardwood forests in the southern Appalachian Mountains (T.D. Schowalter, personal observation). *Heterobasidion* and *Leptographium* management also requires consideration of populations outside the managed stand, because of the spore dispersal capacity of *H. annosum*, and the vectoring of *L. wageneri* by insects (Chapter 9). While practical limits to long distance spread of these pathogens are not known, stand boundaries are not likely to influence these limits.

11.3.3 Harvesting before culmination of stand production

Volume loss varies according to stage of stand development, levels of root disease severity, and bark beetle population levels. Stands with the highest disease severity often have little merchantable volume remaining. The annual mortality volume in these stands often exceeds annual growth resulting in a net loss of volume. Conversely, at low levels of root disease severity, volume accretion may exceed mortality. There is a point along the scale of root disease severity at which stand production will have culminated, i.e. mortality volume equals growth volume. Stands often lose volume rapidly beyond this point. Root systems are large enough to serve as substantial food bases for the pathogens, stems are large and have thick cambium to support large populations of bark beetles, and dying trees are of merchantable diameters.

Stand productivity on sites with severe root disease at the beginning of a rotation may culminate before reaching merchantable volume. Productivity on sites with low root disease severity may culminate with little or no effect of root disease or bark beetles. Setting priorities for harvest can save as much timber volume as more direct root disease control efforts. Limits on harvest activities within a particular area can limit options for managing pathogens and bark beetles. Controlling these organisms in stands that are unlikely to show increased productivity condemns untreated stands to further degradation while awaiting treatment. Stands also may be treated improperly in order to protect other resources, especially wildlife, that would be adversely impacted by harvest. It is important to weigh unrealized growth in a stand with severe root disease but no merchantable volume against mortality volume in a merchantable stand with moderate root disease, to set priorities for stand treatment.

A stand grow simulation model (Stage, 1973) and the recently developed Western Root Disease Model (McNamee *et al*., 1989; Chapter 10) were used to aid in setting stand treatment priorities for 12 000 ha on the Fernan Ranger District in Idaho where watershed protection was a major consideration. A Geographical Information System (GIS) was used to analyze root disease severity, stand volume, species and site productivity information, watershed subdrainage locations, and sensitive soils. This provided a means of integrating stand treatments and watershed effects in order to identify stand treatment priorities. It also provided a unique opportunity to examine potential effects of root pathogen epidemics together with elevated, but non-epidemic, bark beetle populations on watershed quality. The analysis indicated a steady deterioration of watershed quality, due to loss of canopy cover, if root diseases were not controlled. The long-term effect of no treatment far outweighed the short-term impacts of sanitation harvest. These results clearly supported rehabilitation of the watershed through scheduled harvest and planting to non-hosts.

Resource models and Geographical Information Systems are becoming increasingly important tools for land management planning. These models have the potential to reduce negative effects of root diseases and bark beetles through scheduling of stand treatments that are appropriate for meeting management objectives.

11.3.4 Site and stand disturbance

The importance of fire to maintaining the structure and health of conifer forests, especially *Pinus* forests, has lead to widespread use of prescribed fire in the southern US (Wade and Ward, 1976). Controlled use of fire to reduce plant competition, mineralize accumulating litter, and maintain open healthy forests should reduce tree susceptibility to bark beetles and pathogens (Schowalter *et al.*, 1981). However, prescribed burning requires careful application to prevent injury to stems or surficial roots that could increase tree susceptibility to bark beetles or pathogens (Gara *et al.*, 1985). Prescribed fire has not been widely used in western North America where steep topography and fuel accumulation, after decades of fire suppression, complicate fire management.

Thinning can be an effective means of maintaining adequate growing space and resources and disrupting bark beetle spread (Sartwell and Stevens, 1975; Mitchell *et al.*, 1983; Nebeker *et al.*, 1985; Wood *et al.*, 1985; Amman *et al.*, 1988). Thinning at an early age to avoid or reduce the need for thinning later, when the stumps are larger, can minimize stand disturbance. Small stumps provide limited resources to maintain root pathogens. Twenty years after precommercial thinning, mortality rates were low in both thinned and unthinned portions (2.0 and 4.0 trees ha^{-1} $year^{-1}$, respectively) of a sapling *Pinus ponderosa* stand in central Oregon (Filip *et al.*, 1989). Precommercial thinning did not increase mortality to *Armillaria ostoyae*.

Small stumps also contribute little to spread of *Heterobasidion* by spore infection of stump surfaces. Small *Pinus* stumps (<15 cm in diameter) have been shown to be poor substrates for *H. annosum* on the Lassen National Forest in northern California because they reach temperatures lethal to the fungus quickly during the summer (Kliejunas, 1989).

However, thinning is not always an appropriate pest management tool. Although thinning can reduce wood volume killed by bark beetles, total volume removed to meet thinning specifications can exceed volume killed by beetles (Dahlsten and Rowney, 1983). If thinning occurs before trees reach commercial size or if no market exists for the timber cut, the cost of thinning may exceed losses to bark beetles (Wood *et al.*, 1985). Furthermore, accumulating evidence suggests that physical injury, soil compaction, temporary stress due to changed environmental conditions, increased likelihood of windthrow and fire, and woody debris resulting from thinning operations can increase resources for bark beetles and pathogens (Wood *et al.*, 1985).

Increases in *Leptographium* infection of *Pseudotsuga menziesii* can be dramatic following precommercial thinning in young single-species stands in Washington, Oregon and northern California (Harrington *et al.*, 1985; Hansen *et al.*, 1988; Chapter 9). Trees between 10 and 30 years of age are highly susceptible to this fungus, but surviving trees become increasingly resistant to the disease after about 35 years of age. Bark beetles are not known to cause significant problems in young *P. menziesii* stands, but severe infestations of bark beetles occur in *Pinus ponderosa* weakened by *L. wageneri* (Cobb *et al.*, 1974; Goheen and Cobb, 1980; Cobb, 1988).

Multiple partial harvests of a stand result in a mosaic of tree ages (e.g. Franklin *et al.*, 1989). Such uneven-age stand management can be accomplished through individual tree selection (harvest of scattered selected trees) and group selection (harvest of aggregates of trees). Species composition, diameter distribution, stocking density, and cutting frequency each will influence the response of pathogens and insects to uneven-age management. Franklin *et al.* (1989) note that studies of coastal forests of the Pacific Northwest indicate that the diversity of species and functional components favored by uneven-age management promote forest health and resistance to pests. The biological consequences of uneven-age management are only partly understood. Arguments against uneven-age management are often based on pest problems associated with unplanned uneven-age management that resulted from economic selection harvest or salvage and replanting. This evidence does not provide a fair test of uneven-age management, because these practices typically removed vigorous overstory trees and left suppressed subcanopy trees or maintained an uneven-aged monoculture of commercial tree species. The results, however, do indicate bark beetle and pathogen responses to conditions that could be avoided in future uneven-age management. Perhaps the most important clue to success of uneven-age management is whether the uneven-age condition we seek to produce is consistent with the natural ecology of the site. In many cases ecologically sound management may require a mixture of even-age and uneven-age conditions.

Partial retention of existing stands is often a first step in creating an uneven-age condition or an intermediate step toward even-age condition. McGregor *et al.* (1987) demonstrated that commercial thinning can be used successfully to create uneven-aged *Pinus contorta* stands, minimizing losses to *D. ponderosae* while simultaneously providing wildlife cover, watershed protection, and visual quality. However, a variety of pests, notably *Arceuthobium* spp. and *Choristoneura occidentalis*, rapidly spread from the canopy to understory saplings in multi-storied mixed-conifer stands. On most *Pinus* habitat types, the shading effect of partial retention favors regeneration of *Pseudotsuga menziesii, Abies* spp. and *Tsuga heterophylla*, species that often are most susceptible to *C. occidentalis* and root diseases on these habitat types. Tree colonization by *Dendroctonus pseudotsugae, D. ponderosae* and *D. brevicomis* depends on phloem thickness, associated with tree diameter. These bark beetles may find a more constant supply of trees of the appropriate diameters in uneven-aged stands. Given predisposing conditions, such as weakening by root diseases, endemic bark beetle populations may progressively kill the large-diameter trees in stands under uneven-age management. These problems could be alleviated, at least partially, by maintaining trees at low density and high diversity.

Pinus ponderosa on *Pseudotsuga menziesii* habitat types, where *P. menziesii* is removed as an unwanted seed source, may provide the best opportunities for uneven-age management. However, *Heterobasidion annosum* has been an increasingly important problem in *P. ponderosa* stands management by selective harvesting because of the ability of *H. annosum* to infect stump surfaces. The problem in this case is not stand structure but the repeated harvest entries that provide a continuous supply of large stumps.

Group selection is a silvicultural system in which an uneven-aged stand is produced by creating aggregates of even-aged trees within the stand. This may, in effect, be a compromise between even and uneven-age management that somewhat lessens the impact of root disease while meeting other objectives of uneven-age management. Harvested patches generally range from 0.1 to 2 ha in size in most applications, depending on management objectives (Barth, 1990). Group selection has the advantage over single-tree selection in providing for regeneration of shade-intolerant tree species. The size of the clearing must be sufficiently large to meet the light requirements of shade-intolerant seedlings. Perhaps more importantly, the cleared area:edge-length ratio should be large to minimize root disease impact along the edge of the openings while maximizing the area regenerated with disease-resistant tree species. The stumps and injured trees left after harvest are likely to increase disease inoculum and resultant mortality in the uncut portions of the stand adjacent to the openings and could affect susceptible seedlings in the regenerated stand. Group selection can have a further advantage over single-tree selection in making permanent corridors for harvest possible. This can reduce the number of wounded trees by concentrating the damage from all harvest entries to trees along permanent corridors.

11.3.5 Non-silvicultural mitigation

In some situations, creation of healthy and/or diverse stands that limit root disease and bark beetle populations is either impractical or inconsistent with market demand for particular tree species. Tree species preferences generally are based on product and aesthetic values and on costs of production. For example, *Pseudotsuga menziesii* has been preferred over *T. heterophylla* for timber value, large old *Pinus ponderosa* over young *P. ponderosa* or *Quercus* sp. for aesthetic value, and *A. grandis* and *A. concolor* over *P. ponderosa* on *Abies* sites where *Abies* spp. have regenerated naturally but *P. ponderosa* is absent and must be planted. Where silvicultural options for prevention of root disease and bark beetle epidemics cannot or will not be used, a limited number of direct control methods are available (Wood *et al.*, 1985; Hagle and Shaw, 1991).

11.3.5.1 Semiochemicals for bark beetle control

Semiochemicals (chemical signals such as pheromones) can be used to monitor bark beetle populations and, potentially, to reduce or repel bark beetles. Scattered bark beetles must concentrate on a suitable host (using pheromone communication) to ensure that attack densities are sufficient to overcome host resistance and to ensure mating (Chapter 6). This requirement provides a biological basis for strategies that prevent endemic populations from building to outbreak levels.

Endemic bark beetle populations often can be identified by the presence of root pathogens. The associated bark beetle populations can be monitored for beetle abundance using attractive pheromone baits. Suppression of growing populations involves the use of attractive pheromones to lure beetles to traps where they can be destroyed or the use of anti-aggregative pheromones to prevent concentration of endemic populations (Wood *et al.*, 1985). These techniques offer promise as an environmentally acceptable tool (Furniss *et al.*, 1972; Wood *et al.*, 1985; Amman *et al.*, 1989; Lindgren *et al.*, 1989). We note, however, that stand structure can affect the efficacy of pheromone communication (e.g. Fares *et al.*, 1980) and that the fate of beetles deterred from a site and their contribution to proliferation of new infestations remains unknown.

11.3.5.2 Chemical control

Borax (sodium tetraborate decahydrate) treatment to prevent *H. annosum* infection of stump surfaces has been practiced in all *P. ponderosa* timber sales on the Lassen National Forest in California since 1978 (Kliejunas, 1989). Most *H. annosum* root disease patches develop from large stumps following commercial harvests. Field observations in California have indicated that this disease seldom spreads from stumps smaller than about 40 cm in diameter to initiate root disease patches in stands (Kliejunas, 1989). Various studies have indicated that *P. ponderosa* stumps smaller than 30–35 cm in diameter in northern California (Kleijunas, 1989), and 45 cm in diameter in Oregon (Hadfield *et al.*, 1986), contribute little to disease spread.

Fig. 11.4. Fumigation with chloropicrin and other volatile chemicals successfully eliminated *Armillaria ostoyae* from infected stumps of *Pinus ponderosa* (under plastic sheets). Effects on non-target organisms is unknown. (Photo by G.M. Filip.)

Hence, minimum diameter requirements for treating *Pinus* stumps with borax following cutting probably will vary somewhat by location. Similar use of borax treatment to prevent *H. annosum* infection of *Abies* spp. has been recommended (Hadfield *et al.*, 1986).

There is little evidence for long-term effectiveness of chemical control of established root pathogens in stumps or roots (Shaw and Roth, 1978; Filip and Roth, 1987; Hagle and Shaw, 1991). However, the use of chemical fumigants may have some benefit, particularly when used in combination with stump and root removal. Fumigation of stumps (Fig. 11.4) has been effective in reducing *Armillaria ostoyae* in *P. ponderosa* (Filip and Roth, 1977) and *Phellinus weirii* in *Pseudotsuga menziesii* (Thies and Nelson, 1982).

Insecticides also are available for control of bark beetles (Wood *et al.*, 1985). Chemical insecticides may reduce or delay the spread of insects to individual susceptible hosts sufficiently to justify treatments in high-value sites such as developed recreation, administrative, and residential sites. However, bark beetle populations are not reduced by treating individual trees, which remain at risk. Chemical treatment is not economically feasible over large areas, especially because exposure of bark beetles to aerially applied chemicals cannot be ensured. Future availability of many registered pesticides is uncertain, due to safety standards for human and environmental protection. Chemical treatment can cause severe reduction in abundance of non-target organisms, such as bark beetle predators and parasites, soil arthropods and mycorrhizal fungi (Wood *et al.*, 1985).

11.3.5.3 Other direct methods

Removal of bark beetle or pathogen populations should terminate damage. Rapid salvage of dead and dying trees also reduces losses of merchantable timber.

Rapid salvage of dead and infested trees has been used widely in the southern US to remove *D. frontalis* populations (Wood *et al.*, 1985). Cutting infested trees and leaving them exposed on the ground also has been used to reduce *D. frontalis* populations through solar heating and disrupting pheromone communication across canopy gaps. However, the number of new infestations in adjacent untreated areas may increase (Wood *et al.*, 1985). Other studies of paired treated (salvage or cut-and-leave) and untreated areas are not conclusive, because of inadequate replication and non-random assignment of treated and untreated areas. These methods may have short-term benefits but their efficacy in reducing tree mortality to bark beetles over larger areas remains unknown (Coulson and Stark, 1982; Wood *et al.*, 1985). Furthermore, these techniques will not mitigate predisposing conditions nor terminate the spread of root pathogens.

Stump extraction by bulldozing or pulling has been effective in reducing *Armillaria* and *Phellinus* inoculum (Roth *et al.*, 1977; Morrison *et al.*, 1988). Coastal *Pseudotsuga menziesii* stands occasionally are treated in this way to allow regeneration of high-value *P. menziesii* (Wallis, 1976; Thies, 1984). This method of treatment also is feasible in some developed sites to reduce spread of root disease to healthy trees from infected stumps. Effects on non-target organisms, especially beneficial mycorrhizal fungi, are largely unknown.

11.3.5.4 Biological control

As discussed in Chapter 7, bark beetles and pathogens are vulnerable to a variety of predators, parasites and benign competitors. Some of these organisms offer potential benefits as biological control agents (Dahlsten, 1982). For example, parasitic nematodes can reduce the survival and fecundity of infected bark beetles (Thong and Webster, 1975; Kinn, 1980). Antagonistic interactions between *Trichoderma* spp. and *Phellinus weirii* could be exploited to reduce the incidence of *P. weirii* (Goldfarb *et al.*, 1989). Ectomycorrhizal fungi also can suppress root pathogenic fungi and have the additional benefit of improving tree health (Sinclair *et al.*, 1982).

11.3.6 Developed sites

Developed recreation and administrative sites have root disease and bark beetle management opportunities and restrictions that are different in many ways from those in other forest areas. Trees generally have higher individual economic value in developed sites than in other forest settings. The objective of forest insect and disease management in developed sites often is to extend the longevity of the existing stand and to protect people and facilities from injury or damage from tree failure.

Drastic changes in the appearance of forest stands in developed sites is nearly always undesirable. Root pathogen and bark beetle epidemics cause considerable tree loss, as do management activities aimed at control of epidemics. Clearcutting of sites infected with *H. annosum* often is required to control the disease in *Pinus ponderosa* stands in campgrounds in California (West, 1989). Dusting *Pinus* stump surfaces with borax generally is recommended in connection with tree removal. Root disease patches that are clearcut must be replanted with root disease-resistant species, or the inoculum must be killed or removed to allow replanting with susceptible species. Planting disease-resistant species may result in a drastically different forest appearance, such as replacement of *Pinus* stands with *Quercus, Alnus*, and *Acer* spp. in Yosemite National Park (West, 1989). These early successional shrub or hardwood forests may have helped sanitize sites of conifer pathogens under natural conditions, but generally fail to meet aesthetic expectations of park visitors.

11.4 CONCLUSIONS

Bark beetles and root pathogens are natural components of conifer forest ecosystems and function at natural levels to maintain forest health through natural thinning and nutrient cycling. However, fire suppression and management for particular resources on a regional scale has created dense, often stressed forests favorable to bark beetles and/or pathogens. Increased tree diversity, reduced tree density, and controlled use of fire could prevent most bark beetle and pathogen outbreaks. As we try to reverse mistakes that have led to destructive outbreaks, we certainly will find that some of our new approaches also will favor bark beetles, pathogens or other incipient pests. Various suppression options are available for controlling bark beetles and pathogens where silvicultural techniques are ineffective. However, management practices that increase species or age class diversity will maximize natural barriers to pest populations as well as maximizing options for mitigating pest population growth. Careful observation of natural forest patterns and processes and recognition of the important functions of insects and pathogens in forest ecosystems will contribute to management for healthy forests.

REFERENCES

Amman, G.D. and Baker, B.H. (1972). Mountain Pine beetle influence on lodgepole pine structure. *J. Forestry* **70**, 204–209.

Amman, G.D., McGregor, M.D., Schmitz, R.F. and Oakes, R.D. (1988). Susceptibility of lodgepole pine to infestation by mountain pine beetles following partial cutting of stands. *Can. J. For. Res.* **18**, 688–695.

Amman, G.D., Thier, R.W., McGregor, M.D. and Schmitz, R.F. (1989). Efficacy of verbenone in reducing lodgepole pine infestations by mountain pine beetles in Idaho. *Can. J. For. Res.* **19**, 60–64.

Arno, S.F. (1976). The historical role of fire on the Bitterroot National Forest. USDA Forest Serv., Res. Paper INT-187. USDA Forest Serv. Intermountain Res. Sta., Ogden, UT, 29 pp.

Barrett, S.W. and Arno, S.F. (1982). Indian fires as an ecological influence in the northern Rockies. *J. Forestry* **80**, 647–651.

Barrows, K. (1984). Old-growth Douglas-fir forests. *Fremontia* **11(4)**, 20–23.

Barth, R.S. (1990). The group selection method. *In* "Proc. Silviculture/Genetics Workshop" (R. Miller and D. Murphy, eds), pp. 94–105. USDA Forest Serv., Washington, DC.

Bega, R.V. and Smith, R.S., Jr. (1966). Distribution of *Fomes annosus* in natural forests of California. *Plant Dis. Rep.* **50**, 832–836.

Byler, J.W. (1984). Status of disease pests in the interior Douglas-fir and grand fir types. *In* "Proc. Silivicultural Management Strategies for Pests of the Interior Douglas-fir and Grand Fir Forest Types Symposium", pp. 45–50. Washington State University, Pullman, WA.

Byler, J.W., Stewart, C.A. and Hall, L.D. (1985). Establishment report: permanent plots to evaluate the effects of Armillaria root disease in precommercially thinned stands. USDA Forest Serv. Rpt. No. 85-21. USDA Forest Serv. Northern Region, Missoula, MT, 12 pp.

Byler, J.W., Marsden, M.A. and Hagle, S.K. (1987). Opportunities to evaluate root disease incidence and damage using forest inventory and permanent growth plots. *In* "Proc. 34th Annual Western International Forest Disease Work Conference," pp. 52–56. USDA Forest Service, Pacific Northwest Region, Portland, OR.

Byler, J.W., Marsden, M.A. and Hagle, S.K. (1990). The probability of root disease on the Lolo National Forest, Montana. *Can. J. For. Res.* **20**, 987–994.

Chase, T.E. (1989). Genetics and population structure of Heterobasidion annosum with special reference to western North America. *In* "Proc. Symposium on Research and Management of Annosus Root Disease (*Heterobasidion annosum*) in Western North America", pp. 19–25. USDA Forest Serv. Gen. Tech. Rpt. PSW-116. USDA Forest Serv. Pacific Southwest Res. Stn., Berkeley, CA.

Cobb, F.W., Jr. (1988). *Leptographium wageneri*, cause of black-stain root disease: a review of its discovery, occurrence and biology with emphasis on pinyon and ponderosa pine. *In* "*Leptographium* Root Diseases on Conifers," pp. 41–62. American Phytopathological Society Press, St. Paul, MN, 149 pp.

Cobb, F.W., Jr., Parmeter, J.R., Jr., Wood, D.L. and Stark, R.W. (1974). Root pathogens as agents predisposing ponderosa pine and white fir to bark beetles. *In* "Proc. Fourth International Conference on *Fomes annosus*", pp. 8–15. USDA Forest Serv., Washington, DC.

Coulson, R.N. and Stark, R.W. (1982). Integrated management of bark beetles. *In* "Bark Beetles in North American Conifers: A System for the Study of Evolutionary Biology" (J.B. Mitton and K.B. Sturgeon, eds), pp. 315–349. University of Texas Press, Austin, TX.

Coulson, R.N., Hennier, P.B., Flamm, R.O., Rykiel, E.J., Hu, L.C. and Payne, T.l. (1983). The role of lightning in the epidemiology of the southern pine beetle. *Z. Ang. Entomol.* **96**, 182–0193.

Coulson, R.N., Saunders, M.C., Payne, T.L., Flamm, R.O., Wagner, T.L., Hennier, P.B. and Rykiel, E.J. (1985). A conceptual model of the role of lightning in the epidemiology of the southern pine beetle. *In* "The Role of the Host in Population Dynamics of Forest Insects" (L. Safranyik, ed.), pp. 136–146. Canadian Forestry Serv., Pacific Forestry Centre, Victoria, BC.

Dahlsten, D.L. (1982). Relationships between bark beetles and their natural enemies. *In* "Bark Beetles in North American Conifers" (J.B. Mitton and K.B. Sturgeon, eds), pp. 140–182. University of Texas Press, Austin, TX.

Dahlsten, D.L. and Rowney, D.L. (1983). Insect pest management in forest ecosystems. *Environ. Manage.* **7**, 65–72.

Fares, Y., Sharpe, P.J.H. and Magnusen, C.E. (1980). Pheromone dispersion in forests. *J. Theor. Biol.* **84**, 335–359.

Filip, G.M. (1977). An Armillaria epiphytotic on the Winema National Forest, Oregon. *Plant Dis. Rep.* **61(8)**, 708–711.

Filip, G.M. (1979). Root disease in Douglas-fir plantations is associated with infected stumps. *Plant Dis. Rep.* **63(7)**, 580–583.

Filip, G.M. and Goheen, D.J. (1982a). Hazard of root disease in Pacific Northwest recreation sites. *J. Forestry* **80**, 163–164.

Filip, G.M. and Goheen, D.J. (1982b). Tree mortality caused by root pathogen complex in Deschutes National Forest, Oregon. *Plant Dis.* **66**, 240–243.

Filip, G.M. and Roth, L.F. (1977). Stump injections with soil fumigants to eradicate *Armillariella mellea* from young-growth ponderosa pine killed by root rot. *Can. J. For. Res.* **7**, 226–231.

Filip, G.M. and Roth, L.F. (1987). Seven chemicals fail to protect ponderosa pine from Armillaria root disease in central Washington. USDA Forest Serv. Res. Note PNW-RN-460. USDA Forest Serv. Pacific Northwest Res. Stn., Portland, OR, 8 pp.

Filip, G.M. and Schmitt, C.L. (1979). Susceptibility of native conifers to laminated root rot east of the Cascade Range in Oregon and Washington. *Forest Sci.* **30**, 138–142.

Filip, G.M., Goheen, D.J. Johnson, D.W. and Thompson, J.H. (1989). Precommercial thinning in a ponderosa pine stand affected by Armillaria root disease: 20 years of growth and mortality in central Oregon *West. J. Appl. For.* **4(2)**, 58–59.

Franklin, J.F., Perry, D.A., Schowalter, T.D., Harmon, M.E., McKee, A and Spies, T.A. (1989). Importance of ecological diversity in maintaining long-term site productivity. *In* "Maintaining Long-term Productivity of Pacific Northwest Forest Ecosystems" (D.A. Perry, B. Thomas, R. Meurisse, R. Miller, J. Boyle, P. Sollins and J. Means, eds), pp. 82–97. Timber Press, Portland, OR.

Furniss, M.M., Kline, L.N., Schmitz, R.F. and Rudinsky, J.A. (1972). Tests of three pheromones to induce or disrupt aggregation of Douglas-fir beetles in live trees. *Ann. Entomol. Soc. Am* **65**, 1227–1232.

Furniss, M.M., Livingston, R.L. and McGregor, M.D. (1981). Development of a stand susceptibility classification for Douglas-fir beetle. *In* "Proceedings of Hazard-rating Systems in Forest Insect Pest Management", pp. 115–128. USDA Forest Serv. Gen.Tech. Rpt. WO-27, USDA Forest Serv., Washington, DC.

Gara, R.I., Littke, W.R., Agee, J.K., Geiszler, D.R., Stuart, J.D. and Driver, C.H. (1985). Influence of fires, fungi and mountain pine beetles on development of a lodgepole pine forest in south-central Oregon. *In* "Lodgepole Pine, the Species and its Management: Symposium Proceedings", pp. 153–162. Washington State University, Cooperative Extention, Pullman, WA.

Goheen, D.J. and Cobb, F.W., Jr. (1980). Infestation of *Ceratocystis wageneri*-infected ponderosa pine by bark beetles (Coleoptera: Scolytidae) in the central Sierra Nevada. *Can. Entomol.* **112**, 725–730.

Goheen, E.M. and Goheen, D.J. (1989). Losses caused by annosus root disease in Pacific Northwest forests. *In* "Proc. Symposium on Research and Management of Annosus Root Disease (*Heterebasidion annosum*) in Western North America," pp. 66–69. USDA Forest serv. Gen. Tech. Rpt. PSW-116, USDA Forest Serv., Pacific Southwest Res. Sta., Berkeley, CA.

Goldfarb, B., Nelson, E.E. and Hansen, E.M. (1989). *Trichoderma* species from Douglas-fir stumps and roots infested with *Phellinus weirii* in the western Cascade Mountains of Oregon. *Mycologia* **81**, 134–138.

Greig, B.J.W. and Pratt, J.E. (1976). Some observations on the longevity of Fomes annosus in conifer stumps. *Eur. J. For. Pathol.* **6**, 250–253.

Gruell, G.E. (1985). Indian fires in the interior West—a widespread influence. *In* "Proc. Wilderness Fire Symposium," pp. 68–74. USDA Forest Serv. Gen. Tech. Rpt. INT-182, USDA Forest Serv., Intermountain Res. Stn., Ogden, UT.

Hadfield, J.S. and Johnson, D.W. (1977). Laminated root rot. A guide for reducing and preventing losses in Oregon and Washington forests. USDA Forest Serv. Pacific Northwest Region, Portland, OR, 16 pp.

Hadfield, J.S., Goheen, D.J., Philip, G.M., Schmitt, C.L. and Harvey, R.D. (1986). Root diseases in Oregon and Washington Conifers. USDA Forest Serv. Rpt R6-FPM-250-86, USDA Forest Serv. Pacific Northwest Region, Portland, OR, 27pp.

Hagle, S.K. (1985). Monitoring root disease mortality. USDA Forest Serv. Rpt. 85-27, USDA Forest Serv., Northern Region, Missoula, MT, 13 pp.

Hagle, S.K. and Goheen, D.J. (1988). Root disease response to stand culture. *In* "Proc. Future Forests of the Mountain West: a Stand Culture Symposium", pp. 303–309. USDA Forest Serv. Gen. Tech. Rpt. INT-243, USDA Forest Serv., Intermountain Res. Stn., Ogden, UT.

Hagle, S.K. and Shaw, C.G., III 1991). Avoiding and reducing losses from Armillaria root disease. *In* "Armillaria Root Disease" (C.G. Shaw III and G.A. Kile, eds), pp. 157–173. USDA Forest Serv. Agric. Hdbk. 691, USDA Forest Serv., Washington, DC.

Hansen, E.M., Goheen, D.J., Hessburg, P.F., Witcosky, J.J. and Schowalter, T.D. (1988). Biology and management of black-stain root disease in Douglas-fir. *In* "*Leptographium* Root Diseases on Conifers", pp. 63–80. American Phytopathological Society Press, St. Paul, MN.

Harlow, W.M. and Harrar, E.S. (1950). "Textbook of Dendrology" McGraw-Hill Book Co., New York, 55 pp.

Harrington, T.C., Cobb, F.W., Jr. and Lownsbery, J.W. (1985). Activity of *Hylastes nigrinus*, a vector of *Verticicladiella wageneri*, in thinned stands of Douglas-fir. *Can. J. For. Res.* **15**, 519–523.

Hermann, R.K. (1985). The genus *Pseudotsuga*: Ancestral history and past distribution. Oregon State University. Special Publication 2b, Oregon State University, Corvallis, OR, 32 pp.

Hodges, C.S. (1969). Modes of infection and spread of *Fomes annosus*. *Annu. Rev. Phytopathol.* **7**, 247–266.

Hodges, J.D., Elam, W.W., Watson, W.F. and Nebeker, T.E. (1979). Oleoresin characteristics and susceptibility of four southern pines to southern pine beetle (Coleoptera: Scolytidae) attacks. *Can. Entomol.* **111**, 889–896.

Kinn, D.N. (1980). Mutualism between *Dendrolaelaps neodisetus* and *Dendroctonus frontalis*. *Environ. Entomol.* **9**, 756–758.

Kliejunas, J.T. (1989). Borax stump treatment for control of annosus root disease in the eastside pine type forests of northeastern California. *In* "Proc. Symposium on Research and Management of Annosus Root Disease (*Heterobasidion annosum*) in Western North America," pp. 159–166. USDA Forest Serv. Gen. Tech. Rpt. PSW-116, USDA Forest Serv. Pacific Southwest Forest and Range Exp. Stn., Berkeley, CA.

Lindgren, B.C., Borden, J.H., Cushon, G.H., Chong, L.J. and Higgins, C.J. (1989). Reduction of mountain pine beetle (Coleoptera: Scolytidae) attacks by verbenone in lodgepole pine stands in British Columbia. *Can. J. For. Res.* **19**, 65–68.

Lorio, P.L., Jr. (1980). Loblolly pine stocking levels affect potential for southern pine beetle infestation. South. *J. Appl. For.* **4**, 162–165.

Martin, R.E. (1982). Fire history and its role in succession. *In* "Proc. Forest Succession and Stand Development Research in the Northwest", pp. 92–99. Oregon State University, Corvallis, OR.

McGregor, M.D., Amman, G.D., Schmitz, R.F. and Oakes, R.D. (1987). Partial cutting of lodgepole pine stands to reduce losses to the mountain pine beetle. *Can. J. For. Res.* **17**, 1234–1239.

McNamee, P., Sutherland, G., Shaw, C.G. III *et al.*, (1989). Description of a multi-species root disease model developed for silvicultural planning and management in coniferous forests of western North America. *In* "Proc. 7th IUFRO Conference on Root and Butt Rot in Forest Trees" (D.J. Morrison, ed), pp. 320–335. Forestry Canada, Pacific Forest Research Centre, Victoria, BC.

Mitchell, R.G., Waring, R.H. and Pitman, G.B. (1983). Thinning lodgepole pine increases tree vigor and resistance to mountain pine beetle. *Forest Sci.* **29**, 204–211.

Morrison, D.J. (1981). Armillaria root disease. A guide to disease diagnosis, development and management in British Columbia. Canadian Forestry Service, Pacific Forest Research Centre, Victoria, BC, 15 pp.

Morrison, D.J. and Johnson, A.L.S. (1978). Stump colonization and spread of *Fomes annosus* 5 years after thinning. *Can. J. For. Res.* **8**, 177–180.

Morrison, D.J., Wallis, G.W. and Weir, L.C. (1988). Control of Armillaria and laminated root rots: 20-year results from the Skimikin stump removal experiment. Canadian Forestry Service Info. Rpt. BC-P-302, Forestry Canada, Pacific Forestry Centre, Victoria, BC, 16 pp.

Nebeker, T.E., Hodges, J.D., Karr, B.K. and Moehring, D.M. (1985). Thinning practices in southern pines—with pest management recommendations. USDA Forest Serv. Tech. Bull. 1703, USDA Forest Serv., Washington, DC, 36 pp.

Payne, T.L. (1980). Life history and habits. *In* "The Southern Pine Beetle" (R.C. Thatcher, J.L. Searcy, J.E. Coster and G.D. Hertel, eds), pp. 7–28. USDA Forest Serv. Tech. Bull. 1631, USDA Forest Serv., Washington, DC, 266 pp.

Pfister, R.D., Kovalchik, B.L., Arno, S.F. and Presby, R.C. (1977). Forest habitat types of Montana. USDA Forest Serv. Gen. Tech. Rpt. INT-34. USDA Forest Serv. Intermountain Res. St., Ogden, UT, 174 pp.

Roth, L.F., Shaw, C.G. III and Rolph, L. (1977). Marking ponderosa pine to combine commercial thinning and control of Armillaria root rot. *J. Forestry* **75(10)**, 644–647.

Sartwell, C. and Stevens, R.E. (1975). Mountain pine beetle in ponderosa pine: prospects for silvicultural control in second-growth stands. *J. Forestry* **73**, 136–140.

Schmidt, W.C. (1989). Lodgepole pine: an ecological opportunist. *In* "Proc. Symposium on the Management of Lodgepole Pine to Minimize Losses to the Mountain Pine Beetle," pp. 14–20. USDA Forest Serv. Gen. Tech. Rpt. INT-262, USDA Forest Serv., Intermountain Res. Stn., Ogden, UT, 119 pp.

Schmitt, C.L., Goheen, D.J., Goheen, E.M. and Frankel, S.J. (1984). Effects of management activities and dominant species type in pest-caused mortality losses in true fir in the Fremont and Ochoco National Forests. USDA Forest Serv., Pacific Northwest Region, Forest Pest Management, Portland, OR, 34 pp.

Schowalter, T.D., Coulson, R.N. and Crossley, D.A., Jr. (1981). Role of southern pine beetle and fire in maintenance of structure and function of the southeastern coniferous forest. *Environ. Entomol.* **10**, 821–825.

Shaw, C.G., III and Roth, L.F. (1978). Control of Armillaria root rot in managed coniferous forests: a literature review. *Eur. J. For. Pathol.* **8(3)**, 163–174.

Shaw, C.G., III, Roth, L.F., Rolph, L. and Hunt, J. (1976). Dynamics of pine and pathogen as they relate to damage in a forest attacked by Armillaria. *Plant Dis. Rep.* **60**, 214–218.

Sinclair, W.A., Sylvia, D.M. and Larsen, A.O. (1982). Disease suppression and growth promotion in Douglas-fir seedlings by the ectomycorrhizal fungus *Laccaria laccata. Forest Sci.* **28**, 191–201.

Stage, A.R. (1973). Prognosis model for stand development. USDA Forest Serv. Res. Paper INT-137, USDA Forest Serv. Intermountain Res. Stn., Ogden, UT, 32 pp.

Steele, R., Arno, S.F. and Geier-Hayes, K. (1986). Wildfire patterns change in central Idaho's ponderosa pine–Douglas-fir forest. *West. J. Appl. For.* **1**, 16–18.

Stokes, M.A. and Dieterich, J.H. (1980). "Proc. Fire History Workshop." USDA Forest Serv., Gen. Tech. Rpt. RM-81, USDA Forest Serv., Rocky Mountain Res. Stn., Ft. Collins, CO, 142 pp.

Thies, W.G. (1984). Laminated root rot. The quest for control. *J. Forestry* **82(6)**, 345–356.

Thies, W.G. and Nelson, E.E. (1982). Control of *Phellinus weirii* in Douglas-fir stumps by the fumigants chloropicrin, allyl alchohol, Vapam or Vorlex. *Can. J. For. Res.* **12**, 528–532.

Thong, C.H.S. and Webster, J.M. (1975). Effects of the bark beetle nematode, *Contortylenchus reversus*, on gallery construction, fecundity, and egg viability of the Douglas-fir beetle *Dendroctonus pseudotsugae* (Coleoptera: Scolytidae). *J. Invert. Pathol.* **26**, 235–238.

Tkacz, B.M. and Hansen, E.M. (1982). Damage by laminated root rot in two succeeding stands of Douglas-fir. *J. Forestry* **80**, 788–791.

Wade, D.D. and Ward, D.E. (1976). Prescribed use of fire in the South—a means of conserving energy. *Proc. Tall timbers Fire Ecol. Conf.* **14**, 549–558.

Wallis, G.E. (1976). *Phellinus* (*Poria*) *weirii* root rot. Detection and management proposals in Douglas-fir stands. Canadian Forestry Service Forestry Tech. 12, Forestry Canada, Pacific Forestry Centre, Victoria, BC, 16 pp.

West, L. (1989). Management of annosus root disease caused by *Heterobasidion annosum* in coniferous trees in Yosemite National Park. *In* "Proc. Symposium on Research and Management of Annosus Root Disease (Heterobasidion annosum) in Western North America", pp. 167–170. USDA Forest Serv. Gen. Tech. Rpt. PSW-116, USDA Forest Serv. Pacific Southwest Forest and Range Exp. Stn., Berkeley, CA.

Wickman, B.E. (1990). The battle against bark beetles in Crater Lake National Park: 1925-34. USDA Forest Serv. Gen. Tech. Rpt. PNW-259, USDA Forest Serv. Pacific Northwest Forest and Range Exp. Stn., Portland, OR, 40 pp.

Witcosky, J.J., Schowalter, T.D. and Hansen, E.M. (1986). *Hylastes nigrinus*, (Coleoptera: Scolytidae), *Pissodes fasciatus*, and *Steremnius carinatus* (Coleoptera: Curculionidae) as vectors of black stain root disease of Douglas-fir. *Environ. Entomol.* **15**, 1090–1095.

Wood, D.L., Stark, R.W., Waters, W.E., Bedard, W.D. and Cobb, F.W., Jr. (1985). Treatment tactics and strategies. *In* "Integrated Pest Management in Pine–Bark Beetle Ecosystems" (W.E. Waters, R.W. Stark and D.L. Wood, eds), pp. 121–139. John Wiley & Sons, New York.

Worrall, J.J., Parmeter, J.R. and Cobb, F.W. (1983). Host specialization of *Heterobasidion annosum*. *Phytopathology* **73**, 304–307.

PART VI
Synopsis

–12–

Synopsis

G.M. FILIP[1] and T.D. SCHOWALTER[2]
Departments of [1]Forest Science and [2]Entomology, Oregon State University, Corvallis, OR, USA

12.1 SYNTHESIS OF CURRENT KNOWLEDGE

Our understanding of bark beetle–pathogen interactions and their consequences in conifer forests clearly has increased dramatically over the past two decades. Research in this area has contributed to and benefitted from emerging views of species interactions and forest pest management. We are recognizing the need to view forests as ecosystems with natural checks and balances, rather than as collections of independent species, if long-term forest management goals are to be achieved. Forest managers must address potential cumulative effects of management practices on bark beetles and pathogens over long time periods, even under shortened rotation periods. Furthermore, forest health must be sustainable in the long term if forest resources are to remain available to future generations. Forest resources, such as timber and wildlife, must be effectively balanced. Pest management should avoid contributing to pesticide resistance, outbreaks of non-target pests, and environmental degradation.

Management of bark beetle–pathogen complexes has improved over the past two decades as a result of two new concepts: the importance of species interactions in controlling pest epidemiologies, and integrated pest management (IPM) methodology. Interactions among bark beetles, pathogenic fungi and conifers are a major concern to forest managers in North America. Loss of timber volume to bark beetles and root pathogens has been estimated at 0.9–1.1 billion m^3 $year^{-1}$ in western North America (Chapter 9). Consequently, suppression of these mortality agents has been a priority. However, these organisms together with fire have been instrumental (through natural thinning, nutrient cycling and selection for pest and fire tolerance) in maintaining the structure and productivity of these forests (Chapter 1).

We have made progress in identifying important mortality agents (Chapters 2 and 3). New systematics techniques, such as cytogenetics and enzyme electrophoresis, have permitted discrimination among pathogenic and non-pathogenic species and varieties. Morphological characteristics, such as bark beetle mycangia and sticky conidia, that facilitate interaction between bark beetles and fungi have been identified.

Abiotic and biotic factors often predispose conifers to attack by mortality agents (Chapter 4). Water stress is a major predisposing factor. Both drought and soil saturation can disrupt plant physiological processes and increase susceptibility to insects and pathogens. Soil com-

BEETLE–PATHOGEN INTERACTIONS IN CONIFER FORESTS
ISBN 0-12-628970-0

paction, nutrient imbalances, lightning strike, windthrow, atmospheric pollution, and road building also have been shown to predispose conifers to attack by insects and pathogens. Although wildfire traditionally has been viewed as a disturbance, creating damaged trees susceptible to bark beetles and pathogens, fire suppression in western North America has resulted in dense forests predisposed to bark beetles and pathogens.

Stress is not inevitably bad for trees and good for the invading organisms. Tree physiological processes mediate the relationship between environmental conditions and pest epidemiology (Chapter 5). Even healthy trees undergo physiological changes during seasonal and ontogenetic growth and development that increase susceptibility to invading organisms. The timing, intensity and duration of stress, as well as the proximity and size of bark beetle or pathogen populations, determine the risk of attack by these organisms (Chapters 5 and 6). Biotic factors often exacerbate plant stress and increase susceptibility to less virulent organisms. Root pathogens increase tree susceptibility, and often attractiveness, to bark beetles. On the other hand, relatively innocuous stem rusts, mistletoes, defoliators and some bark beetles and fungi can increase tree susceptibility to tree-killing organisms (Chapter 4).

Over the past 20 years we have learned much about the mechanisms of tree defense against bark beetles and pathogenic fungi. Conifer defense mechanisms include both a preformed resin system and/or an induced wound response (Chapters 5 and 8). Oleoresin exudation pressure and flow rate have been correlated with conifer susceptibility to attack by bark beetles, especially for *Pinus* in the western and southern US. Trees with low OEP or flow rate tend to be more susceptible to attack by bark beetles. The induced wound response includes localized autolysis of parenchyma cells, tissue necrosis, secondary resinosis by adjacent secretory and parenchyma cells, and formation of wood periderm to physically isolate the lesion. These two defense mechanisms function to limit the number of organisms that can successfully penetrate the bark and to isolate or kill those that do reach the subcortical habitat. However, these defenses constitute a substantial drain of energy that could otherwise be channeled into growth and reproduction. Consequently, defensive capability often is reduced during periods of physiological stress.

Stand factors can affect susceptibility. Conifers with large crowns produce more photosynthate and are more capable of defense. Dense stands are characterized by reduced light, temperature, leaf area, growth rate, and distance between trees. All of these factors can facilitate host discovery and colonization by bark beetles and pathogenic fungi (Chapters 5 and 6).

Host colonization by bark beetles consists of four phases: dispersal, host selection, concentration, and establishment (Chapter 6). Bark beetles often can travel many kilometers, depending on physiological and environmental conditions, but most travel much shorter distances. Long-distance dispersal by some individuals facilitates outbreeding.

Bark beetles can be divided into two functional groups on the basis of host selection and concentration mechanisms. Secondary, or non-aggressive, bark beetles, such as *Dendroctonus valens* and *Scolytus ventralis*, are attracted to weakened trees by host chemical cues. Aggregation is not required to overcome defenses of such trees. These beetles typically are restricted to scattered susceptible hosts but have relatively stable populations. The primary, or aggressive, bark beetles, such as *D. frontalis* and *D. ponderosae*, use primarily visual cues to locate tree boles. Once on the bark, beetles respond to feeding, pressure and/or tactile stimuli and either initiate boring or resume flight. At large population levels, pheromones in combination with host tree volatiles attract large numbers of beetles, resulting in mass attack and exhaustion of tree defenses. These species are capable of attacking and killing more resistant trees in forest stands. However, populations of these species are unstable because of the dependence on large populations to attack living trees (Chapter 6).

Colonization begins with boring activity initiated by chemical and/or physical stimuli. Drought-induced acoustic signals also may trigger the tunneling response. Penetration of the bark provides entry to pathogenic *Ophiostoma* spp. and *Leptographium* spp. that may accelerate tree decline and death, perhaps through toxin production, mycelial plugging of tracheids, gas release into tracheids, and/or blockage of pit openings. The establishment phase begins when tree defenses are exhausted and beetle oviposition becomes possible.

At least 100 species of microorganisms associated with bark beetles have been discovered, and more likely await discovery. Dissemination is the primary benefit fungi receive from this association. Mycangia are specialized structures that provide nutrients to mutualistic microorganisms, but mycangia are not necessary for vectoring of many fungi. *Ophiostoma* and other pathogenic fungi are transported externally on dispersing beetles or on phoretic mites. Microorganisms can be both beneficial and detrimental to the bark beetles. Some bark beetles may feed on associated fungi, but most larvae mine in advance of fungal penetration. The suitability of the subcortical environment as a food base is enhanced by some microorganisms, such as yeasts and nitrogen-fixing bacteria. Nitrogen-fixing bacteria may improve the nutritional status of the phloem for bark beetles. Some bacteria and fungi convert host tree terpenes into beetle aggregation pheromones, thereby facilitating mass attack and successful colonization.

In addition to the microorganisms, bark beetles vector or provide entry for a large number of associated invertebrates, including mites, nematodes, microorganisms and protozoans (Chapter 7). Many of the microorganisms and invertebrates carried by bark beetles into their galleries are mutualists that increase the nutritional suitability of the subcortical resources or prey on bark beetle parasites and predators. A large number of predators and competitors are attracted by bark beetle pheromones or by volatiles emanating from the dying tree. These associates interact with each other and with the bark beetles and microorganisms within the gallery system. Even indirect interactions, such as nematophagous mites feeding on bark beetle parasites, or saprophagous mites feeding on or carrying host spores, can influence bark beetle and pathogen epidemiologies.

Under natural conditions, bark beetles and pathogenic fungi do not threaten the long-term productivity of the forest (Chapter 9). These organisms, together with fire, function as natural thinning agents and accelerate nutrient cycling from decomposing trees, contributing to forest diversity, health and long-term productivity and stability. Humans and their activities have become part of this ecosystem. Bark beetles and pathogenic fungi become viewed as destructive agents warranting suppression when we introduce the concept of economic value to timber resources. Although the ultimate effect of bark beetle and pathogen attack is usually tree mortality, our response to this mortality depends on forest management objectives. For example, mortality in stands managed for timber production could be detrimental to management goals to the extent that the number of high yield crop trees is reduced or altered harvest or salvage scheduling is required. On the other hand, this mortality might contribute to timber management goals if the pattern of natural thinning and increased light and soil fertility enhance long-term productivity and resistance of surviving trees. Mortality in stands managed for wildlife or biodiversity could contribute to these objectives if it increased stand diversity and wildlife habitat, or it could be detrimental if it threatened habitat for endangered species.

Development of complex mathematical simulation models has improved our ability to predict and evaluate the effects of bark beetle and pathogen outbreaks (Chapter 10). These models also can indicate priorities for future research. A model developed for western North America has been used to simulate effects of root diseases and associated bark beetles.

Currently, the model incorporates effects of combinations of three root diseases and four bark beetle functional groups. Continued testing and refinement of this model will improve its use for forest management decisions.

Tree mortality due to bark beetles and pathogenic fungi should be rare events in forests managed to facilitate natural interactions and balances among species. Prevention of outbreaks through forest management practices that mimic natural processes is a key to reducing losses without disrupting natural processes in integrated ecosystems (Chapter 11). This goal can be achieved through management of tree species composition, density, age or size distribution (stand structure), and disturbance-related factors such as tree injury and stump density.

12.2 RECOMMENDATIONS FOR FUTURE RESEARCH

Despite our advanced understanding of species interactions in forests, this synthesis has demonstrated the limits of our knowledge. Major questions remain to direct future research on interactions between bark beetles and pathogenic fungi in conifer forests.

Failure to meet forest management objectives often indicates inaccurate identification of target species. Funding for bark beetle and fungal systematics typically corresponds to concern during outbreaks. While much has been learned through the use of modern morphological techniques, both bark beetles and pathogenic fungi continue to challenge those who wish to distinguish species or ecologically significant varieties (biotypes). Among bark beetles, morphologically indistinguishable populations (cryptic and sibling species) respond differently to host factors and management strategies. Perhaps new chemosystematic techniques, such as cuticular hydrocarbons or DNA analysis, or behavioral chemicals (semiochemicals), will distinguish incipient species (Chapter 2). Similarly, fungal systematists still contend with questions of hyphal interfertility and life history relationships between sexual and asexual stages among the Deuteromycotina, Ascomycotina and Basidiomycotina. The difficulty in distinguishing pathogenic and non-pathogenic species of *Armillaria* and *Leptographium* is an obvious example (Chapter 3). Again, new techniques in cytogenetics and molecular systematics are certain to reveal new information about the relationships among insects and fungi.

We need more information on factors predisposing conifer forests to outbreaks of bark beetles and pathogenic fungi. For example, what degree of stress is necessary to increase tree susceptibility to mortality agents? What combinations of stress, resource concentration, and incipient populations of bark beetles and pathogenic fungi are sufficient to trigger outbreaks? What are the likely consequences of global climate change for trees, insects, and fungi? How will forest management practices stabilize forests or exacerbate effects of climate change? Because forests respond to changing conditions over long time periods, often with significant lag time because of mediating interactions (Franklin *et al.*, 1989; Perry and Borchers, 1990), research on effects of predisposing factors must cover sufficiently long time periods. This will require agency commitments to longer term research support than currently available. At the same time, critical scientific methods for evaluating effects of specific factors must be balanced against the need for information on interactive effects among multiple predisposing factors. Assessing interactions among many factors will require use of multivariate analytical techniques, as well as multiple analysis of variance (ANOVA) techniques.

Our current understanding of interactions among the invertebrates and microorganisms in this system inspire awe at their complexity. Apparently innocuous interactions, such as between bark beetles and nematophagous mites, or between fungi and the saprophagous mites phoretic on bark beetles, influence the reproductive and developmental success of the beetles. The tarsonemid mites phoretic on bark beetles only recently were found to be the principle vectors of the pathogenic *Ophiostoma minus* (Bridges and Moser, 1986). Because funding for work on these interactions traditionally has been limited to outbreak periods, we still know little about changes in interactions that trigger the shift from endemic to epidemic populations. Further intricacies in interactions critical to regulating bark beetle and pathogen epidemiologies likely await discovery.

The physical and biochemical defenses of conifers still are imperfectly understood. The roles of oleoresin production and flow rate, terpene composition, lesion development, and resource partitioning patterns underlying these tree defenses require clarification. Evaluation of the relative importances of the oleoresin system and the induced response to wounding will improve our understanding of tree defense physiology. The mechanisms by which fungal growth or metabolites interfere with tree defensive mechanisms and lead to tree death also require further investigation.

We have only begun to appreciate the variety of effects of these interactions on forest ecosystems. Bark beetles and pathogenic fungi regularly cause extensive tree mortality and loss of forest resources (Filip and Goheen, 1982, 1984; Leuschner and Berck, 1985). However, bark beetle and pathogen roles in thinning and decomposition processes contribute to soil structure and fertility, and to long-term forest productivity and biodiversity (Franklin *et al.*, 1989; Schowalter *et al.*, 1992). Research has addressed non-destructive effects only during the past decade. Experimental evaluation of the importance of insects, pathogens and fire to long-term forest health lags far behind advancing concepts on potential roles. Few entomologists or pathologists have been integrated in long-term multidisciplinary research on nutrient cycling and forest health. Experimental studies on these long-term effects will require funding over longer periods.

We have seen considerable advance in computer technology and modeling software. However, even the most complex simulation models and decision-support systems still rely on untested assumptions, especially regarding the roles of tree physiological processes and stand structure and composition. Knowledge about these roles will form the links between environmental conditions and tree susceptibility to bark beetles and pathogenic fungi. In addition, models have yet to incorporate effects of bark beetles and fungi on soil fertility and biodiversity into evaluations of "pest" epidemiology and treatment efficacy. Furthermore, model structure and data requirements necessarily represent a compromise between scientific validity and management utility. Data requirements for model initialization often cannot be collected conveniently by forestry personnel. Few models have been evaluated for the costs, in terms of accuracy and reliability of predictions, of using inadequate data to initiate model simulations (Schowalter *et al.*, 1982).

Finally, we have recognized that our forest management practices, especially fire suppression and tree species selection, have created forests susceptible to bark beetles and pathogenic fungi over broad geographic areas and that this susceptibility is likely exacerbated by problems of atmospheric pollution and global climate change. Nevertheless, few critical experiments have demonstrated the effects of particular management practices on bark beetle and pathogen responses. For example, is the reduced tree mortality to bark beetles following thinning due to increased resource availability for remaining trees, to increased distance between hosts, and/or to changes in ambient temperature and moisture? How will bark beetles and pathogens respond to reintroduction of periodic fire? Similarly, few critical experiments have evaluated the long-term or region-wide consequences of pest activity or suppression tactics (Honnold and Wood, 1990)

A large proportion of the pest control measures in current use have not been evaluated for efficacy, in terms of resource protection, through appropriate controlled comparisons among replicated treatments (Waters and Stark, 1980; Wood *et al.*, 1985; Honnold and Wood, 1990). Typically, an infested (or treated) stand is compared to an uninfested (or untreated) stand. Inadequate replication and non-random assignment of treatments among study plots preclude convincing statistical evaluation. Random assignment of treatments often is especially difficult where insect and disease-caused damage dictates the type of treatment employed (e.g. plots containing numerous root disease centers cannot receive commercial thinning treatment).

Data necessary to evaluate effects of pest activity and pest suppression must be obtained experimentally, with adequate replication and random assignment of treatments among experimental plots. Application of results will require that replicate plots be distributed across large areas rather than contained within individual stands. For example, Schowalter and Turchin (1993) evaluated the effects of stand density and composition on initiation of *Dendroctonus frontalis* infestations by introducing a standard number of brood beetles (two infested trees) into replicate treatment plots (Fig. 12.1) distributed over 100 000 ha in each of 2 years. Replication and equal exposure to bark beetle populations among plots were ensured using this method.

Rigorous scientific testing of hypotheses using experimental methods will require greater collaboration between researchers (including entomologists, pathologists and forest ecologists) and forest managers. Such scientific evaluation of management practices is absolutely necessary to justify pest management decisions, especially in sensitive areas or areas set aside to protect ecosystem integrity, such as designated wilderness (Honnold and Wood, 1990). In fact, areas set aside as preserves will become increasingly important as natural laboratories for evaluation of natural interactions as demands for resources from managed forests increase (Schowalter, 1988; Filip, 1990; Honnold and Wood, 1990; Wickman, 1990).

The traditional emphasis in North American forestry has been on timber values. Increasing concern over sustainable forest health, preservation of biodiversity, protection of long-term site productivity, and the fate of future forests subject to changing atmospheric quality and global climate already is altering our management objectives and approaches. Our research efforts must be directed toward a better understanding of the roles played by bark beetles, pathogenic fungi, environmental factors, and their interactions, in meeting these objectives. Perhaps by incorporating or mimicking these natural roles into our forest management, we can accomplish our management objectives more efficiently.

REFERENCES

Bridges, J.R. and Moser, J.C. (1986). Relationship of phoretic mites (Acari: Tarsonemidae) to the bluestaining fungus, *Ceratorcystis minor*, in trees infested by southern pine beetle (Coleoptera: Scolytidae). *Environ. Entomol.* **15**, 951–953.

Filip, G.M. (1990). Effects of tree harvesting on Armillaria root disease in an old-growth mixed-conifer stand in northeastern Oregon. *Northw. Environ. J.* **6**, 412–413.

Filip, G.M. and Goheen, D.J. (1982). Tree mortality caused by root pathogen complex in Deschutes National Forest, Oregon. *Plant Dis.* **66**, 240–243.

Filip, G.M. and Goheen, D.J. (1984). Root diseases cause severe mortality in white and grand fir stands in the Pacific Northwest. *Forest Sci.* **30**, 138–142.

Fig. 12.1. Experimental design for testing effects of stand treatments on bark beetle infestation development. Trees containing mature beetles ready to emerge were cut into 1.3 m sections and distributed evenly among replicate treatment plots. These methods ensured that treatments were equally exposed to beetle populations, rather than allowing naturally distributed beetles to discover some treatment plots and not others.

Honnold, D.L. and Wood, D.L. (1990). Pest management in wilderness and roadless areas. *Northw. Environ. J.* **6**, 177–193.

Franklin, J.F., Perry, D.A., Schowalter, T.D., Harmon, M.E., McKee, A. and Spies, T.A. (1989). Importance of ecological diversity in maintaining long-term site productivity. *In* "Maintaining Long-term Productivity of Pacific Northwest Forest Ecosystems" (D.A. Perry, B. Thomas, R. Meurisse, R. Miller, J. Boyle, P. Sollins and J. Means, eds), pp. 82–97. Timber Press, Portland, OR.

Leuschner, W.A. and Berck, P. (1985). Impacts on forest uses and values. *In* "Integrated Pest Management in Pine–Bark Beetle Ecosystems" (W.E. Waters, R.W. Stark and D.L. Wood, eds), pp. 105–120. John Wiley & Sons, New York.

Perry, D.A. and Borchers, J.G. (1990). Climate change and ecosystem responses. *Northw. Environ. J.* **6**, 293–313.

Schowalter, T.D. (1988). Forest pest management: a synopsis. *Northw. Environ. J.* **4**, 313–318.

Schowalter, T.D. and Turchin, P. (1993). Southern pine beetle infestation development: interaction between pine and hardwood basal areas. *Forest Sci.* (in press).

Schowalter, T.D., Coulson, R.N., Turnbow, R.H. and Fargo, W.S. (1982). Accuracy and precision of procedures for estimating populations of the southern pine beetle (Coleoptera: Scolytidae) by using host tree correlates. *Environ. Entomol.* **75**, 1009–1016.

Schowalter, T.D., Caldwell, B.A., Carpenter, S.E., Griffiths, R.P., Harmon, M.E., Ingham, E.R., Kelsey, R.G., Lattin, J.D. and Modenke, A.R. (1992). Decomposition of fallen trees; effects of initial conditions of heterotroph colonization rates. *In* "Tropical Ecosystems: Ecology and Management" (K.P. Singh and J.S. Singh, eds), pp. 373–383. Wiley Eastern Ltd., Bombay.

Waters, W.E. and Stark, R.W. (1980). Forest pest management: concept and reality. *Annu. Rev. Entomol.* **25**, 479–509.

Wickman, B.E. (1990). Old-growth forests and history of insect outbreaks. *Northw. Environ. J.* **6**, 401–403.

Wood, D.L., Stark, R.W., Waters, W.E., Bedard, W.D. and Cobb, F.W., Jr. (1985). Treatment tactics and strategies. *In* "Integrated Pest Management in Pine–Bark Beetle Ecosystems" (W.E. Waters, R.W. Stark and D.L. Wood, eds), pp. 121–139. John Wiley & Sons, New York.

–Appendix 1–

Scientific and Common Names of Trees and Shrubs[a]

Scientific name	Common name
Abies alba Mill.	Silver fir (Europe)
A. balsamea (L.) Mill.	Balsam fir
A. concolor (Gord. & Glend.) Lindl. ex Hildebr.	White fir
A. grandis (Dougl. ex D. Don) Lindl.	Grand fir
A. lasiocarpa (Hook.) Nutt.	Subalpine fir
Acer circinatum Pursh	Vine maple
A. macrophyllum Pursh	Bigleaf maple
Alnus rubra Bong.	Red alder
Calocedrus decurrens Torr.	Incense cedar
Chamaecyparis lawsoniana (A. Murr.) Parl.	Port-Orford cedar
Glycine max (L.) Merr.	Soybean
Heteromeles arbutifolia (Lindl.) M. J. Roem.	Toyon
Larix occidentalis Nutt.	Western larch
Liquidambar styraciflua L.	Sweetgum
Picea abies (L.) Karst)	Norway spruce
P. engelmannii Parry ex Engelm.	Engelmann spruce
P. mariana (Mill.) B.S.P.	Black spruce
P. rubens Sarg.	Red spruce
P. sitchensis (Bong.) Carr.	Sitka spruce
Pinus banksiana Lamb.	Jack pine
P. contorta Dougl. ex Loud. var. *contorta*	Shore pine
P. contorta var. *latifolia* Engelm.	Lodgepole pine
P. echinata Mill.	Shortleaf pine
P. elliottii Engelm. var *elliottii*	Slash pine
P. flexilis	James Limber pine
P. jeffreyi Grev. & Balf.	Jeffrey pine
P. lambertiana Dougl.	Sugar pine
P. monophylla Torr. & Frem.	Singleleaf pinyon pine
P. monticola Dougl. ex D. Don	Western white pine
P. palustris Mill.	Longleaf pine
P. pinaster Ait.	Maritime pine
P. ponderosa Dougl. ex Laws.	Ponderosa pine
P. radiata D. Don	Monterey pine
P. resinosa Ait.	Red pine
P. rigida Mill	Pitch pine
P. strobus L.	Eastern white pine
P. sylvestris Ait.	Scots pine
P. taeda L.	Loblolly pine

Pseudotsuga menziesii (Mirb.) Franco var. *menziesii*	Douglas-fir (coastal)
P. menziesii var. *glauca* (Beissn.) Franco	Douglas-fir (interior)
Quercus alba L.	White oak
Q. kelloggii Newb.	California black oak
Sequoia sempervirens (D. Don) Endl.	Redwood
Sideroxylon sessiliflorum (Poiret) Aubrev.	Tambalacoque
Thuja plicata Donn ex D. Don	Western red-cedar
Tsuga heterophylla (Raf.) Sarg.	Western hemlock

[a]From Little, E.L., Jr. (1979). "Checklist of United States Trees (Native and Naturalized)." USDA Agric. Handbook 541. USDA, Washington, DC, 375 pp.

–Appendix 2–

Scientific and Common Names of Insects[a]

Scientific name	Common name
Adelges piceae (Ratzeburg)	Balsam wooly adelgid
Choristoneura pinus Freeman	Jack pine budworm
C. fumiferana (Clemens)	Spruce budworm
C. occidentalis Freeman	Western spruce budworm
Coloradia pandora Blake	Pandora moth
Conophthorus banksianae McPherson	Jack pine cone beetle
C. ponderosae Hopkins	Ponderosa pine cone beetle
C. resinosae Hopkins	Red pine cone beetle
Corythuca arcuata (Say)	Oak lace bug
Cryptoxyleborus naevus Schedl	
Dendroctonus adjunctus Blandford	Roundheaded pine beetle
D. brevicomis LeConte	Western pine beetle
D. frontalis Zimmerman	Southern pine beetle
D. jeffreyi Hopkins	Jeffrey pine beetle
D. mexicanus Hopkins	
D. micans (Kugelann)	Great spruce bark beetle
D. ponderosae Hopkins	Mountain pine beetle
D. pseudotsugae Hopkins	Douglas-fir beetle
D. punctatus LeConte	Allegheny spruce beetle
D. rhizophagus Thomas & Bright	
D. rufipennis (Kirby)	Spruce beetle
D. simplex LeConte	Eastern larch beetle
D. terebrans (Olivier)	Black turpentine beetle
D. valens LeConte	Red turpentine beetle
D. vitei Wood	
Dryocoetes autographus Ratzeburg	
D. confusus Swaine	Western balsam bark beetle
Epilachna varivestis Mulsant	Mexican bean beetle
Euwallacea fornicatus (Eichhoff)	
Gnathotrichus primus (Bright)	
G. retusus (LeConte)	
G. sulcatus (LeConte)	
Hylastes macer LeConte	
H. nigrinus (Mannerheim)	
Hypothenemus curtipennis (Schedl)	
Indocryphalus intermedius (Sampson)	
Ips acuminatus (Gyllenhal)	

I. amitinus (Eichhoff)	
I. apache Lanier	
I. avulsus (Eichhoff)	Small southern pine engraver
I. borealis Swaine	
I. calligraphus (Germar)	Sixspined ips
I. confusus (LeConte)	Pinyon ips
I. grandicollis (Eichhoff)	Eastern fivespined ips
I. hoppingi Lanier	
I latidens LeConte	
I. lecontei Swaine	Arizona fivespined ips
I. mexicanus (Hopkins)	Monterey pine ips
I. paraconfusus Lanier	California fivespined ips
I. pertubatus (Eichoff)	
I. pini (Say)	Pine engraver
I. plastographus (LeConte)	
I. sexdentatus Boerner	Pine engraver beetle
I. typographus L.	European spruce bark beetle
Microcorthylus castaneus Schedl	
Monarthrum bicallosum (Schedl)	
Monochamus titillator (F.)	Southern pine sawyer
Orgyia pseudotsugata (McDunnough)	Douglas-fir tussock moth
Pissodes fasciatus LeConte	
Polygraphus rufipennis (Kirby)	Four-eyed spruce beetle
Pseudohylesinus nebulosus (LeConte)	Douglas-fir pole beetle
Scolytus abietis Blackman	
S. laricis Blackman	Larch engraver
S. multistriatus (Marsham)	Smaller European elm bark beetle
S. piceae (Swaine)	Spruce engraver
S. praeceps LeConte	
S. scolytus (Fabricius)	
S. subscaber LeConte	
S. tsugae (Swaine)	Hemlock engraver
S. unispinosus LeConte	Douglas-fir engraver
S. ventralis LeConte	Fir engraver
Spondylis upiformis Mannerheim	
Steremnius carinatus (Mannerheim)	
Synanthedon sequoiae (Hy. Edwards)	Sequoia pitch moth
Tomicus minor (Hartig)	Smaller pine shoot beetle
Trypodendron lineatum (Olivier)	Striped ambrosia beetle
Xyleborus dispar (Fabricius)	European shothole borer
X. ferrugineus (Fabricius)	
Xyloterinus politus (Say)	

[a]From Stoetzel, M.B. (1989). "Common Names of Insects and Related Organisms." Entomological Society of America, Lanham, MD, 198 pp.

–Appendix 3–

Scientific Names of Mites

Alobates pennsylvanica
Cercoleipus coelonotus (Kinn)
Dendrolaelaps isodentatus (Hurlbutt)
D. neocornutus (Hurlbutt)
D. neodisetus (Hurlbutt)
Ereynetes scutulis (Hunter)
E. sinescutulis Hunter and Rosario
Heterotarsonemus lindquisti Smiley
Histiogaster arborsignis Woodring
Histiostoma media (Woodring and Moser)
H. varia Woodring and Moser
Iponemus gaebleri (Schaarschmidt)
Macrocheles boudreauxi Krantz
Mexecheles virginiensis (Baker)
Paracarophaenax ipidarius (Redikortsev)
Paraleius leontonychus (Berlese)
Pleuronectocelaeno drymocoetes Kinn
Proctolaelaps dendroctoni Lindquist and Hunter
P. fiseri Samsinak
P. hystricoides Lindquist and Hunter
P. hystrix (Vitzthum)
Pyemotes giganaticus Cross, Moser and Rack
P. parviscolyti (Cross and Moser)
P. scolyti (Oudemans)
Pygmephorellus bennetti Cross & Moser
Tarsonemus ips Lindquist
T. krantzi Smiley and Moser
T. subcorticalis Lindquist
Trichouropoda austalis Hirschmann
T. hirsuta Hirschmann
Tyrophagus putrescentae (Schrank)
Vulgarogamasus lyriformis (McGraw and Farrier)

–Appendix 4–

Scientific Names of Pathogens and Common Names of Resulting Diseases[a]

Scientific name	Common name
Alternaria alternata (Fr.) Keissler	
Arceuthobium abietinum Engelm. ex Munz (Fir)	True fir dwarf mistletoe
A. campylopodum Engelm.	Western dwarf mistletoe
Armillaria calvescens Berube & Dessureault	Armillaria root disease
A. cepistipes Velenovski	Armillaria root disease
A. gallica Marxmuller & Romagn.	Armillaria root disease
A. gemina Berube & Dessureault	Armillaria root disease
A. mellea (Vahl.:Fr.) Kummer	Armillaria root disease
A. ostoyae (Romang.) Herink	Armillaria root disease
A. sinapina Berube & Dessureault	Armillaria root disease
A. tabescens (Scop.:Fr.) Dennis, Orton & Hora	Armillaria root disease
Bacillus cereus Frankland & Frankland	
Beaveria bassiana (Bals.) Vuill.	
Bursaphelenchus xylophilus (Steiner & Buhrer) Nickle	Pine wilt nematode
Ceratocystiopsis proteae Wingfield, Van Wyk & Marasas	
C. ranaculosus Perry and Bridges	
Ceratocystis coerulescens (Munch) Bakshi	
C. fagacearum (Bretz.) Hunt	Oak wilt
C. fimbriata Ellis & Halst.	
C. laricicola Redfern & Minter	
Cronartium comandrae Pk.	Comandra rust
C. ribicola Fisch.	White pine blister rust
Cryptoporus volvatus (Pk.) Shear	Gray saprot
Cytospora abietis Sacc.	Fir canker
Diplodia pinea (Desm.) Kickx	Shoot dieback of pine
Entomocorticium dendroctoni Whitney	
Fomitopsis pinicola (Swartz:Fr.) Karst.	Brown crumbly rot
Fusarium subglutinans (Wollenw. & Reink.) Nelson	Pitch canker
Hansenula capsulata Wickerham	
H. holstii Wickerham	
Heterobasidion annosum (Fr.) Bref.	Annosus root disease
H. araucariae Buchanan	
Inonotus circinatus (Fr.) Gilb.	
I. tomentosus (Fr.) Teng	Tomentosus root disease
Leptographium procerum (Kendr.) Wingf.	Procerum root disease
L. terebrantis Barras and Perry	

L. wageneri var. *ponderosum* (Harrington & Cobb) Harrington & Cobb	Black-stain root disease on hard pines
L. wageneri var. *wageneri* (Kendr.) Wingf.	Black-stain root disease on pinyon pine
L. wageneri var. *pseudotsugae* Harrington & Cobb	Black-stain root disease on *Pseudotsuga*
Ophiostoma clavigerum (Robins.-Jeff. & Davids.) Harrington	
O. dryocoetidis (Kendr. & Molnar) De Hoog & Scheffer	
O. huntii (Robins.-Jeff. & Davids.) Harrington	
O. ips (Rumb.) Nannf.	
O. minus (Hedgc.) H. & P. Sydow	Blue stain fungus
O. montium (Rumb.) von Arx	
O. nigrocarpum (Davids.) De Hoog	
O. polonicum Siem.	
O. ulmi (Buism.)	Dutch elm disease
O. wageneri (Goheen & Cobb) Harrington	Black-stain root disease (teleomorph on *Pinus ponderosa*)
Pezizella chapmanii Whitney & Funk	
Phaeolus schweinitzii (Fr.) Pat.	Schweinitzii root disease
Phellinus weirii (Murr.) Gilb.	Laminated root rot
Phoradendron bolleanum (Seem.) Eich.	True mistletoe
Phytophthora cinnamomi Rands	Littleleaf disease
P. lateralis Tucker & Milbrath	Port-Orford-cedar root disease
Pichia pini (Holst) Phaff	
Spiniger meineckellum (Olson) Stalpers	Annosus root disease (anamorph)
Trichosporium symbioticum Wright	

[a]Scientific names of polypore fungi from Gilbertson, R.L. and Ryvarden, L. (1986) and (1987). "North American Polypores," Vols 1 and 2." Fungiflora A/S, Oslo, Norway, 885 pp. Scientific names of fungi causing black-stain root disease from Harrington, T.C. and Cobb, F.W., Jr., eds (1988). "Leptographium Root Diseases on Conifers." The American Phytopathological Society, St. Paul, MN, 149 pp.

Index